BLAZAR DEMOGRAPHICS AND PHYSICS

COVER ILLUSTRATION:

The family of multifrequency blazar spectra for γ-ray-detected sources averaged in five luminosity bins. The characteristic double-peaked spectrum is produced by synchrotron radiation at low frequencies and possibly Compton scattering at high frequencies. The lines (mainly to guide the eye) represent a phenomenological analytic model that describes the run of spectral shapes from the lowest luminosity (BL Lac objects) to the highest (flat-spectrum quasars). Adapted from Fossati et al. (1998), reproduced by the kind permission of the author, referenced in MNRAS, 299, figure 12, page 445.

A SERIES OF BOOKS ON RECENT DEVELOPMENTS IN
ASTRONOMY AND ASTROPHYSICS

Publisher

THE ASTRONOMICAL SOCIETY OF THE PACIFIC
390 Ashton Avenue, San Francisco, California, USA 94112-1722
Phone: (415) 337-1100   Fax: (415) 337-5205
E-Mail: catalog@aspsky.org   Web Site: www.aspsky.org

ASP CONFERENCE SERIES - EDITORIAL STAFF
Managing Editor: D. H. McNamara   LaTeX-Computer Consultant: T. J. Mahoney
Associate Managing Editor: J. W. Moody   Production Manager: Enid L. Livingston

PO Box 24453, 211 KMB, Brigham Young University, Provo, Utah, 84602-4463
Phone: (801) 378-2111   Fax: (801) 378-4049   E-Mail: pasp@byu.edu

ASP CONFERENCE SERIES PUBLICATION COMMITTEE:
Alexei V. Filippenko   Geoffrey Marcy
Ray Norris   Donald Terndrup
Frank X. Timmes   C. Megan Urry

A listing of all other ASP Conference Series Volumes and IAU Volumes
published by the ASP is cited at the back of this volume

ASTRONOMICAL SOCIETY OF THE PACIFIC
CONFERENCE SERIES

Volume 227

BLAZAR DEMOGRAPHICS AND PHYSICS

Proceedings of a conference held at
Space Telescope Science Institute, Baltimore, Maryland, USA
12-14 July 2000

Edited by

Paolo Padovani
Space Telescope Science Institute, Baltimore, Maryland, USA

Affiliated with
Astrophysics Division, Space Science Department, European Space Agency

On leave from
Dipartimento di Fisica, II Università di Roma "Tor Vergata", Italy

and

C. Megan Urry
Space Telescope Science Institute, Baltimore, Maryland, USA

© 2001 by Astronomical Society of the Pacific. All Rights Reserved

No part of the material protected by this copyright notice may be reproduced or utilized in any form or by any means – graphic, electronic, or mechanical including photocopying, taping, recording or by any information storage and retrieval system, without written permission from the publisher.

Library of Congress Cataloging in Publication Data
Main entry under title

Card Number: 2001087623
ISBN: 1-58381-059-5

ASP Conference Series - First Edition

Printed in United States of America by Sheridan Books, Chelsea, Michigan

Contents

Preface viii

Conference participants ix

Conference photograph xii

Part 1. Properties across Classification and Luminosities

Issues in Blazar Research 3
    P. Padovani and C. M. Urry

Images and Polarization of Compact Jets in Blazars 10
    A. P. Marscher and S. G. Jorstad

The Parsec-Scale Structure of TeV Blazars 18
    B. G. Piner, P. G. Edwards, and S. Fodor

Host Galaxies of Blazars and Radio Galaxies 22
    R. Scarpa

Deep NOT Imaging of Radio Selected BL Lac Objects 32
    T. Pursimo, K. Nilsson, A. Sillanpää, L. O. Takalo, and J. Heidt

Intrinsic Differences in the Inner Jets of High and Low Optically Polarized Radio Quasars 36
    M. L. Lister and P. S. Smith

Spectral Energy Distributions of Blazars: Facts and Speculations 40
    L. Maraschi and F. Tavecchio

The HST View of the Nuclei of Radio Galaxies: Implications for the Radio-loud AGN Unification Models 50
    M. Chiaberge, A. Celotti, A. Capetti, and G. Ghisellini

Blazars in Low-Luminosity and Radio-Weak AGN? 56
    H. Falcke, S. Markoff, and P. L. Biermann

Connection between Superluminal Ejections and γ-Ray Flares in Blazars ... 69
    S. Jorstad, A. Marscher, M. Aller, and H. Aller

The Emission Line Properties of DXRBS Blazars 73
    H. Landt, P. Padovani, E. S. Perlman, and P. Giommi

On the Relation between Radio and Non-Radio Elliptical Galaxies ... 77
    R. Scarpa and C. M. Urry

Part 2. Jet Physics

Blazar Jets: The Spectra 85
    G. Ghisellini

Jets in Quasars 95
    M. Sikora

On the Energy Content of Blazar Jets 105
    A. Celotti

Beaming in Jets with Evolving Relativistic Features ... 108
    Z. Abraham

Time Scales of Blazar Variability 112
    S. J. Wagner

Size-Luminosity-Scaling and Inverse Compton Seed Photons in Blazars . 116
    M. Georganopoulos, J. G. Kirk, and A. Mastichiadis

Chandra Detection of an X-ray Jet in 3C 371 122
    R. M. Sambruna, C. M. Urry, R. Scarpa, F. Tavecchio, L. Maraschi, and J. E. Pesce

Continuous Long-look Observations of TeV Blazars Using ASCA 127
    C. Tanihata, T. Takahashi, J. Kataoka, and C. M. Urry

Spectral Energy Distribution of Low Power FR I Radio Galaxies 131
    E. Trussoni, A. Capetti, A. Celotti, and M. Chiaberge

BL Lacs at the Blue End of the Blazar Sequence 135
    L. Costamante, G. Ghisellini, A. Wolter, G. Tagliaferri, G. Fossati, P. Padovani, and P. Giommi

A Polarization Flare in 3C 273: A Clue to Jet Physics 140
    L. L. Cross, B. J. Wills, J. H. Hough, and J. A. Bailey

Simultaneous Optical and X-Ray Observations of BL Lacertae 144
    R. Nesci, E. Massaro, F. Montagni, S. Sclavi, T. Balonek, M. Caler, C. Tremonti, F. D'Alessio, S. Catalano, A. Frasca, E. Marilli, G. Tagliaferri, G. Ghisellini, M. Ravasio, P. Giommi, L. Chiappetti, T. Kato, M. Uemura, O. M. Kurtanidze, M. G. Nikolashvili, M. T. Carini, J. C. Noble, G. Tosti, G. Nucciarelli, and J. Mattox

The Physics of Blazar Optical Emission Regions. I. Alignment of Optical Polarization and the VLBI Jet 150
    M. J. Yuan, H. Tran, B. Wills, and D. Wills

The Physics of Blazar Optical Emission Regions. II. Magnetic Field Orientation, Viewing Angle, and Beaming 154
    M. J. Yuan, H. Tran, B. Wills, and D. Wills

On the AGN Storage Ring Central Engine Model for Blazars 158
    H. D. Greyber

Contents   vii

Part 3. New Surveys, Number Counts, Luminosity Functions

Deep Blazar Surveys 163
    P. Padovani

The BLEIS Project 176
    I. Cagnoni, A. Celotti, and D. Poccecai

The Reddest BL Lacs? 180
    S. Antón and I. Browne

Hidden BL Lacertae Objects Near and Far 184
    J. T. Stocke

Radio Properties of REX BL Lacs and Galaxies 190
    A. Wolter, A. Caccianiga, T. Maccacaro, R. Della Ceca, I. M. Gioia,
    F. Cavallotti, and M. Minoia

The EMSS Radio Loud Quasar Sample 196
    A. Wolter and A. Celotti

Surveys and the Blazar Parameter Space 200
    E. S. Perlman, P. Padovani, H. Landt, J. T. Stocke, L. Costamante,
    T. Rector, P. Giommi, and J. F. Schachter

How Many 'Flavors' Can a BL Lac Have? 208
    M. J. Marchã

Discovery of Hidden Blazars inside Quasars 212
    F. Ma and B. J. Wills

Demographics of Blazars 218
    G. Fossati

Part 4. Evolution

The Cosmological Evolution of BL Lacertae Objects 227
    P. Giommi, A. Pellizzoni, M. Perri, and P. Padovani

The Cosmological Evolution of BL Lacs: The REX Point of View 238
    A. Caccianiga, T. Maccacaro, A. Wolter, R. Della Ceca, and I. M. Gioia

Extragalactic Radio Source Evolution and Unification: Clues to the
Demographics of Blazars 242
    C. A. Jackson and J. V. Wall

The Connection between BL Lacs and Flat-Spectrum Radio Quasars .. 252
    V. D'Elia and A. Cavaliere

Author Index 259

Preface

The idea for this workshop came from a number of people. We recall suggestions from Maria Marchã (Lisbon Observatory) and John Stocke (U. Colorado), a few years ago, to convene the key experts to try to understand the "HBL/LBL" distinction (i.e., why the different types of spectral energy distributions arise and whether/how they are tied to jet power). To maximize attendance with a very attractive venue, we decided to forego Portugal or Colorado in favor of Baltimore in July, always a "warmly" hospitable place!

More than fifty experts braved the Baltimore summer to tackle not only the physics of blazars but, for the first time, their demographics. Workers in the field know this has been a puzzling and sometimes contentious issue, and one that has resisted an easy solution.

The timing was ideal. After years during which our knowledge of blazars was based mostly on relatively shallow and small samples, new, deeper samples have recently been identified from large surveys at various wavelengths. A highlight of the meeting was the realization that real progress on blazar demographics is now possible.

The demographics are important because blazar type has an underlying physical significance. This was discussed extensively, with comprehensive talks on jet physics, unified schemes, evolution. Here, too, new data (e.g., from Chandra) are making rapid advances possible.

The number densities and physical significance of blazar types have implications beyond blazars. Workshop participants seemed persuaded that the same observational biases that preferentially select different blazar types must also be relevant to other AGN classifications, notably the radio-loud/radio-quiet division. Clearly, the understanding of blazars will illuminate the much larger subject of jet formation and evolution in AGN.

We thank the Space Telescope Science Institute for hosting this meeting and for financial support. Quindarian Gryce has our profound thanks for having organized the logistics beautifully, and Sharon Toolan has our eternal gratitude for helping us assemble the Proceedings.

Paolo Padovani and Meg Urry

Space Telescope Science Institute

Conference Participants

Zulema Abraham	zulema@iagusp.usp.br
Mark Allen	mga@stsci.edu
Sonia Anton	sac@jb.man.ac.uk
Thomas Balonek	tbalonek@mail.colgate.edu
Katherine Blundell	kmb@astro.ox.ac.uk
Alessandro Caccianiga	caccia@oal.ul.pt
Ilaria Cagnoni	ilale@sissa.it
Alfonso Cavaliere	cavaliere@roma2.infn.it
Annalisa Celotti	celotti@sissa.it
Teddy Cheung	tcheung@stsci.edu
Marco Chiaberge	chiab@sissa.it
Luigi Costamante	costa@merate.mi.astro.it
Lara Cross	lcross@astro.as.utexas.edu
Valerio D'Elia	delia@roma2.infn.it
Heino Falcke	hfalcke@mpifr-bonn.mpg.de
Giovanni Fossati	gfossati@mamacass.ucsd.edu
Markos Georganopoulos	markos@mickey.mpi-hd.mpg.de
Gabriele Ghisellini	gabriele@merate.mi.astro.it
Paolo Giommi	giommi@napa.sdc.asi.it
Howard Greyber	hgreyber@yahoo.com
Carole Jackson	cjackson@mso.anu.edu.au
Hermine Landt	landt@stsci.edu
Matthew Lister	lister@sgra.jpl.nasa.gov
Feng Ma	feng@astro.as.utexas.edu
Duccio Macchetto	macchetto@stsci.edu
Laura Maraschi	maraschi@brera.mi.astro.it
Maria Marcha	mmarcha@sintra.oal.ul.pt
Svetlana Jorstad	jorstad@rjet.bu.edu
Alan Marscher	marscher@bu.edu
John Mattox	mattox@bu.edu
Giridhar Nandikotkur	giridhar@iastate.edu
Roberto Nesci	roberto.nesci@uniroma1.it
Paolo Padovani	padovani@stsci.edu
Eric Perlman	perlman@pha.jhu.edu
Glenn Piner	B.G.Piner@jpl.nasa.gov
Tapio Pursimo	tpursimo@utu.fi
Rita Sambruna	rms@astro.psu.edu
Riccardo Scarpa	rscarpa@eso.org
Marek Sikora	sikora@camk.edu.pl
John Stocke	stocke@casa.colorado.edu

Chiharu Tanihata — tanihata@astro.isas.ac.jp
Fabrizio Tavecchio — fabrizio@brera.mi.astro.it
Edoardo Trussoni — trussoni@to.astro.it
Meg Urry — cmu@stsci.edu
Stefan Wagner — swagner@lsw.uni-heidelberg.de
Richard L. White — rlw@stsci.edu
Beverley Wills — bev@astro.as.utexas.edu
Anna Wolter — anna@brera.mi.astro.it
Juntao Yuan — juntao@astro.as.utexas.edu

Conference Photograph

Conference Photograph

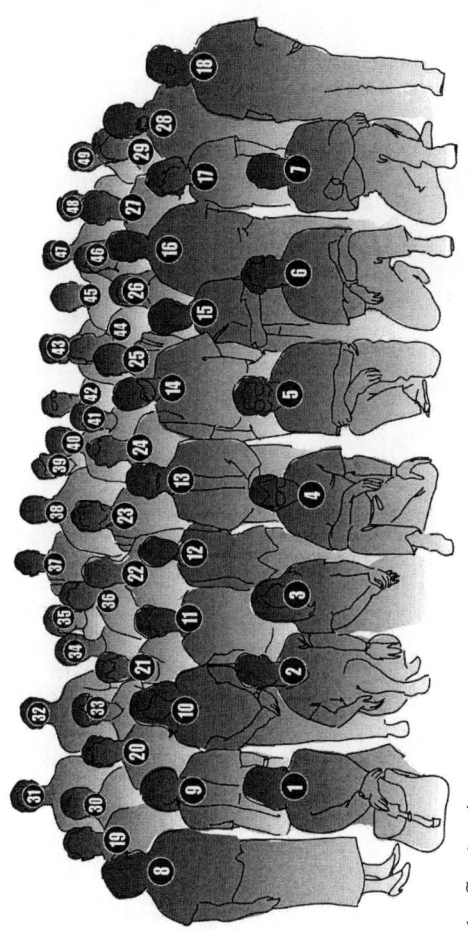

1. Sonia Anton; 2. Ilaria Cagnoni; 3. Svetlana Jorstad; 4. Alan Marscher; 5. Thomas Balonek; 6. Chris O'Dea; 7. Matthew Lister; 8. Meg Urry; 9. Laura Maraschi; 10. Katherine Blundell; 11. Anna Wolter; 12. Elena Pian; 13. Edoardo Trussoni; 14. Gabriele Ghisellini; 15. Zulema Abraham; 16. Glenn Piner; 17. Lara Cross; 18. Alfonso Cavaliere; 19. Paolo Padovani; 20. Riccardo Scarpa; 21. Chiharu Tanihata; 22. Roberto Nesci; 23. Fabrizio Tavecchio; 24. Marco Chiaberge; 25. Paolo Giommi; 26. Rita Sambruna; 27. Juntao Yuan; 28. Heino Falcke; 29. Mark Allen; 30. Howard Greyber; 31. Stefan Wagner; 32. Duccio Macchetto; 33. Teddy Cheung; 34. Hermine Landt; 35. Maria Marcha; 36. Luigi Costamante; 37. John Mattox; 38. Giridhar Nandikotkur; 39. Markos Georganopoulos; 40. Tapio Pursimo; 41. Valerio D'Elia; 42. Beverley Wills; 43. Marek Sikora; 44. Eric Perlman; 45. Feng Ma; 46. Carole Jackson; 47. Alessandro Caccianiga; 48. Annalisa Celotti; 49. Giovanni Fossati; Missing: John Stocke, Richard L. White.

Part 1

Properties across Classification and Luminosities

Issues in Blazar Research

Paolo Padovani

Space Telescope Science Institute, 3700 San Martin Drive, Baltimore, MD, 21218, USA

Affiliated to the Astrophysics Division, Space Science Department, European Space Agency

On leave from Dipartimento di Fisica, II Università di Roma "Tor Vergata," Via della Ricerca Scientifica 1, I-00133 Roma, Italy

C. Megan Urry

Space Telescope Science Institute, 3700 San Martin Drive, Baltimore, MD, 21218, USA

Abstract. As an overview of blazar research, we briefly discuss some of the themes of this conference, including the various types of blazars, blazar samples and selection effects, unified schemes, evolution, and jet physics.

1. Introduction

The initial motivation for this conference was to try to solve the seemingly simple problem "which blazars are more numerous," HBL-like or LBL-like. We thought that by convening in the same room advocates from the various factions we could perhaps agree on an answer. As this question had broad implications for our understanding of blazars, we expanded the scope of the conference to include also jet physics. The timing was good. After years during which our knowledge of blazars was based mostly on relatively shallow and small samples, new, deeper samples are now being assembled out of the large surveys that are available at various wavelengths.

More than fifty astronomers responded to our call. Many of them were the "usual suspects" but also a good number were young scientists, relatively new to the field, a sign that blazars continue to arouse interest.

The conference program addressed some of the outstanding key issues in blazar research. It is impossible to do justice to the whole conference in a short paper. Instead we highlight some of the most fundamental themes, referring to the single papers for a more in-depth discussion. We hope in this way to give a glimpse of the "big picture" underlying the various efforts.

One demographic issue we clearly solved, at least, was that of nationality. We had participants from all over the world and we were curious to see "which blazar pundits are more numerous." The answer should have been obvious to

Blazar Demographics

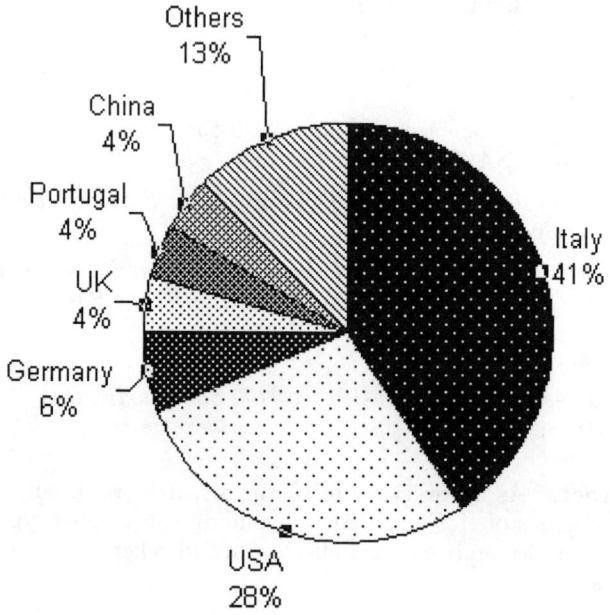

Figure 1. The distribution of nationalities amongst the attendants to the blazar conference.

anybody overhearing some of the conversations during the coffee breaks but it comes out explicitly in Fig. 1: blazars are Italian!

2. Types of Blazars

There is general agreement that blazars are radio-loud active galactic nuclei with relativistic jets pointed at us, which means their numbers and jet physics are relevant to all radio-loud active galactic nuclei (AGN). Blazars include both BL Lac objects, which have very weak line emission, and more luminous flat-spectrum radio quasars (FSRQ), which have normal, strong emission lines. Notwithstanding that nomenclature is the bane of this field, the conference necessarily began with definitions of common terms for types of blazars, RBL and XBL, LBL and HBL, red and blue. Because BL Lacs found in radio and X-ray surveys had strikingly different properties, they were further designated radio-selected BL Lac (RBL) or X-ray-selected (XBL) BL Lac. It was eventually realized that most RBL were radio bright and X-ray weak because their synchrotron component peaked at relatively low frequency (infrared-optical), while most XBL were radio weak and X-ray bright because their peaks fell at ultraviolet-X-ray frequencies. Hence the more precise terms: low-frequency peaked (LBL) and high-frequency peaked (HBL) BL Lacs, depending on whether the value of the two-point radio/X-ray spectral index α_{rx} is greater than or less than ~ 0.8, respectively. These definitions extend naturally to all blazars, both BL Lacs and quasars. Finally, to avoid the over-use of these undescriptive acronyms, we often employ the more colorful terms "red" and "blue" blazars for LBL and HBL, respectively.

Several participants emphasized that blazars show a clear continuity of properties with radio galaxies and steep-spectrum radio quasars (Scarpa, Maraschi, Anton, Wolter, Marchã, Ma). Radio powers and emission line strengths rise smoothly from BL Lacs to quasars—the very definition of a BL Lac as having equivalent width emission lines weaker than 5 Å is completely arbitrary (Landt). In a similar way, the radio powers and host galaxy magnitudes of the "parent population" of blazars, radio galaxies of Fanaroff-Riley (FR) types I and II, overlap. The host galaxies of blazars are uniformly luminous giant ellipticals, regardless of intrinsic nuclear power (Scarpa, Pursimo). They also have similar properties (morphologies, luminosities, and sizes) to the host galaxies of FR I and IIs or to luminous ellipticals without central radio sources.

The spectral energy distributions (SED) of blazars were interpreted as a combination of synchrotron emission and higher-energy radiation, possibly from Compton scattering (Maraschi). The observed SEDs follow an interesting trend with luminosity, such that high luminosity blazars have synchrotron and Compton peaks at lower frequencies. This can be explained as a cooling sequence, although there are concerns that it is induced at least in part by strong selection effects in the limited samples studied.

3. Blazar Samples and Selection Effects

The importance of selection biases was a recurring theme of this conference. The unification paradigm first arose a decade ago to explain the very strongest selection biases, that blazar radiation is strongly beamed in our direction (but maybe not as strongly as we think? Abraham), causing AGN appearance to change dramatically with orientation. Now experts in the field are focusing on more subtle effects stemming from the SED shapes (implying strong K-corrections) and from variability (especially important at gamma-ray energies where detectors/telescopes are not yet very sensitive).

Most of our current knowledge of blazars is based on a handful of relatively small and high-flux-limit samples. As a result, we have only probed the intrinsically most luminous sources. The new, deeper surveys currently underway are changing that. Participants at this meeting described several major new efforts, including the BLEIS, DXRBS, FIRST, Parkes 0.25 Jy, REX, RGB, and Sedentary surveys (Cagnoni, Padovani, Caccianiga, Giommi). Many of these new surveys employ multiple flux limits, usually radio and X-ray, and in some cases also optical. Because of the steepness of the source counts (more faint sources), such surveys find objects mostly at the survey vertex in the flux-flux(-flux) plane (or volume). Extension of the results of any survey to the whole blazar population is not always possible and in any case always requires understanding how such flux limits bias the derived sample (Padovani).

The selection biases can be very strong. The distinction between HBL and LBL, which previously looked like different types of blazar, has now been shown to be far more arbitrary because many intermediate blazars have been found (e.g. in the RGB, DXRBS, and REX surveys). There must therefore be a broad distribution of peak synchrotron frequencies between the far-infrared and the X-ray bands. These new surveys have also found a previously unknown class of blue FSRQ, although there are clearly none as blue as the most extreme BL Lacs (Perlman, Padovani, Wolter, Costamante). We still do not know how these blue FSRQ relate to their red relatives, although there are some ideas (Georganopoulos, Perlman). Simulations presented at the conference showed clearly that the content of new blazar samples, even the deeper ones, is affected more strongly by the sample flux limit(s) than by intrinsic properties (Fossati, Giommi). *In the simplest terms, the surveys find what they can rather than what exists.*

4. Unification of Blazars and Radio Galaxies

The original unification ideas, equating BL Lac objects with FR I galaxies and quasars with FR IIs, seem to hold up well to newer, more extensive data. The radio counts of BL Lac objects from the $\sim$ 20-times deeper DXRBS survey match well the predictions from beaming models matched to the 34-object 1 Jy sample (i.e. starting from FR I luminosity function). Similarly, the luminosity function of flat-spectrum radio quasars has now been extended a factor $\gtrsim$ 20 lower in luminosity by the DXRBS, and it too matches well predictions based on beaming FR IIs into high-luminosity flat-spectrum radio quasars based on the 2 Jy sample (Padovani). Given the overlapping properties of FR Is and FR IIs, and of BL Lacs and quasars, it seems likely there is one over-arching "grand unification" scheme relating all radio galaxies to all blazars, although we still do not know what this scheme is. Bulk Lorentz factors $\sim$ 10 still explain, at zeroth order, the number statistics, the SED (Chiaberge), superluminal motion (Marscher), and the correlation between core-dominance parameter and optical polarization (Yuan), although there might be some inconsistencies among different methods (Chiaberge).

A few intriguing details remain unresolved, however. Recent work with HST implies that nuclear obscuration (e.g. a torus) may be very rare in FR I radio galaxies, implying that few have hidden broad-emission-line regions (Chiaberge). This would make them physically different from higher-power radio sources, where the presence of an obscuring torus has been inferred from spectropolarimetric observations (broad lines are seen in the polarized light from FR II radio galaxies, presumably scattered from a hidden broad-line region). It would further imply that broad lines should be extremely rare in BL Lac objects, yet they have been observed, including in BL Lac itself. On the other hand, the spectral energy distribution of (at least a couple) FR Is is consistent with that of BL Lacs of the LBL type (Trussoni).

We now fold evolution into the unification picture. The evolutionary properties of BL Lac objects have been puzzling for a number of years. It appeared that the evolution of different samples was quite different, the radio-selected 1 Jy sample evolving in the same sense as quasars (more/brighter at high redshift) and the X-ray-selected EMSS sample evolving in the opposite sense. Participants at the conference brought some welcome clarity to this mess. In general, BL Lac objects evolve little or not at all, regardless of the survey in which they were found (Padovani, Caccianiga, Stocke). There is, however, an intriguing dependence of evolution (as measured by $\langle V_e/V_a \rangle$, the ratio of "enclosed" to "available" volume) on spectral type, with the extreme HBL (very low α_{rx} or equivalently very high f_x/f_r) still showing negative evolution (fewer/dimmer at high redshift; Giommi, Stocke). This could possibly be explained, for example, by the dependence of beaming on apparent luminosity (objects that appear less luminous on average, like HBL, are less beamed), which could cause the (large!) K correction to depend artificially and systematically on apparent luminosity (Giommi).

In contrast, quasars clearly evolve in the positive sense (Padovani, Jackson), and the highest luminosity BL Lacs might have similar evolution. In fact, a thorough analysis of radio sources spanning the full FR I/II power range suggests there is a smooth dependence of evolution on radio power, with the low-power sources showing little or no evolution and the high-power sources showing large, positive evolution—in other words, luminosity-dependent evolution, similar to that seen in radio-quiet quasars (Jackson). An evolutionary connection between FSRQ and BL Lacs has also been suggested (Stocke, D'Elia). This reinforces the previously mentioned "grand unification" of radio sources.

One can in principle extend the unification idea to even lower radio power, into the so-called "radio-quiet" regime. New, deeper radio surveys like the FIRST are finding a unimodal distribution of the radio-to-optical flux ratio—i.e. there do not appear to be two populations of radio sources, one radio-quiet and one radio-loud —in contrast to what has been inferred from large optical samples (e.g. PG and LBQS). A close look at conventionally radio-quiet AGN indeed shows they have some blazar-like properties: flat-spectrum radio cores, large variability, high brightness temperatures, and in a handful of cases, superluminal expansion velocities (Falcke). *Thus AGN might all have similar nuclear engines, producing relativistic radio-emitting jets that in most cases fail to form large, bright radio sources.* Given the demonstrated influence of (multiple) flux limits on BL Lac samples, it is clearly past time to consider similar effects in quasar samples—perhaps the radio-quiet/radio-loud "division" is artificial too.

5. Jet Physics

The observed multiwavelength variability of blazars is complex and thus interpretation in terms of underlying jet physics is not so straightforward. For the high energy synchrotron radiation, both soft and hard lags are seen, in the same object, within the same sequence of short flares (Wagner). Soft lags can be explained by synchrotron cooling providing electron acceleration is rapid, while possibly hard lags can be explained by slow acceleration. In either case, the observed symmetry of flares implies the light crossing time is the dominant time scale, and the absence of prominent plateaus suggests that the electron injection time must be comparable to it. Interestingly, the handful of blazars well studied with long ASCA observations show a break in the structure function at ~ 1 day, similar to the radio and optical light curves of intra-day-variable AGN

(Tanihata, Wagner). This rapid variability offers the strongest evidence for a characteristic time scale in any AGN. There appears to be a connection between X-ray and optical variability and γ-ray and radio variability, at least in some sources (Nesci, Jorstad).

Jets clearly carry considerable kinetic power, sufficient to power extended radio lobes in FR II sources. How this energy is converted to radiation is not completely clear but it must be a relatively low-efficiency process (so as not to tap all the kinetic energy), and it may involve internal shocks, in the manner expected to hold in gamma-ray bursts (Ghisellini, Sikora, Cross). Far less clear is how the energy is extracted from the black hole and transported to the radiative zone. Intrinsic differences might exist in the inner jets of radio sources (Lister).

Most of the jet kinetic energy is probably carried by protons. If carried instead by electrons, there would have to be a large number of cold electrons (with minimum energy $\gamma_{min} \sim 1$). The bulk relativistic outflow in the jet ($\Gamma_{bulk} \sim 10$) would allow these electrons to Comptonize optical/UV photons to soft X-ray energies (Sikora). The absence of such a "bulk Compton bump" has been used to rule out such a low γ_{min}; however, recent interpretations of soft X-ray spectra suggest such features may exist (Celotti). On the other hand, electron-positron jets are easier to slow down, and would probably be more consistent with the relatively slow jets observed, for example, in TeV BL Lacs (Marscher, Piner). Whether or not protons dominate the kinetic energy budget, pairs certainly have to be present, if only because they are formed easily in high-energy-density environments, through proton-proton collisions, proton-induced cascades, and $\gamma\gamma$ collisions (Sikora).

Chandra observations of radio jets will undoubtedly provide critical new information. Already there are some suprises. X-ray jets extend to very large scales, many tens of kiloparsecs in projected distance from the nucleus, up to half a Megaparsec (de-projected) in the most extreme example, PKS 0637−752. This emission is difficult to explain via synchrotron or synchrotron self-Compton models—the energy requirements are too large and/or the spectral shape implies multiple populations of electrons. A viable alternative is that relativistic electrons in the jet inverse-Compton scatter photons in the ambient cosmic microwave background radiation field, provided the jet still has relativistic bulk outflow on such large spatial scales (Tavecchio, Sambruna, Ghisellini). Further Chandra (and HST) observations will test this important hypothesis.

6. Blazar Demographics

The original goal of the conference was to address the demographics of blazars. A seemingly prosaic question—whether there are more "red" or "blue" blazars—actually has profound physical meaning since red blazars might be more powerful than blue, on average. That is, the demographics issue translates to the fundamental question: *What kind of jets does nature make, powerful or weak?* This clearly bears on the radio-quiet issue as well: if radio-quiet quasars are much more numerous, and if (weak) jets are common in the nuclei of radio-quiet AGN, then weak jets have to be more common.

An answer to the demographics question proved elusive, largely due to the overwhelming influence of selection effects but also to the fact that most new samples are still not completely identified. Before the information content of those samples can be interpreted, extensive simulations will also be needed. Preliminary results presented at the conference make clear that quite broad blazar populations—quite a range of jet physical states—are allowed by the current data. Physical limits (e.g. to peak synchrotron frequencies) are still not well constrained by any single sample. With joint constraints from various surveys sampling different regions of flux-flux space and improved simulations, however, it should be possible to definitively answer the demographics question. Thus interesting work still lies ahead.

Acknowledgments. We acknowledge the Space Telescope Science Institute, via the AGN Journal Club, the Visitor Program, and the Director Discretionary Research Fund, for financial support. We thank Quindairian Gryce for expert organization of the meeting and Sharon Toolan for critical assistance with the proceedings. Finally, we enthusiastically thank our blazar colleagues for their valuable participation in this meeting.

Images and Polarization of Compact Jets in Blazars

Alan P. Marscher, Svetlana G. Jorstad

*Institute for Astrophysical Research, Boston University,
725 Commonwealth Ave., Boston, MA 02215*

Abstract. We discuss the results of imaging of compact jets in blazars with the VLBA since its completion in January 1995. Our own work has concentrated on high frequencies—mainly 22 and 43 GHz—and, since 1997, polarized intensity imaging. The apparent speeds of the γ-ray bright blazars are considerably faster than in the general population of bright compact radio sources. The cores of γ-ray blazars are only weakly polarized, with electric-vector position angles (EVPAs) usually within $40°$ of the local direction of the jet. The EVPAs of the jet components are usually within $20°$ of the local jet direction.

1. Introduction

We now know that the defining feature of blazars—a spectrum dominated by rapidly variable, nonthermal emission—is the result of the presence of a highly relativistic jet that flows from the nucleus almost directly into our line of sight. The variability provides us with valuable insight into the structure and physics of jets. Fortunately, at radio frequencies we can directly image these jets, at least as they appear projected onto the sky plane.

The advent of the Very Long Baseline Array (VLBA) has enabled us to obtain high-quality images of compact jets in blazars at resolutions as high as 0.1 milliarcseconds. This corresponds to 0.07 pc in Mkn 501 (redshift $z = 0.034$), 0.14 pc in BL Lac ($z = 0.069$), 0.6 pc in 3C 279 ($z = 0.538$), and 0.9 pc in PKS 0528+134 ($z = 2.06$). [These numbers are calculated for a Hubble constant $H_0 = 65$ km s^{-1} Mpc^{-1} and either $q_0 = 0.1$ or a flat universe with $\Omega(\text{baryons}) = 0.3$ and $\Omega(\Lambda) = 0.7$.] The fact that the VLBA is operational throughout the entire year allows us to monitor the images to detect and follow changes in the jets such as superluminal motion of bright spots. (For a graphic example, see the paper on the radio galaxy 3C 120 by Gómez et al. 2000.) Furthermore, the circularly polarized feeds of the VLBA antennas permit mapping of the polarized as well as the total intensity.

Multi-epoch data from the VLBA therefore provide a rich amount and variety of information that has opened new doors toward understanding the fascinating, complex, and exotic phenomenon of blazar jets. This is only now becoming apparent to the astronomical community at large simply because monitoring programs of objects that vary on timescales of months to years require several years to bear fruit.

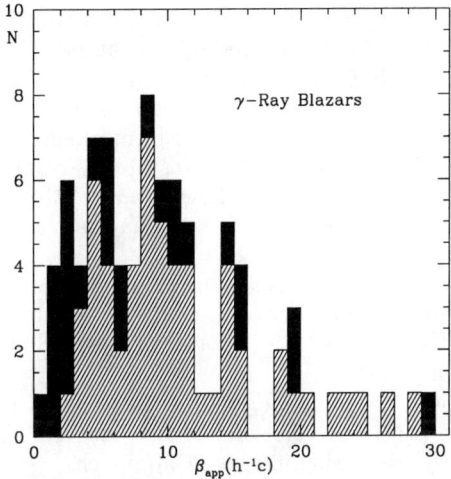

Figure 1. Histogram of apparent velocities of jet components of γ-ray bright blazars with proper motions detected (85 components in 33 sources). The cross-hatched regions are quasars while the solid regions are BL Lac objects.

Here we will discuss some of the results that are now emerging from VLBA studies of blazars. We will focus on our work (with many collaborators) at 22 and 43 GHz, especially with γ-ray bright blazars.

2. γ-ray Bright Blazars

In order to explore the properties of the jets of γ-ray bright blazars at the highest possible resolution, we have monitored 42 EGRET-detected blazars with the VLBA, mostly at 22 and 43 GHz. Starting in 1997, near the end of the project, observations were carried out in dual-polarization mode so that polarized intensity images could be made. In the jets of the γ-ray blazar sample, we have found 43 stationary components in 27 sources and 85 moving components in 33 sources.

The apparent velocity distribution of the moving components is given in Figure 1. The apparent speeds range from sub-c (Mkn 421) to $> 40c$ (0235+164). The peak of the apparent velocity distribution is at $11c$, but there is a long high-velocity tail: about 30% (13) of the sources contain components (18 in all) whose speeds exceed $20c$. The distribution of apparent velocities of the general population of bright, flat-spectrum radio sources is quite different (Pearson et al. 1998), heavily weighted toward low speeds.

In general, for blazars in which more than one superluminal component is detected, the apparent velocities are not the same. For example, in PKS 0528+134 the speeds range from 4 to $13c$. If the separation from the core is deprojected in a crude fashion (e.g. by assuming that the fastest component is moving at the optimal angle for superluminal motion—the result does not depend strongly on the details of the deprojection scheme), there is a positive correlation of apparent speed with linear distance from the core for the sample.

One of the main findings of this study is that standing features are quite common close to the core in compact blazar jets. The most obvious interpretation is that these are "recollimation" shocks from pinch-mode instabilities of a jet confined by an external medium. The instability results from pressure imbalances that inevitably occur if the jet flow is variable (see Gómez et al. 1997). Stationary features with steeper spectra are found farther from the core, often at sites where the jet bends.

Many of the jets of the γ-ray blazars are strongly bent. Examples are shown in Figure 2. Furthermore, a number of the jets of γ-ray blazars are quite broad within a few mas of the core (see Fig. 4 for examples), which suggests that they are being viewed nearly end-on.

Since compact jets are dynamic, one needs to add the dimension of time to receive the full impact of the glorious VLBA images. The reader is referred to Jorstad et al. (2001), in which images at all epochs, as well as separation of jet components from the core vs. time, are plotted. Still more appropriate are movies of the jets made by interpolation between closely spaced (no longer than 3-month intervals) epochs of observation. Such animations can be found for selected objects at web sites http://www.bu.edu/astronomy/people/marscher.html and ../astronomy/research/apm.

Unfortunately, until we can do 43 and 86 GHz VLBI that include one or more space-based antennas—as in the proposed Japanese VSOP 2 and NASA ARISE missions—we cannot directly image the radio cores: the resolution of an Earth-based interferometer is just too low. However, if the suggestion of Daly & Marscher (1988) is right, then the core is a system of standing conical shocks. These have been simulated with 2D+ relativistic hydrodynamical codes by Gómez et al. (1997). Immediately downstream of the conical shocks is a thin but rather long rarefaction, also in the shape of a cone.

The polarization properties of the blazars whose images are shown in Figures 2 and 3 are representative of the entire sample. The core polarizations are very low—all less than 6% and most less than 3%, with 25% (9 blazars) less than 1%. The electric-vector position angles (EVPAs) have a rather broad distribution relative to the local jet direction (see Figure 4, top panels). The EVPAs of the cores therefore tend to favor orientations somewhat oblique to the jet axis. The substantial number of oblique EVPAs might suggest further bending closer to the central engine (so that the fields would be transverse to the inner jet direction), which can be tested with higher frequency or space-VLBI observations at 22 or 43 GHz. The distribution of the EVPAs of the jet components (see Fig. 4, bottom panels) is more skewed, favoring magnetic fields that are more transverse to the jets, although still usually oblique. There is no statistical difference between the polarization properties of γ-ray bright BL Lac objects and quasars, but the number of BL Lac objects in our sample is too small for a meaningful comparison. In the general population of compact radio sources the jet components in BL Lac objects have magnetic fields that are preferentially oriented transverse to the local jet axis (Cawthorne et al. 1993; Lister, these proceedings).

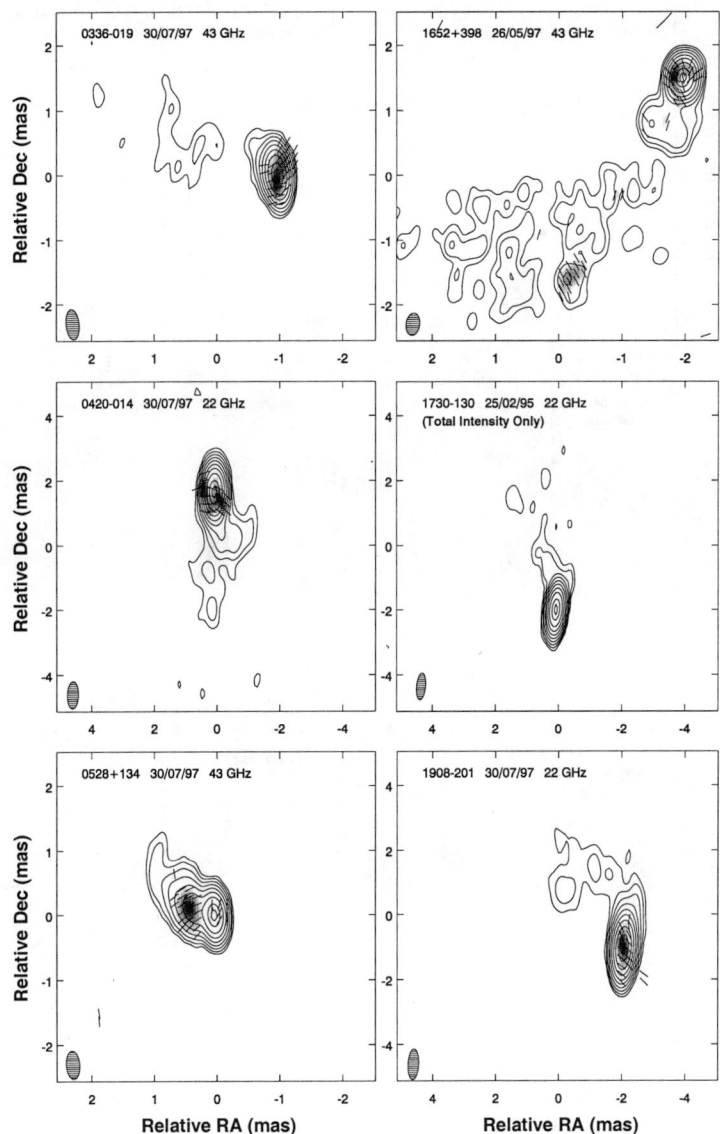

Figure 2. VLBA images of 6 γ-ray blazars with strongly bent jets. The contours correspond to factors of 2 in total intensity, with the highest at 64%. The gray-scale shows polarized intensity, while the lines indicate the direction of the electric vectors. The maximum total intensities in Jy beam^{-1} (fractional polarizations of the cores) are: 0336−019: 1.73 (1.9%); 0420−014: 3.00 (0.5%); 0528+134: 1.74 (0.7%); 1652+398 (Mkn 501): 0.23 (1.2%); 1730−130: 7.81 (—); 1908−201: 2.41 (1.4%).

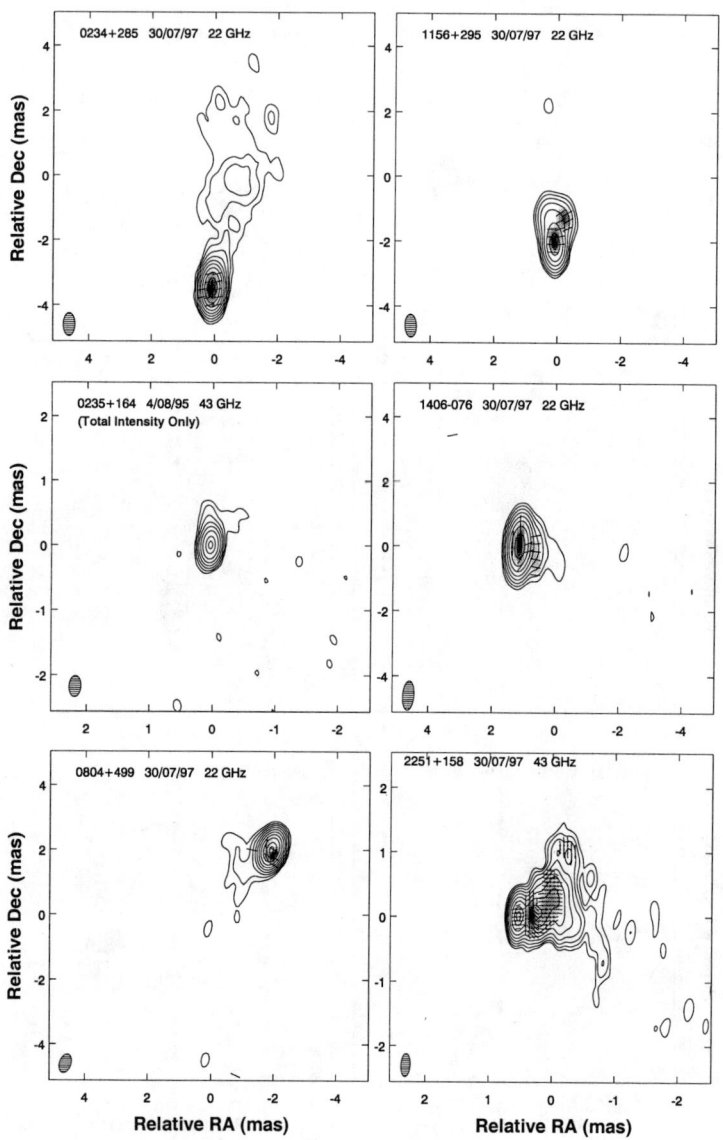

Figure 3. VLBA images of 6 γ-ray blazars with broad jets near the core. The contours correspond to factors of 2 in total intensity, with the highest at 64%. The gray-scale shows polarized intensity, while the lines indicate the direction of the electric vectors. The maximum total intensities in Jy beam^{-1} (fractional polarizations of the cores) are: 0234+285: 1.48 (4.6%); 0235+164: 0.52 (—); 0804+499: 1.10 (0.5%); 1156+295: 0.58 (2.2%); 1406−076: 0.90 (1.8%); 3C 454.3: 1.14 (2.0%).

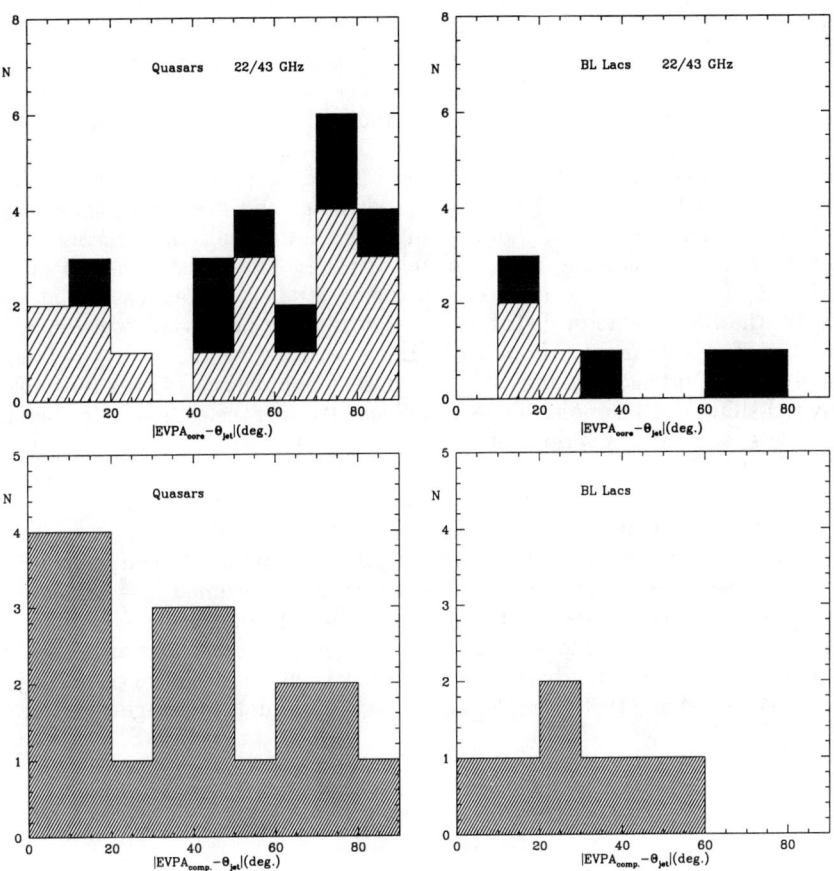

Figure 4. *Upper panels*: Distribution of polarization E-vector position angles of the cores in γ-ray blazars: *left*: quasars; *right*: BL Lac objects. For both panels, the cross-hatched regions correspond to measurements at 22 GHz and the solid regions at 43 GHz. *Lower panels*: The same for jet components in γ-ray blazars.

3. Demographics

It is clear from the results of our VLBA study that γ-ray sources are different from the general population. They are more highly beamed, which as indicated by the high apparent superluminal velocities of the sample. We hope that our work will clear up the confusion wrought by conflicting conclusions of previous studies based on smaller samples or more heterogeneous data (see Jorstad et al., these proceedings, for a brief discussion.)

Another source of confusion arises when inferences are drawn about a sample without considering all the selection effects involved. As the population simulation by Lister & Marscher (1997, 1999) shows, a flux-limited sample of blazars contains a mixture of sources with high luminosities at high redshifts, sources with low luminosities at low redshifts, objects with relatively low luminosities and high levels of relativistic beaming, and everything in between. Statements that are made without taking this heterogeneity into account range from somewhat misleading to completely wrong. One must consider (1) the luminosity function of the phenomenon being studied, in our case relativistic jets, (2) the Lorentz factor distribution—a decreasing power-law reproduces the statistics of superluminal radio sources (Lister & Marscher 1997), (3) the random distribution of orientation angles for the parent population, (4) the (cosmologically red-shifted and Doppler blue-shifted) spectrum of the emission, (5) bending of the jets, and (6) the various observational selection criteria. Time variability might be a factor as well, but it seems to have the effect mainly of swapping one highly variable source for another in terms of meeting the flux limit.

According to Lister & Marscher (1999), the current samples are too small to use them to discriminate between γ-ray emission caused by the synchrotron self-Compton mechanism and the somewhat more highly beamed inverse Compton scattering of photons external to the jet. However, the future GLAST γ-ray mission should provide enough information to sort this out. Nevertheless, it is clear that the jets of γ-ray blazars are different from those typical of the objects in the Pearson et al. (1998) samples: they have much higher average bulk Lorentz factors.

4. Conclusions

The association of γ-ray emission with extremely relativistic jets in blazars is demonstrated by the high mean apparent velocities and the coincidence of γ-ray flares with superluminal ejections and/or polarization events (Jorstad et al., these proceedings). The properties of the γ-ray blazars are consistent with a population that is more strongly beamed than the general population of strong, compact radio sources (Lister & Marscher 1999). The stronger beaming of the γ-rays occurs in the case of Compton scattering of photons from outside the jet because of the blue-shifting of these seed photons in the jet frame. In the SSC case, the extra beaming comes from the K-correction owing to the steeper spectral index (typically ~ -1) of the γ-ray emission relative to the radio emission (~ 0).

The polarization of the cores and jets of both quasars and BL Lac objects favors magnetic fields that lie at oblique angles to the jet direction. This contrasts with the general population of compact radio sources, for which nearly parallel (quasars) or transverse (BL Lacs) fields dominate.

The images shown here reveal cores that are nearly unresolved despite the high frequencies of observation. In order to explore the nature of the cores of blazar jets and to understand the production of energy and acceleration of relativistic electrons in the jets, we need the angular resolution provided by even higher-frequency (e.g. 86 GHz) VLBI with antennas in space, as well as multiwavelength monitoring missions and campaigns. In addition, continued development of theoretical models, in particular 3D relativistic MHD simulations, is required in order to interpret properly the results of such observations.

Acknowledgments. This work was supported in part by NASA through CGRO Guest Investigator grants NAG5-7323 and NAG5-2508, and by U.S. National Science Foundation grant AST-9802941. The VLBA is an instrument of the National Radio Astronomy Observatory, a facility of the National Science Foundation operated under cooperative agreement by Associated Universities, Inc.

References

Cawthorne, T. V., Wardle, J. F. C., Roberts, D. H. & Gabuzda, D. C. 1993, ApJ, 416, 519

Hartman, R. C., et al. 1999, ApJS, 123, 79

Daly, R. D. & Marscher, A. P. 1988, ApJ, 334, 539

Gómez, J. L., Marscher, A. P., Alberdi, A., Jorstad, S. G., & García-Miró, C. 2000, Science, 289, 2317

Gómez, J. L., Martí, J. M., Marscher, A. P., Ibáñez, J. M., & Alberdi, A. 1997, ApJ, 482, L33

Jorstad, S. G., Marscher, A. P., Mattox, J. P. Wehrle, A. E., Bloom, S. D., & Yurchenko, A. V. 2001, ApJ, submitted

Lister, M. L. & Marscher, A. P. 1997, ApJ, 476, 572

Lister, M. L. & Marscher, A. P. 1999, Astroparticle Phys., 11, 65

Marchenko, S. G., Marscher, A. P., Mattox, J. R., Hallum, J., Wehrle, A. E., & Bloom, S. D. 2000, in Proceedings of the 5th Compton Symposium, eds. M. L. McConnell & J. M. Ryan, AIP Conf. Proc., 510, 357

Marscher, A. P. 1980, ApJ, 235, 386

Marscher, A. P. & Gear, W. K. 1985, ApJ, 298, 114

Marscher, A. P., Gear, W. K., & Travis, J. P. 1992, in Variability of Blazars, eds. E. Valtaoja & M. Valtonen (Cambridge: Cambridge Univ. Press), 85

Meier, D. L. & Koide, S. 2000, in Astrophysical Phenomena Revealed by Space VLBI, ed. H. Hirabayashi, P. G. Edwards, & D. W. Murphy (Sagamihara: ISAS), 31

Pearson, T. J., et al. 1998, in Radio Emission from Galactic and Extragalactic Compact Sources, ASP Conf. Ser. Vol. 144, eds. J. A. Zensus, G. B. Taylor & J. M. Wrobel, (San Francisco: ASP), 17

Blazar Demographics and Physics
ASP Conference Series, Vol. 227, 2001
Paolo Padovani and C. Megan Urry, eds.

The Parsec-Scale Structure of TeV Blazars

B. G. Piner

Dept. of Physics and Astronomy, Whittier College, 13406 E. Philadelphia St., Whittier, CA 90608

P. G. Edwards

ISAS, Yoshinodai 3-1-1, Sagamihara, Kanagawa 229-8510, Japan

S. Fodor

Dept. of Physics and Astronomy, Whittier College, 13406 E. Philadelphia St., Whittier, CA 90608

Abstract. We are studying the parsec-scale jet structures of TeV blazars and candidates with VLBI. Results to date show that, despite the high Doppler factors inferred from γ-ray observations, components in the VLBI jets of TeV blazars are significantly slower than those in the EGRET blazars. This suggests a deceleration of the TeV blazar jets between the γ-ray emitting regions and the VLBI scales.

1. Introduction

Detectable fluxes of TeV ($\sim 1\times 10^{12}$ eV) γ-rays have been reported for six AGNs; however, at present only Mrk 421 and Mrk 501 have been independently confirmed by multiple TeV telescopes as sources at these energies. Short-timescale variability of the γ-ray emission implies high Doppler factors in the γ-ray emitting regions (see Catanese & Weekes 1999 and references therein). We are undertaking VLBI observations of these TeV detected sources (with the exception of 3C 66A) to determine if high Doppler factors are also evident in the parsec-scale jet structure.

2. Observations

Mrk 421 and Mrk 501 both have a significant number of archival VLBI observations. For these sources we have combined this archival VLBI data (from the U.S. Naval Observatory Radio Reference Frame Image Database and from the VLBA 2 cm survey) with our own VLBA and VSOP data. All archival data has been independently imaged and model fit by us, starting from the calibrated (u, v) data. The other three sources have fainter VLBI fluxes (~ 200 mJy for PKS 2155−304 and ~ 100 mJy for 1ES 2344+514 and 1ES 1959+650) and had not been previously studied with VLBI. We are monitoring these sources with the NRAO VLBA at 15 GHz.

2.1. Mrk 421

Analysis of 30 VLBI images of Mrk 421 at 15 epochs is presented by Piner et al. (1999). The apparent speeds of three components were measured, and all were found to be subluminal (see Table 1). These measured speeds differed from apparent superluminal speeds obtained from 3 epochs of VLBI data from the early 1980s by Zhang & Bååth (1990).

2.2. Mrk 501

For Mrk 501, we have imaged a total of five epochs of VLBI data: three from the USNO RRFID, one from the VLBA 2 cm survey, and one VSOP observation (Edwards et al. 2000). A total of four jet components can be identified in these observations, their speeds are listed in Table 1. We have an additional four epochs of VLBA 2 cm survey data yet to be included, which will bring the total number of epochs for this source to nine.

2.3. 1ES 2344+514

We observed 1ES 2344+514 four times with the VLBA between 1999 October and 2000 March. All four epochs have been imaged, an image from the latest epoch is shown in Figure 1. A total of three components were identified by model fitting, at each of the four epochs. The measured apparent speeds for these components are given in Table 1.

2.4. PKS 2155−304

PKS 2155−304 is a well-known X-ray selected BL Lac object. We have observed this source with the VLBA at a total of 3 epochs. Only the first epoch has been reduced to date, the image from this epoch is shown in Figure 1.

2.5. 1ES 1959+650

As with PKS 2155−304, 1ES 1959+650 has been observed by the VLBA at 3 epochs, with the first epoch reduced. The image from this epoch is shown in Figure 1. The parsec-scale morphology of this source is intriguing. The jet begins in a southward direction from the presumed core as seen in the uniformly weighted image in Figure 1. In the naturally weighted image, and in a 5 GHz VLBA image (Rector 2000 private communication), continuing emission to the south is not seen, but diffuse emission is present to the north of the presumed core. The jet may start out to the south nearly aligned with the line-of-sight, and then turn back across the line-of-sight toward the north.

3. Discussion

All of these sources are morphologically similar on parsec-scales. They all belong to the 'blue' blazar class, and all have jets that break up into extended, diffuse structure after several milliarcseconds in the VLBI images. All of the VLBI components have relatively slow apparent speeds, with the fastest measured speed being $1.3c$. These speeds are significantly slower than speeds measured for the lower energy 'red' γ-ray blazars by Marscher et al. (2000), as shown by

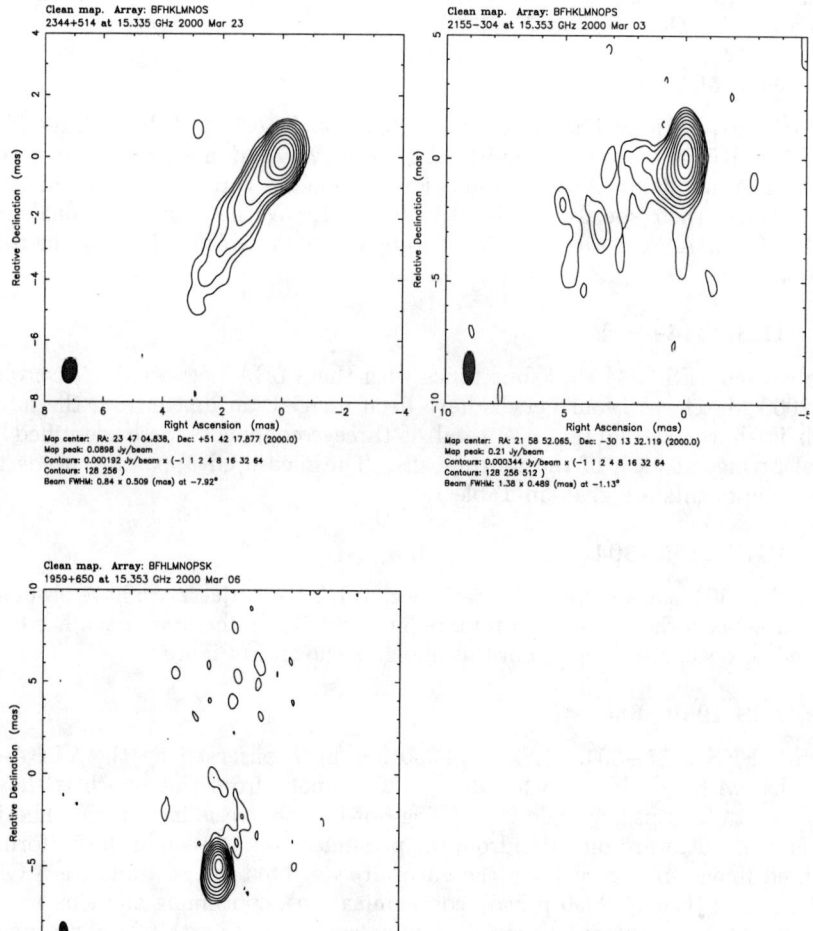

Figure 1. 15 GHz VLBA images of 1ES 2344+514, PKS 2155−304, and 1ES 1959+650. These panels show the uniformly weighted images.

Table 1. Apparent Component Speeds

Source	Comp.[a]	Speed[b] (c)
Mrk 421	C4	0.2 ± 0.3
	C5	0.3 ± 0.1
	C6	-0.1 ± 0.1
Mrk 501	C1	-0.2 ± 0.2
	C2	1.1 ± 0.2
	C3	0.2 ± 0.2
	C4	0.0 ± 0.2
1ES 2344+514	C1	1.3 ± 0.8
	C2	0.5 ± 0.8
	C3	-0.2 ± 0.8

[a] Components are numbered C1, C2, ... from the outermost component inward.
[b] for $H_0 = 65$ km s^{-1} Mpc^{-1}.

a Kolmogorov-Smirnov test applied to the two speed distributions. We consider the most likely explanation to be that the jets are slowing down between the γ-ray emitting regions and the VLBI scales in these TeV sources. A mechanism to decelerate TeV blazar jets is described by Marscher (1999). If the jet is electron-positron with a sufficiently flat electron index, then most of the energy will be carried in the highest energy electrons that lose energy very efficiently to synchrotron and inverse Compton emission, producing the X-rays and TeV γ-rays, and dumping most of the energy and momentum close to the base of the jet. If deceleration of TeV blazar jets is confirmed by additional work, it would be an important argument for light (electron-positron) jets in these sources.

Acknowledgments. We thank Ken Kellermann for providing us with data from the VLBA 2 cm survey. This research has made use of the United States Naval Observatory (USNO) Radio Reference Frame Image Database (RRFID). The VLBA is operated by NRAO with Associated Universities Inc. and is funded by the National Science Foundation.

References

Catanese, M. & Weekes, T. C. 1999, PASP, 111, 1193
Edwards, P. G., Piner, B. G., Unwin, S. C., et al. 2000, in Astrophysical Phenomena Revealed by Space VLBI, eds. H. Hirabayashi, P. G. Edwards & D. W. Murphy, (Sagamihara: ISAS), 235
Marscher, A. P. 1999, Astroparticle Phys., 11, 19
Marscher, A. P., Marchenko-Jorstad, S. G., Mattox, J. R., et al. 2000, in Astrophysical Phenomena Revealed by Space VLBI, eds. H. Hirabayashi, P. G. Edwards & D. W. Murphy, (Sagamihara: ISAS), 39
Piner, B. G., Unwin, S. C., Wehrle, A. E., et al. 1999, ApJ, 525, 176
Zhang, F. J. & Bååth, L. B. 1990, A&A, 236, 47

Host Galaxies of Blazars and Radio Galaxies

Riccardo Scarpa

European Southern Observatory, Alonso de Cordova, Casilla 19001, Santiago, Chile

Abstract. Our view of unification schemes for AGN is rapidly changing thanks to the improvement of available observational data. Here, the current status of unification of BL Lac objects, radio galaxies, and non-radio ellipticals galaxies is reviewed, discussing the global photometric and structural properties for several types of elliptical galaxies. In particular the luminosity, size, ellipticity, isophotal twist, isophotal shape, and isophotal displacement for active and non-active ellipticals are presented. Results show that non-active ellipticals, ellipticals hosting powerful radio sources, and ellipticals hosting BL Lac nuclei are all indistinguishable. This supports a broad unification scheme in which all ellipticals are similar, and therefore host a massive black hole at their center. When conditions are favorable (i.e. gas is available for accretion), nuclear activity starts and we observe either a radio galaxy or a blazar depending on orientation.

1. Introduction

Proving the correctness of unified schemes is of major importance for our understanding of the true nature of AGN, and is among the most lively topics in modern astrophysics. In the current paradigm for radio-loud AGN, BL Lac objects have relativistically outflowing jets oriented nearly along the line of sight (Urry & Padovani 1995). Strong relativistic beaming of the jet emission then alters the observed properties of these blazars, so that radio-loud AGN pointing at different angles are seen as quasars or radio galaxies—this is the so-called "unification" picture.

Preliminary studies on host galaxy properties soon identified blazars with radio galaxies (Schwartz & Ku 1983; Perez-Fournon & Biermann 1984; Ulrich 1989; Browne 1989). More specifically, BL Lacs were identified with low-luminosity Fanaroff & Riley I (FR I; Fanaroff & Riley 1974) radio galaxies, while flat spectrum radio quasars (FSRQ) were linked to powerful FR II radio galaxies.

Subsequent studies have supported this hypothesis (Urry & Padovani 1995, and references therein). However, as is often the case in nature, the picture seems not so simple and some authors have proposed more complex scenarios in which the BL Lacs' parent population includes only some FR Is (e.g. Wurtz, Stocke, & Yee 1996), or a mix of FR I and FR II (Kollgaard et al. 1996). Indeed, the extended radio morphologies of BL Lacs can be of both FR I and II types (Kollgaard et al. 1992), and some of the line emission from FR IIs is weak enough to be BL Lac-like (Laing et al. 1994). Thus, it may be appropriate to unify blazars more generally with radio galaxies.

Our understanding of unification is, however, changing thanks to the Hubble Space Telescope (HST), which has given an unprecedented detailed view of galaxies and AGN host galaxies (Disney et al. 1995; Bahcall et al. 1997; Hooper, Impey, & Fultz 1997; Falomo et al. 1997, 2000; Malkan, Gorjian, & Tam 1998; Urry et al. 1999, 2000; McLeod, Rieke, & Storrie-Lombardi 1999; McLure et al. 1999; Scarpa et al. 2000a,b).

It is now well established (Scarpa et al. 2000a; Urry et al. 2000; Kotilainen et al. 1998a,b) that blazars are systematically hosted by elliptical galaxies, as is the case for radio galaxies. Moreover, it seems now clear that most galaxies, if not all, host a massive black hole in their center, with mass proportional to the bulge mass (Kormendy & Richstone 1995; Magorrian et al. 1998). If this is correct, than the most obvious picture is that all ellipticals have the potential to experience a phase of nuclear activity (see Scarpa & Urry, these proceedings), and become radio galaxies or blazars depending on their orientation. The differences between FR I and FR II, and between BL Lacs and FSRQ can then be due, for example, to differences in the stage of evolution and/or accretion rate, rather than to basic structural differences. As we shall show, there is now good observational evidence to support this picture, which is simple and for this reason attractive. We compare results from the HST survey of BL Lac objects (Scarpa et al. 2000a; Urry et al. 2000; Falomo et al. 2000), with similar data for radio galaxies (both FR I and FR II; Govoni et al. 2000; Ledlow & Owen 1995), and control samples of non-radio elliptical galaxies (Fasano & Vio 1991; Fasano & Bonoli 1989). In particular, in Falomo et al. (2000), a full two-dimensional analysis of the host galaxy of 30 low redshift BL Lacs was presented, providing for the first time a statistically large sample of host galaxies with reliable determination of ellipticity, isophotal shape, isophotal displacement, and isophotal twist. This allows an unprecedentedly detailed comparison of the properties of extremely active galaxies with similar properties for non-active ellipticals.

2. Comparing Radio-Quiet Ellipticals, Radio Galaxies, and BL Lac Hosts

2.1. Size and Luminosity

In Urry et al. (2000) the luminosity and size distributions of BL Lac hosts and radio galaxies are fully discussed. It is found that both classes share the same luminosity and size, this being true on average and differentially across the whole redshift range explored ($0 \lesssim z \lesssim 0.5$; Fig. 1). Virtually without exception, these galaxies are giant ellipticals, with average absolute magnitude $M_R \sim -23.5$ and average effective radius $R_e \sim 10$ kpc, comparable to brightest cluster members and at least one magnitude brighter than M* galaxies.

Plotting the surface brightness μ_e at the effective radius r_e, versus the effective radius (the Kormendy relation), it is found that active and non-active ellipticals follow the same relation (Fig. 1). This is true both in the optical and near-infrared (Scarpa et al. 2000b).

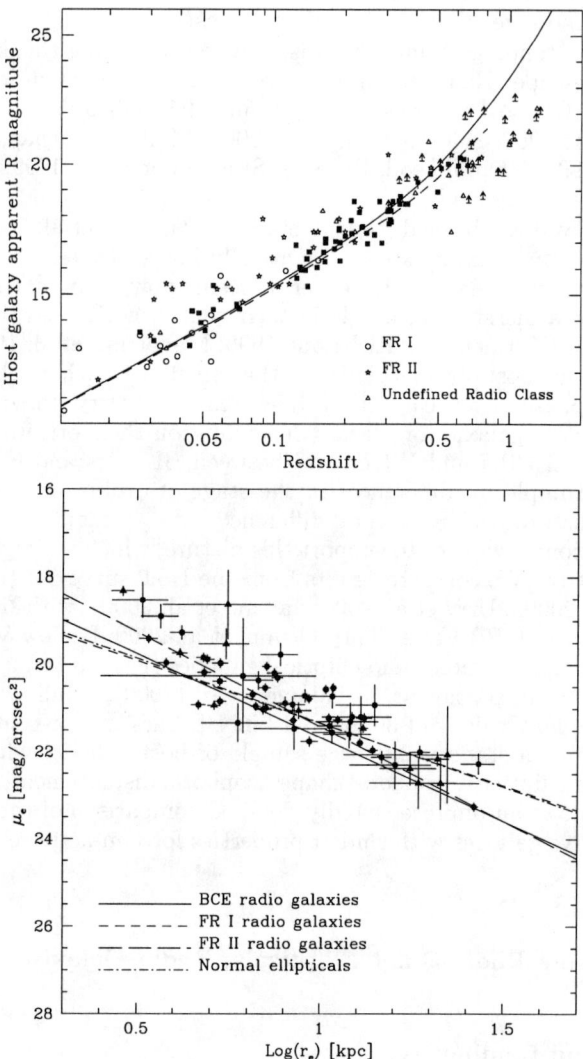

Figure 1. **Upper Panel:** Apparent luminosity for BL Lac host galaxies (solid squares) and radio galaxies from the 2 Jy sample (Morganti et al. 1993), as a function of distance. The solid line is the locus for a galaxy of constant magnitude, while the dashed line is the locus for a galaxy of constant luminosity undergoing passive evolution. It is seen that radio galaxies (irrespective of the radio morphology) and BL Lacs' hosts have the same luminosity at all redshifts. **Lower Panel:** Kormendy relation in the R band for BL Lacs host galaxies (solid symbols) compared to several kind of elliptical galaxies. In all cases, the relations are consistent with each other.

2.2. Ellipticity Distribution

The distribution of galaxy ellipticity at the effective radius for various samples of ellipticals is present in Fig. 2. Data are for 200 radio-quiet ellipticals taken from Fasano & Vio (1991), 79 radio galaxies from Govoni et al. (2000), and for 30 low z BL Lacs from Falomo et al. (2000). It is found that the three samples have statistically similar distribution of ellipticity with average around $<\epsilon> = 0.2$.

Interestingly, the distribution for BL Lac hosts is slightly shifted toward smaller values of ellipticity (but consistent with being drawn from the same population; $P_{KS} \sim 0.1$), which suggests that BL Lacs are preferentially seen face-on, as expected in unified models.

2.3. Isophotal Displacement and Nuclear De-Centering

Isophote displacement with respect to the galaxy center may represent either a global distortion of the host galaxy as due to recent tidal interactions (e.g. Aguilar & White 1986), or a consequence of gravitational microlensing (Ostriker & Vietri 1985). To quantify this asymmetry we plot in Fig. 2 the dimensionless parameter $\delta \equiv \sqrt{(X - X_o)^2 + (Y - Y_o)^2}/r$, where X, Y are the centers a given isophote, X_o, Y_o is the location of the nucleus, and r is the semi-major axes of the isophote under consideration. Data refer to low-redshift BL Lacs and radio galaxies, and give for each object the value of δ at the effective radius of the galaxy. Both samples display a modest deviation (2σ) of δ from zero, suggestive of recent gravitational interaction. However, what is most relevant for unification schemes is that both samples have the same distribution.

Interestingly, in BL Lacs no systematic displacement of the galaxy with respect to the host galaxy is observed, contrary to the expectation of microlensing (Ostriker & Vietri 1985, 1990). This result reinforces the unification scenario.

2.4. Isophotal Twisting

Isophotal twisting is another indicator of recent galaxy interaction (Kormendy 1982) and has been used to show that active galaxies have peculiar structure (Colina & de Juan 1995). To test whether isophotal twisting is a specific property of active galaxies, the total variation of position angles $\Delta\theta$ over the whole range of observed surface brightness is presented in Fig. 2. Data for BL Lacs are restricted to host galaxies having ellipticities larger than 0.15, so that the position angles of the isophotes can be measured precisely. Data for a sample of 43 isolated non-radio ellipticals are taken from Fasano & Bonoli (1989), while data for radio galaxies are from Govoni et al. (2000). In all cases the same method to derive $\Delta\theta$ was used.

A strict similarity between different types of ellipticals is observed. BL Lac host galaxies, if any, seem to be less affected by isophotal twisting than the other ellipticals, again consistent with them being seen face-on.

2.5. Isophotal Shapes: Disky and Boxy

The amplitude of the fourth cosine component, C_4, of the Fourier fit to the isophotes (e.g. Bender & Saglia 1998) can be used to check for small deviations from purely elliptical isophotes. A significant positive value of C_4 corresponds to *disky*-shaped isophotes while a negative value indicates a *boxy*-shaped structure.

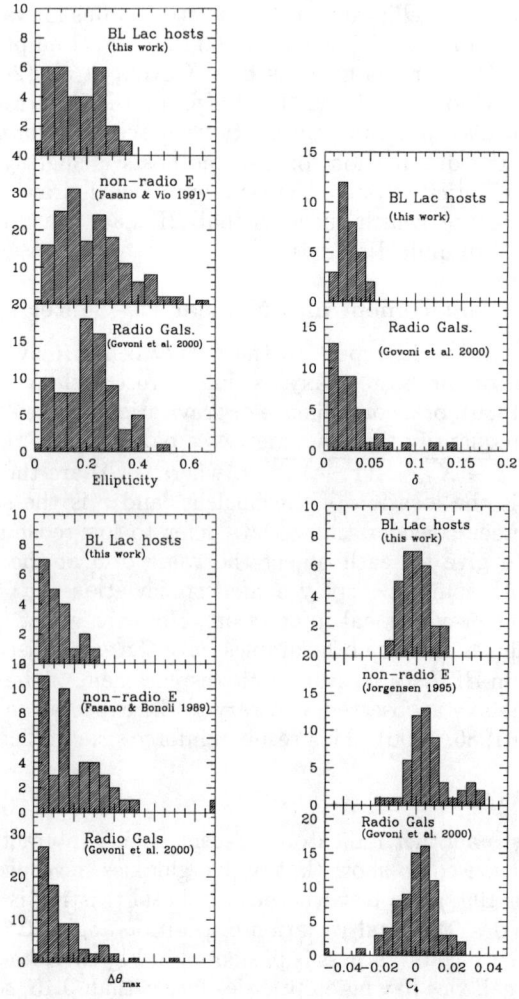

Figure 2. Structural parameters for several types of elliptical galaxies. Data for BL Lac objects are for the 30 sources at $z < 0.2$ from the HST survey (Falomo et al. 2000). **Upper Left:** Ellipticity distribution at the effective radius. **Upper Right:** Isophotal displacement, i.e. variation of the position of the center of the isophotes with respect to the galaxy center, normalized to the radius of the isophote. **Lower Left:** Distribution of isophotal twisting, i.e. total variation of position angle in degrees from 0 to 90 of the isophotes major-axes. **Lower Right:** Distribution of C_4. A positive value of C_4 corresponds to *disky*-shaped isophotes while a negative value indicates a *boxy*-shaped structure.

BL Lac host galaxies have in 80% of cases C_4 amplitude (at the effective radius) smaller than 1% with no object exhibiting $|C_4|$ larger than 3% (Fig. 2). The distribution of C_4 is symmetric around zero suggesting that there is no systematic deviation of isophotal shape from a simple ellipse. Similar distributions are observed in a sample of low-redshift radio galaxies analyzed in the same way ($<C_4>_{RG}= 0.04\%$ ±1.2%; Govoni et al. 2000), a sample of radio galaxies in rich clusters (Ledlow & Owen 1995), as well as normal ellipticals (Jorgensen et al. 1995). Therefore, if some boxiness or diskiness is present in giant ellipticals, this occurs regardless of the nuclear activity.

2.6. Host Galaxy Integrated Color and Color Gradient

Two well-known characteristics of elliptical galaxies are their homogeneity in integrated colors and the existence of a color gradient in the sense that that makes galaxies get bluer going from the center to the edge.

In Kotilainen et al. (1998b), Urry et al. (1999) and Scarpa et al. (2000b) integrated R–H color and color profiles for a dozen BL Lac host galaxies were presented. Results show that the average BL Lac host galaxy color is R–H=2.3± 0.3 mag, consistent with the value reported for a sample of 12 nearby normal elliptical galaxies, R–H=2.3±0.2 mag (Peletier et al. 1990), and with $\langle R-H \rangle =$ 2.3±0.3 mag derived for a number of 3C radio galaxies (Lilly & Longair 1982; see Scarpa et al. 2000b for details on this value). Similar colors ($\langle R-H \rangle = 2.4$ mag) were also reported by De Vries et al. (1998) for a different sample of radio galaxies of various morphological type at $\langle z \rangle \sim 0.3$.

Comparing color profiles, it is again found that BL Lac host galaxies are indistinguishable from other, less active ellipticals. An average color profile for eight host galaxies was presented in Scarpa et al. (2000b). This has slope of $\Delta(R-H)/\Delta \log r = -0.2 \pm 0.1$ mag (Fig. 3). These data offer strong evidence that BL Lac host galaxies become bluer away from the galaxy center.

Color gradients of the same sign have been found in several studies of normal elliptical galaxies (Boroson et al. 1983; 1987; Cohen 1986; Peletier et al. 1990; Munn 1992). A value of $\Delta(V-K)/\Delta \log r = -0.16 \pm 0.18$ mag was found by Peletier et al. (1990). Given the median redshift of the BL Lac sample ($z = 0.2$), the R-band profiles are roughly equivalent to V-band data at $z = 0$; therefore, assuming H and K profiles are similar (since the red stars evolve only slowly), the V–K gradient measured by Peletier et al. is roughly equivalent to the R–H color gradient presented by Scarpa et al. (2000b). The fact that the two values are similar therefore underscores the similarity of BL Lac host galaxies and normal ellipticals. Specifically, the stellar populations of active and non-active ellipticals must be similar not only in an integrated sense, but differentially across the galaxy.

3. Summary and Conclusions

A detailed comparison of several properties of elliptical galaxies of different types shows that these galaxies have very homogeneous properties. In every respect studied, galaxies hosting very active nuclei appear to be indistinguishable from non-active ellipticals. Given the high precision of the HST data, this result is largely free of confusion from the bright nucleus. Had there been dust lanes

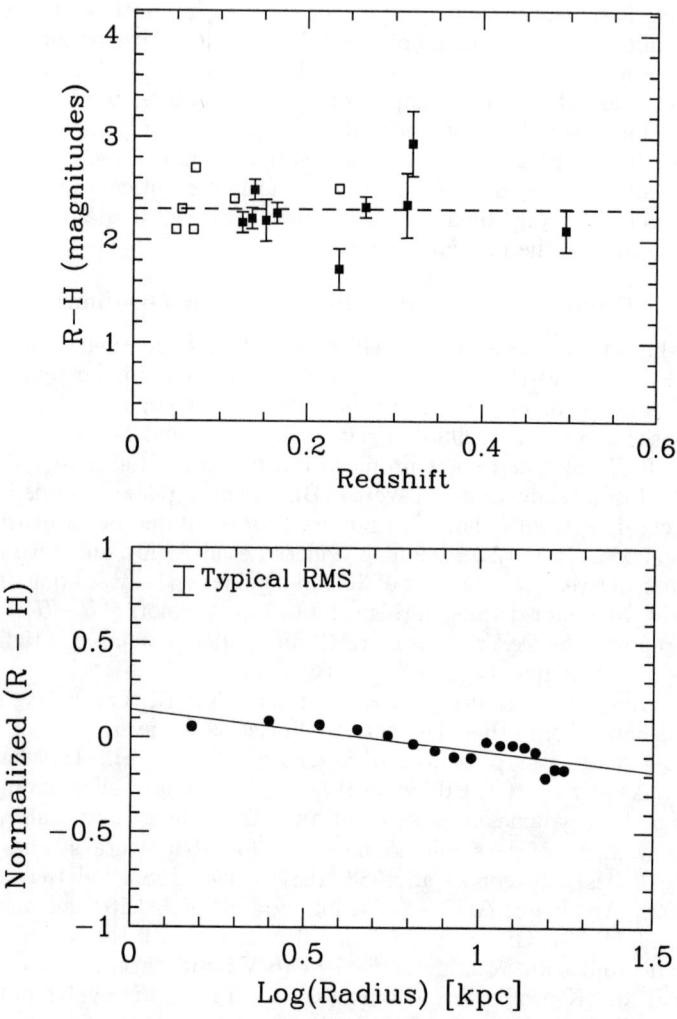

Figure 3. Integrated R–H color and color gradient for BL Lac host galaxies from NICMOS and WFPC2 images (Scarpa et al. 2000b; solid symbols), and from Kotilainen et al. (1998b, open squares). **Upper Panel:** Integrated rest-frame (that is extinction and K-corrected) R–H color for BL Lacs host galaxies. The dashed line is the color expected for a normal, non-active elliptical galaxy. **Lower Panel:** Average R–H color profile for eight BL Lac host galaxies. Profiles were combined after first subtracting their average color, then computing the weighed average of the points on intervals of 1 kpc. The intrinsic dispersion of the points in each bin, due to object to object color variations, is shown at the top-left corner of the figure. The best fit slope (*solid line*) is $\Delta(R-H)/\Delta \log r = -0.2 \pm 0.1$ mag, fully consistent with results for non-active elliptical galaxies.

or other morphological peculiarities beyond ∼ 0.2 arcsec from the center, they would have been readily apparent. It may be concluded that to first order, the host galaxy "knows" nothing about the active nucleus it harbors. This result argues against AGN and galaxies being separate entities, and instead supports a picture in which all galaxies can be AGN but with a relatively low duty cycle. This implies all galaxies (at least, elliptical galaxies) have super-massive black holes at their centers, as indeed appears to be the case in many nearby ellipticals (Richstone et al. 1998; van der Marel 1999).

One thing that remains to be understood is why basically only giant ellipticals are seen in (radio-loud) activity? If ellipticals are all the same, as seems to be the case, then it is natural to expect that small ellipticals (like M32) should also experience a phase of activity. But we do not observe such hosts of radio-loud AGN. The answer to this question may rest in the proposed correlation between bulge mass and central black hole mass (Kormendy & Richstone 1995; Magorrian et al. 1998). If this correlation holds, then it is plausible to expect that only the most massive black holes can produce the enormous energies required to form and sustain an AGN jet. Giant ellipticals are often surrounded by smaller companion galaxies, so it is plausible to imagine that every once in a while one of them is captured and somehow funneled toward the central black hole. In order not to alter the global structure, color, and color gradient of the main galaxy, in the vast majority of cases the nuclear activity must be triggered by the merging of very small companion galaxy.

Present data therefore support a global unification picture for elliptical galaxies and AGN, which is simple and appealing. In this scenario, a single process of galaxy formation is required for all ellipticals, which has to end up producing galaxies with a massive black hole in their center. Then, every time there is accretion of gas, the galaxy becomes active, otherwise it remains quiet. As shown in Scarpa & Urry (these proceedings), the probability that a given galaxy is active seems to be proportional to the square of its luminosity, or, assuming a constant mass-to-light ratio, the square of the mass. This proportionality must reflect both the probability of initiating the nuclear activity and the length of the active period.

Whatever the truth, all observational facts consistently supports a picture in which all galaxies have the potential of being active with a certain probability and duty cycle. This in turn suggests the formation and evolution of galaxies occurred hand in hand with the growth of super-massive black holes at their centers.

Acknowledgments. This work was done in collaboration with Meg Urry at STScI. Additional collaborators include Renato Falomo, Matt O'Dowd, Joe Pesce, and Aldo Treves.

References

Aguilar, L. A. & White, S. D. M. 1986, ApJ, 307, 97
Bahcall, J. N., Kirhakos, S., Saxe, D. H., & Schneider, D. P. 1997, ApJ, 479, 642

Bender, R. & Saglia, R. P. 1998, in Galaxy Dynamics, ASP Conf. Ser. Vol. 182, eds. D. R. Merritt, M. Valluri, & J. A. Sellwood, (San Francisco: ASP), 113

Boroson, T. A., Thompson, I. B., & Shectman, S. A. 1983, AJ, 88, 1707

Boroson, T. A. & Thompson, I. B. 1987, AJ, 93, 33

Browne, I. W. A. 1989, in BL Lac Objects, eds. L. Maraschi, T. Maccaro, & M. H. Ulrich, (Berlin: Springer-Verlag), 401

Cohen, J. G. 1986, AJ,92, 1039

Colina, L. & de Juan, L. 1995, ApJ, 448, 548

de Vries, W. H., O'Dea, C. P., Baum, S. A., & Perlman, E. 1998, ApJ, 503, 156

Disney, M. J., Boyce, P. J., Blades, J. C., et al. 1995, Nature, 376, 150

Fanaroff, B. & Riley, J. M. 1974, MNRAS, 167, 31

Falomo, R., Urry, C. M., Pesce, J. E., Scarpa, R., Giavalisco, M., & Treves, A. 1997, ApJ, 476, 113

Falomo, R., Urry, C. M., Pesce J. E., Scarpa, R., & Treves, A. 2000 ApJ, in press (astro-ph/0006388)

Fasano, G. & Bonoli, C. 1989, A&AS, 79, 291

Fasano, G. & Vio, R. 1991, MNRAS, 249, 629

Govoni, F., Falomo R., Fasano, G., & Scarpa, R. 2000, A&A, 353, 507

Hooper, E. J., Impey, C. D., & Foltz, C. B. 1997, ApJ, 480, L95

Jorgensen, I., Franx, M., & Kjaergaard, P. 1995, MNRAS, 273, 1097

Kollgaard, R. I., Palma, C., Laurent-Muehleisen, S. A., & Feigelson, E. D. 1996, ApJ, 465, 115

Kollgaard, R. I., Wardle, J. F. C., Roberts, D. H., & Gabuzda, D. C. 1992, AJ, 104, 1687

Kormendy, J. 1982, in Morphology and Dynamics of Galaxies, 12th Advanced Course of the Swiss Society of Astronomy and Astrophysics, eds. L. Martinet & M. Mayor (Sauveryny: Geneva Obs.), 113

Kormendy, J. & Richstone, D. 1995, ARA&A, 33, 581

Kotilainen, J. K., Falomo, R., & Scarpa, R. 1998a, A&A, 332, 503

Kotilainen, J. K., Falomo, R., & Scarpa, R. 1998b, A&A, 336, 479

Laing, R. A. 1994, in The Physics of Active Galaxies, ASP Conf. Ser. Vol. 54, eds. G. V. Bicknell, M. A. Dopita, & P. J. Quinn, (San Francisco: ASP), 227

Ledlow, M. J. & Owen. F. N. 1995, AJ, 109, 853

Lilly, S. J. & Longair, M. S. 1982, MNRAS, 199, 1053

Malkan, M. A., Gorjian, V., & Tam, R. 1998, ApJS, 117, 25

Magorrian, J., Tremaine, S., Richstone, D., Bender, R., Bower, G., Dressler, A., Faber, S. M., Gebhardt, K., Green, R., Grillmair, C., Kormendy, J., & Lauer, T. 1998, AJ, 115, 2285

McLeod, K. K., Rieke, J. H., & Storrie-Lombardi, L. J. 1999, ApJ, 511, L67

McLure, R. J., Kukula, M. J., Dunlop, J. S., Baum, S. A., O'Dea, C. P., & Hughes, D. H. 1999, MNRAS, 308, 377

Morganti, R., Killen, N. E. B., & Tadhunter, C. N. 1993, MNRAS, 263, 1023
Munn, J. A. 1992, ApJ, 399, 444
Ostriker, J. P. & Vietri, M. 1985, Nature, 318, 446
Ostriker, J. P. & Vietri, M. 1990, Nature, 344, 45
Peletier, R. F., Valentijn, E. A., & Jameson, R. F. 1990, A&A, 233, 62
Perez-Fournon, I. & Biermann, P. 1984, A&A, 130, L13
Richstone, D., et al. 1998, Nature, 395, 14
Scarpa, R., Urry, C. M., Falomo, R., Pesce, J. E., & Treves, A. 2000a, ApJ, 532, 740
Scarpa, R., Urry, C. M., Padovani, P., Calzetti, D., & O'Dowd, M. 2000b, ApJ, 544, 258
Schwartz, D. A. & Ku, W. H. M. 1983, ApJ, 266, 459
Ulrich, M. H. 1989, in BL Lac Objects, eds. L. Maraschi, T. Maccacaro, & M. H. Ulrich, (Berlin: Springer-Verlag), 45
Urry, C. M. & Padovani, P. 1995, PASP, 107, 803
Urry, C. M., Falomo, R., Scarpa, R., Pesce, J. E., Treves, A., & Giavalisco, M. 1999, ApJ, 512, 88
Urry, C. M., Scarpa, R., O'Dowd, M., Falomo, R., Pesce, J. E., & Treves, A. 2000, ApJ, 532, 861
van der Marel, R. 1999, AJ, 117, 744
Wurtz, R., Stoke, J. T., & Yee, H. K. C. 1996, ApJS, 103, 109

Deep NOT Imaging of Radio Selected BL Lac Objects

T. Pursimo, K. Nilsson, A. Sillanpää, L. O. Takalo

Tuorla Observatory, University of Turku, 21500 Piikkiö, Finland

J. Heidt

Landessternwarte, Königstuhl, D-69117 Heidelberg, Germany

Abstract. We present deep optical imaging of the radio selected BL Lac sample obtained with the Nordic Optical Telescope (NOT). In eleven out of 25 observed objects we could detect underlying fuzz. Elliptical host galaxy was preferred in all low redshift objects. Our results, including luminosity and half light radius are similar to the recent ground based and HST results on host galaxy studies. Finally, we found no evidence of a correlation between the half light radius and rest frame wavelength between 4000 and 6000 Å.

1. Introduction

Optical imaging is a powerful tool to study the relationship between BL Lac objects and other classes of active galactic nuclei (AGN). Deep imaging can be used in studies of the morphology of host galaxies, close environments and clustering properties. Based on the unified scheme BL Lacs are FRI-radio galaxies seen along the jet (see review by Urry & Padovani 1995). This predicts that the host galaxies of BL Lacs and FRIs are similar.

Typically BL Lac host galaxies are slightly larger and brighter than field ellipticals ($< M_{\text{host}} > \sim M_* - 1$, $r_e \sim 10$ kpc) (see review by Heidt 1999). In previous studies most host galaxies have been found to have elliptical morphology, however there are some exceptions when disc-type galaxy has been preferred (e.g. OQ530 Abraham et al. 1991). A large number of peculiar BL Lac host galaxies or a significant difference between the assumed parent population, would challenge the unified model. In this study properties of the host galaxies of the 1 Jy sample BL Lacs (Stickel et al. 1994) were studied. Recently, two other BL Lac host galaxy studies have been published: a sample of EMSS/Slew objects by Falomo & Kotilainen (1999, hereafter FK99) and HST snapshot survey (Scarpa et al. 2000; Urry et al. 2000; Scarpa, these proceedings).

2. Observations and Data Reduction

The observations were carried out at the 2.56m NOT mostly with under subarcesecond seeing. The large majority of the images were obtained using the ALFOSC instrument. Altogether 25 objects out of 37 were observed from the

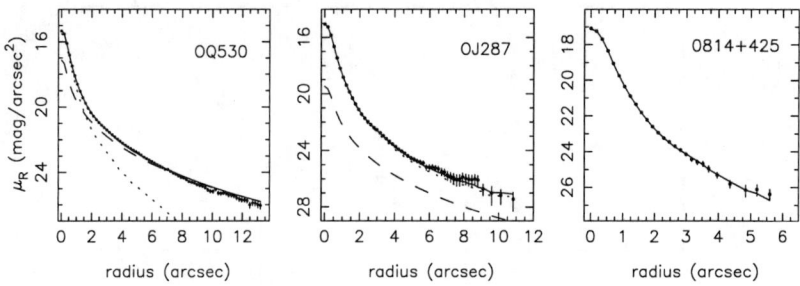

Figure 1. Examples of surface brightness profiles for clearly resolved, marginally resolved and unresolved objects. The dotted line shows the scaled PSF, dashed line the best fit elliptical galaxy model and solid line the sum of the two components.

1 Jy sample (Stickel et al. 1994). Data analysis was made using PROFIT package, which was kindly provided by R. Scarpa. See more details of the observations and data analysis from Pursimo et al. (2000). In addition, we present B-band data for three low redshift 1 Jy objects (OQ 530, Mkn 501 and AP Lib). Unfortunately the nights were nonphotometric.

Analysis of the radial profiles in the present data, EMSS/Slew and HST snapshot data, were carried out in a similar way. The adopted cosmology is $H_0 = 50$ km/s/Mpc and $q_0 = 0$. It should be noted that the present 1 Jy observations and the FK99 data form a homogeneous sample including the instrumentation, analysis and astronomical corrections.

Table 1. R-band and B-band Half Light Radius

object	$r_e['']$	$r_e[kpc]$	$r_e['']$	$r_e[kpc]$
	R-band		B-band	
OQ530	2.32	8.3	2.99	10.7
Mkn 501	17.21	15.7	13.62	12.5
AP Lib	6.72	8.9	6.58	8.7

3. Results

We were able to resolve the host galaxy in eleven objects, of which in one case (OJ287) the detection is marginal. Figure 1 shows examples of a well resolved (OQ530), marginally resolved (OJ287) and unresolved object (0814+425). All objects with $z \lesssim 0.4$ were resolved and all objects with $z \gtrsim 0.7$ or unknown redshift were unresolved. These results are very similar to those of the HST snapshot survey. Table 2 summarises the average host galaxy properties of the present study and other recent studies. There is a striking similarity between the average host galaxy magnitude in our 1 Jy NOT, HST and EMSS/Slew samples. These results also agree well with the values of a typical FRI host galaxy.

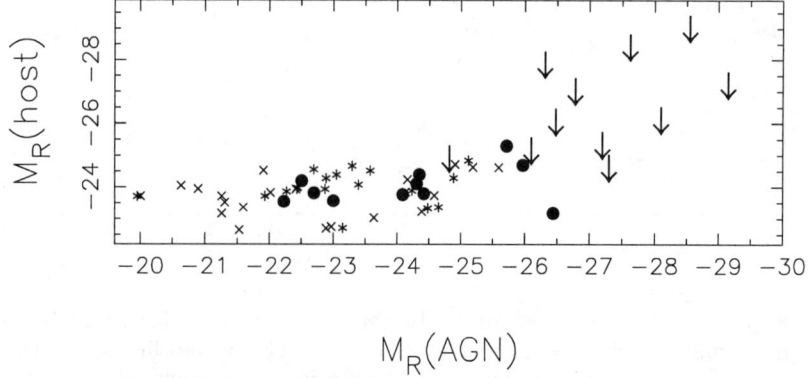

Figure 2. Absolute brightness of the host galaxy vs. the AGN-nucleus. Resolved 1 Jy objects are indicated with filled dots and upper limits with an arrow. The results of FK99 are shown with asterisks (C-EMSS sample) and crosses.

The host galaxy morphologies agree well with the recent HST and FK99 results. In all well resolved objects, including OQ530, elliptical host is preferred. Only in one case (S5 2007+777) a disc-type model gave slightly better fit than an elliptical model, however neither failed to give good fit along the whole extracted profile.

Table 3. The average brightness and size of a BL Lac host galaxy based on various imaging studies. Column two gives the number of observed objects and the next column resolved objects with known redshift. The bottom line refers to a study of FRI type radio galaxies.

Sample	obs	res.	$< M_{host} >$	$< r_e >$	Ref.
1 Jy	34	7	-23.90 ± 0.32	12.4 ± 5.1	Stickel et al. 1993
$z < 0.5$	50	41	-23.7 ± 0.7	10.3 ± 8.3	Wurtz et al. 1996
HST deep	9	7	-23.85 ± 0.54	9.3 ± 6.5	Urry et al. 1999
EMSS/Slew	52	42	-23.92 ± 0.59	7.9 ± 4.9	FK99
HST snap	110	65	-23.71 ± 0.56	8.5 ± 5.6	Urry et al. 2000
1 Jy	25	11	-23.82 ± 0.60	14.8 ± 8.7	Pursimo et al. 2000
FR I	79	79	-24.11 ± 0.69	14.0 ± 9.7	Govoni et al. 1999

The correlation between the AGN-nucleus and the host galaxy has been reported for QSOs (e.g. McLure et al. 1999). In Figure 2 we show the absolute magnitude of the host galaxy vs. AGN-nucleus. There is a weak indication of a trend, however, it is not statistically significant. Also the two BL Lac samples, 1 Jy and EMSS/Slew, have indistinguishable distributions. Again, ground based results are consistent with the HST data.

Urry et al. (2000) found a tendency that the size of the galaxy increases with the redshift. Figure 3 shows the relation between the best fit half light

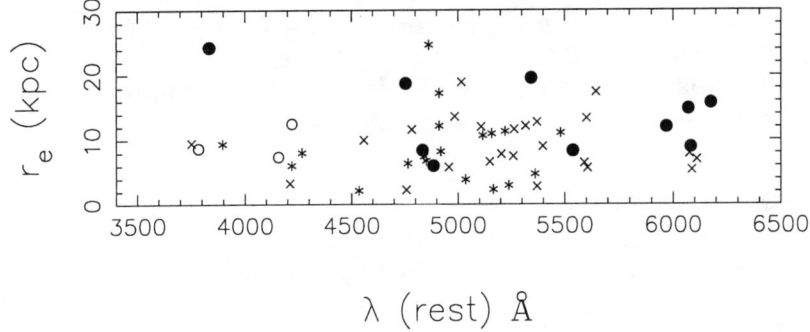

λ (rest) Å

Figure 3. Best fit elliptical galaxy model half light radius vs. the rest-frame wavelength. The symbols are the same as in Fig. 2, in addition the new B-band data is shown by circles.

radius as a function of objects rest-frame wavelength. There is no evidence of any correlation between 4000–6000 Å. The B-band data of the low-z objects show similar values as the values of the more distant R-band data.

Our results suggest that the optical properties of nearby BL Lacs are independent on the peak frequency of the blazar nucleus. These results combined with the FK99 data agree well with the results of the HST snapshot survey.

Acknowledgments. Thanks to Riccardo Scarpa for the use of the PROFIT package. This work was partly supported by the Academy of Finland.

References

Abraham, R. G., et al. 1991, MNRAS, 252, 482
Falomo, R. & Kotilainen, J. K. 1999, A&A, 352, 85 (FK99)
Govoni, F., et al. 2000, A&A, 353, 507
Heidt, J. 1999, in ASP Conf. Ser. Vol. 159, BL Lac Phenonemon, eds. L. O. Takalo & A. Sillanpää, (San Francisco: ASP), 367
McLure, R., et al. 1999, MNRAS, 308, 377
Pursimo, T., et al. 2000, A&A, submited
Scarpa, R., et al. 2000, ApJ, 532, 740
Stickel, M., et al. 1993, A&AS, 98, 393
Stickel, M., et al. 1994, A&AS, 105, 211
Urry, C. M. & Padovani, P. 1995, PASP, 107, 803
Urry, C. M., et al. 1999, ApJ, 512, 88
Urry, C. M., et al. 2000, ApJ, 532, 816
Wurtz, R., et al. 1996, ApJS, 103, 109

Intrinsic Differences in the Inner Jets of High and Low Optically Polarized Radio Quasars

Matthew L. Lister

Jet Propulsion Laboratory, California Institute of Technology, MS 238-332, 4800 Oak Grove Drive, Pasadena, CA 91109-8099

Paul S. Smith

Steward Observatory, The University of Arizona, Tucson, AZ 85721

Abstract. We report on a high-resolution polarization study with the VLBA at 22 and 43 GHz of a sample of 18 high- and low-optically polarized, compact radio-loud quasars (HPQs and LPRQs, respectively). The polarization level of the unresolved parsec-scale radio core at 43 GHz and the overall optical polarization of the source are well correlated, suggesting a common (possibly co-spatial) origin. The electric vectors of the polarized 43 GHz radio cores are roughly aligned with the inner jet direction, indicating magnetic fields perpendicular to the flow. Similar orientations are seen in the optical, suggesting that the polarized flux at both wavelengths is due to one or more strong, transverse shocks located very close to the base of the jet. The LPRQs tend to have less luminous radio cores than the HPQs, and jet components with magnetic fields predominantly parallel to the jet. The components in HPQ jets, on the other hand, tend to have perpendicular magnetic field orientations. These differences cannot be entirely due to differences in jet orientation, and suggest that LPRQs may represent a quiescent phase of blazar activity, in which the inner jet flow does not undergo strong shocks.

1. Introduction

Traditionally one of the best means of identifying blazars has been their strong and variable polarization at optical wavelengths. Early optical polarization surveys by Stockman, Moore, & Angel (1984) and Berriman et al. (1990) revealed that the vast majority of optically-selected quasars have fractional polarizations below 3%. A small subset of flat-spectrum, radio loud sources, however, have polarizations that greatly exceeded this value, with m_{opt} ranging up to $\sim 46\%$ in some cases (Mead et al. 1990). These sources (i.e. blazars) are thought to contain highly relativistically beamed outflows in the form of jets that are pointed nearly directly at us.

A lingering question in blazar research has been why many compact, flat-spectrum, superluminal quasars display nearly all of the characteristics of blazars, yet remain weakly polarized in the optical regime. Some authors (e.g. Valtaoja et al. 1992) have suggested that their jets lie farther away from the line of sight, while others speculate that these low-optically polarized radio-loud quasars (LPRQs) are in a temporary quiescent phase of blazar activity (e.g. Fugmann 1988). In an attempt to resolve this issue, we have conducted a high-resolution polarization study with the VLBA at 22 and 43 GHz, in which we examine the magnetic fields of LPRQs and their high-optically polarized counterparts (HPQs) on parsec scales. We present here the main conclusions of our study, in which we find strong evidence for intrinsic differences in the jets of LPRQs and HPQs. Our findings are described in greater detail in Lister & Smith (2000).

2. Sample Selection and Observations

Our LPRQ sample consists of nine quasars whose optical polarization has never exceeded 3% at three or more epochs. We assembled a complementary sample of nine HPQs with a similar distribution in radio luminosity, redshift, spectral index, and optical magnitude. Our full source list is given in Table 1 of Lister & Smith (2000).

The radio observations of the LPRQs were carried out with the VLBA in dual polarization mode at 22 and 43 GHz on UT 1999 January 12–14. The HPQ data were obtained as part of several other 43 GHz VLBA studies. We also obtained optical polarization data with the Steward Observatory 60-inch telescope at Mt. Lemmon on UT 1999 February 12–14.

3. Discussion

We find a strong correlation (99.995% confidence, according to Kendall's tau) between the level of optical polarization and that of the unresolved VLBI core component at 43 GHz (Figure 1). There is no analogous correlation between m_{opt} and the total source polarization at 43 GHz, which suggests that the correlation of Fig. 1 is due to a shared radio and optical (and possibly co-spatial) emission mechanism in the core. The electric vectors in the optical, as well as those of the cores at 43 GHz, are mainly aligned with the inner jet direction, indicating magnetic fields perpendicular to the jet. Such a configuration may arise from one or more strong transverse shocks near the base of the jet (e.g. Laing 1980).

The optical polarization level is also positively correlated (above 95% confidence) with 43 GHz core luminosity, the ratio of core-to-extended 43 GHz flux density, and the misalignment angle of the jet between parsec and kiloparsec scales. These correlations are all consistent with the predictions of the relativistic beaming model (e.g. Impey, Lawrence, & Tapia 1991), in which the optical polarization of AGN increases as the beamed synchrotron jet component becomes more prominent with respect to the underlying optical continuum.

In addition to having weaker and less polarized cores, the jets of LPRQs have distinctly different polarization characteristics than their HPQ counterparts. In Figure 2 we plot the distribution of misalignment angle between the electric vector and local (downstream) jet direction for the polarized components in HPQ and LPRQ jets. Assuming that the observed electric polarization vectors are perpendicular to the true magnetic field (valid for high observing frequencies and

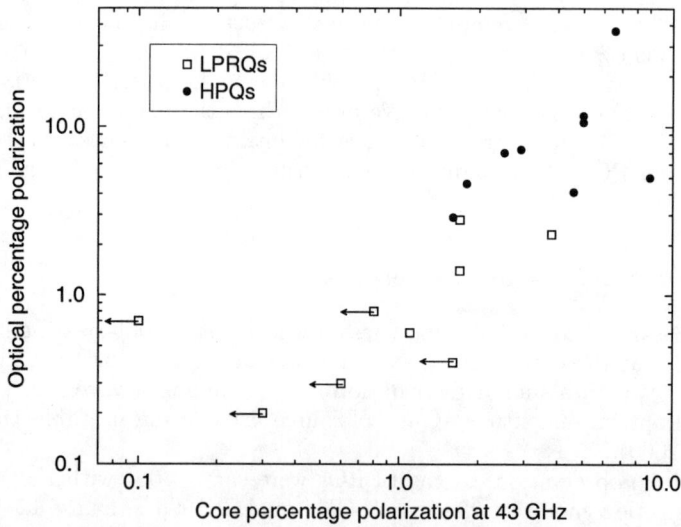

Figure 1. Total optical percentage polarization plotted against that of the unresolved radio core at 43 GHz. Upper limits are given for sources with no detected core polarization.

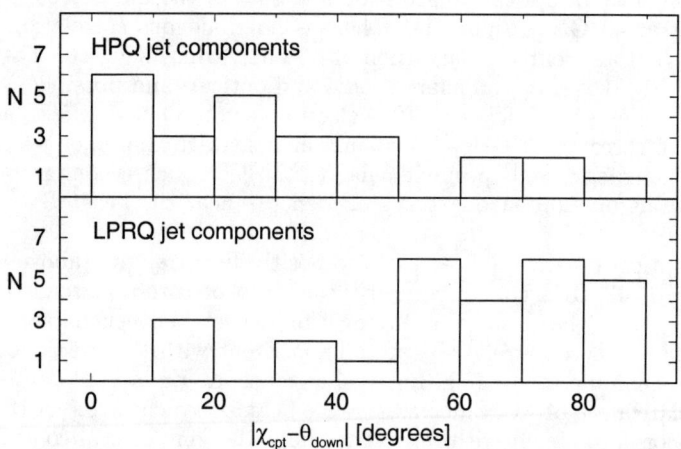

Figure 2. Distribution of 43 GHz electric polarization vector misalignment of polarized jet components with respect to the local downstream jet direction ($\theta_{\rm down}$) for HPQs (top) and LPRQs (bottom).

low source opacities), the magnetic fields in HPQ jets lie primarily perpendicular to the local jet direction. Those of the LPRQs lie mainly parallel to the jet, with the two distributions differing at the 99.989% confidence level according to a Kolmogorov-Smirnov test. The magnetic fields in the LPRQs also tend to take on more perpendicular orientations as you move down the jet. This trend is not seen in the HPQs. These differences cannot be explained by jet orientation alone, and argue strongly in favor of intrinsic differences between LPRQs and HPQs.

4. Conclusions

Our polarization-sensitive VLBI observations have revealed important differences in the parsec-scale magnetic field properties of high- and low-optically polarized radio quasars. The more weakly polarized cores and aligned jet magnetic field orientations of LPRQs suggest that they may be in a quiescent phase, in which their jets undergo weaker shocks. We are currently gathering polarization data on the well-studied Pearson-Readhead AGN sample (Pearson & Readhead 1988) with which we plan to investigate this scenario in greater detail.

Acknowledgments. We thank the Director of Steward Observatory for access to the 60" telescope, and Jim Grantham and Bob Peterson for the maintenance of this facility. We also thank Gary Schmidt for maintaining and for allowing us to use the Two-Holer polarimeter. PSS acknowledges partial support of this research from a Lucas Junior Faculty Award.

This research was performed in part at the Jet Propulsion Laboratory, California Institute of Technology, under contract to NASA.

References

Berriman, G., Schmidt, G. D., West, S. C., & Stockman, H. S. 1990, ApJS, 74, 869
Fugmann, W. 1988, A&A, 205, 86
Impey, C. D., Lawrence, C. R., & Tapia, S. 1991, ApJ, 375, 46
Laing, R. A. 1980, MNRAS, 193, 439
Lister, M. L. & Smith, P. S. 2000, ApJ, in press
Mead, A. R. G., Ballard, K. R., Brand, P. W. J. L., Hough, J. H., Brindle, C., & Bailey, J. A. 1990, A&AS, 83, 183
Pearson, T. J. & Readhead, A. C. S. 1988, ApJ, 328, 114
Stockman, H. S., Moore, R. L., & Angel, J. R. P. 1984, ApJ, 279, 485
Valtaoja, E., Teräsranta, H., Urpo, S., Nesterov, N. S., Lainela, M., & Valtonen, M. 1992, A&A, 254, 80

Spectral Energy Distributions of Blazars: Facts and Speculations

Laura Maraschi, Fabrizio Tavecchio

Osservatorio Astronomico di Brera, Via Brera 28, 20121 Milano, Italy

Abstract. We discuss the present knowledge about the Spectral Energy Distributions (SEDs) of blazars within a unified approach emphasizing overall similarities. The properties of the average SEDs of different samples of blazars suggest that more powerful sources contain on average less energetic particles. Detailed studies of TeV emitting blazars show that the energy of the particles emitting the bulk of the power increases during flares. A framework for a general theoretical understanding is proposed. We present recent results on the SEDs of a group of blazars with emission lines, allowing to estimate both the luminosity in the jet and the luminosity of the accretion disk. Implications for the origin of the power carried by relativistic jets are considered.

1. Introduction

It is now generally accepted that the blazar "phenomenon" (highly polarized and rapidly variable radio/optical continuum) is due to a relativistic jet pointing close to the line of sight. This concept, introduced by Blandford and Rees in 1978, represented a fundamental breakthrough. It took, however, a long time to gather sufficient data to apply it consistently and quantitatively to individual objects and to object classes. Taxonomy has confused the issues, since observational definitions were based (as usual in astronomy) on analogies with known objects rather than on well defined intrinsic properties. Moreover variability could alter the "classification" of the same object. After decades of studies to distinguish between objects with or without emission lines (flat spectrum quasars vs. BL Lacs) or between radio bright and X-ray bright BL Lacs, we think that presently the most productive approach is to assume that all blazars contain relativistic jets and ask in what way these jets differ in different objects and, eventually, why.

Therefore here and in the following we will assume that Quasars with Flat Radio Spectrum (FSQs, which include OVVs and HPQs) and BL Lac objects are essentially "similar" objects in the sense that the nature of the central engine is similar apart from some basic scales. The obvious parameters are the central black hole mass, angular momentum and the accretion rate (Blandford 1990). We do not know yet what governs the phenomenology but we assume that it is some combination of these three parameters. The goal is to understand the role of these fundamental parameters starting from a physical comprehension of

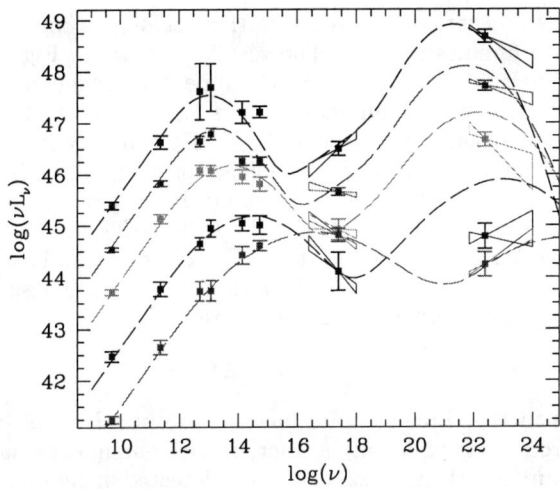

Figure 1. Average SEDs of blazars with different radio luminosity (from Fossati et al. 1998).

the phenomenology. We will therefore take a particular "point of view" and not attempt a review of all the work carried out in this field.

2. The Average SEDs of Different Samples of Blazars

It was noted early on that the SEDs of blazars exhibited remarkable systematic properties (Landau et al. 1986; Sambruna et al. 1996). The subsequent discovery by the Compton Gamma Ray Observatory of gamma-ray emission from blazars (a summary can be found in Mukherjee et al. 1997) was a major step forward, showing that in many cases the bulk of the luminosity was emitted in this band and questioning the importance of previous studies of the SEDs at lower frequencies.

Although a large fraction of the sky was surveyed with the EGRET instrument on board CGRO, its sensitivity was just sufficient to detect the brightest or flaring sources. Therefore the results did not allow to define a complete sample.

A simple approach was taken by Fossati et al. (1998) in order to explore systematically the properties of the blazar SEDs up to gamma-ray energies. The procedure was to construct "average" SEDs of known complete samples of blazars (FSRQs from the 2 Jy radio sample, BL Lacs from the 1 Jy radio sample and BL Lacs from the Slew Survey X-ray sample) using published fluxes at 6 wavelengths in the radio to UV interval plus average fluxes and spectral indices in the X-ray and GeV range.

Clear systematic differences emerged between the three samples, in particular concerning the slopes of the X-ray and gamma-ray emission. Moreover, the average luminosities of the three samples were different. Since in each sample there are objects with different spectral properties (e.g. X-ray to radio flux ratios) we thought that if luminosity was an important parameter it would be useful to bin the objects not according to their belonging to a specified sample, but according to luminosity only. The result is shown in Fig. 1 (from Fossati et al. 1998). Indeed, the resulting SEDs appear homogenous and systematic trends are evident. For each luminosity class the derived SEDs show two very broad components peaking between 10^{13}–10^{17} Hz and between 10^{21}–10^{24} Hz respectively. Analytic curves have been plotted for comparison (see Fossati et al. 1998 for a complete account). The underlying simple assumptions were: i) that the peak frequencies are inversely related to the radio-luminosity; ii) that the ratio of the two peak frequencies in each SED is constant; and iii) that the height of the second peak is proportional to the radio luminosity. These "analytic" laws seem to represent the average SEDs quite closely.

2.1. Biases

All the sources in the three samples have radio optical and X-ray data, most have also infrared data, while only a fraction have gamma-ray data. Within the three original samples, the fraction of objects detected in gamma-rays are 19/50, 9/34 and 8/48 respectively.

Thus it is still an open issue whether the γ-ray properties of the average SEDs are representative of the whole population or are significantly biased. Clearly, since most sources are close to the detection limit and many are known to be strongly variable (e.g. Mukherjee et al. 1997), EGRET preferentially detected those that were in an active state. In an interesting but rarely quoted paper, Impey (1996) showed via MonteCarlo simulations that, assuming all FSQ in the 1 Jy sample have the same average $L_\gamma/L_{\rm radio}$ ratio, the properties of the *gamma-ray detected* sample can be reproduced if the intrinsic average $\langle L_\gamma/L_{\rm radio}\rangle$ is about 1/10 of the actually observed one, with a "normal" scatter of about 1 order of magnitude (the scatter could be due to variability). We did not apply any correction to the SEDs because the estimate though interesting is rather uncertain. Moreover the analysis of Impey refers to FSQ and it would not be justified to assume the same correction factor for all luminosity classes. There can be little doubt, however, that the average gamma ray fluxes in Fig. 1 are overestimated.

In order to assess the possible differences between blazars detected and non detected in gamma-rays we obtained *Beppo*SAX observations of "non gamma-ray" HPQs. A good example is 1641+399 (3C 345): the source is clearly detected up to the PDS range with a hard X-ray luminosity comparable to that in the first peak (Tavecchio et al., in preparation). Thus the hypothesis that all blazars emit gamma-rays at an average level not much different than those actually detected with EGRET (within an order of magnitude) finds support.

2.2. The Unified Framework for the SEDs of Blazars

Figure 1 strongly suggests that the same radiation mechanisms operate in all blazars. The SEDs define a spectral sequence (we will call red and blue the objects at the different extremes of the sequence), implying that the jet properties change with continuity, therefore there are no qualitative differences among jets of different power. Beamed synchrotron and inverse Compton emission from a single population of electrons account very well for the observed SEDs except in the radio to mm range where effects of selfabsorption and inhomogeneity are important (see also Kubo et al. 1998). This model predicts that variability of the two spectral components should be correlated especially at frequencies near the peaks. Given the difficulty of getting adequate data, it is remarkable that this has been verified at least in some well studied objects. In the following we will assume that this emission model holds in general.

There are however quantitative differences between jets producing the SEDs depicted in Fig 1. Adapting model spectra to the observed SEDs of about 50 sources with gamma-ray information (but in many cases rather poor broad band data) Ghisellini et al. 1998 derived the physical parameters of the jets including seed photons of internal (SSC) as well as external (EC) origin. The model fits suggest that the observed trends are due to i) an increasing importance of external seed photons with increasing jet luminosity; and ii) a decreasing "critical" energy of the radiating electrons with increasing (total) radiation energy density. The latter dependence is physically plausible since the radiation energy density determines the energy losses of relativistic particles. If the critical electron energies were determined by a balance between injection/acceleration and cooling processes the latter dependence could be understood.

This "unified" theoretical scheme, attractive as it is, needs to be tested in many respects. One can think of at least two ways of doing so: i) determine the physical parameters in individual objects with good data; and ii) understand the mechanisms of particle acceleration and injection from detailed variability studies.

More indirectly, if the shapes of blazar SEDs are "biunivocally" related to luminosity (of course some scatter will have to be allowed for) there must be predictable consequences on number counts and luminosity functions of different types of objects. Even if FSQs and BL Lacs contain "similar" jets (at least close to the nucleus) as suggested by the continuity of the SEDs, we still need to understand the differences in emission line properties. Also in this respect continuity could hold in the sense that the accretion rate may decrease continuously along the sequence but the emission properties of the disk may not simply scale with the accretion rate.

3. Studies of Individual Objects

The detailed study of the spectral and variability characteristics of single blazars is an important approach complementary to the study of samples described above. In the following we will focus on observations obtained with the Italian-Dutch satellite *Beppo*SAX.

Unfortunately in the case of red blazars the study of the synchrotron component is difficult, because of the position of the Synchrotron peak, which falls in the poorly covered IR–FIR range. Furthermore the study of the gamma-ray component in the MeV–GeV region of the spectrum has been difficult in the last few years due to the loss of efficiency of EGRET and is now impossible after reentry of CGRO.

We show in Fig. 2 the case of 3C 279, observed with *Beppo*SAX in January 1997 simultaneously with CGRO (Hartman et al. 2000) and close in time (December 1996) with ISO (Haas et al. 1998). The source was found to be in a

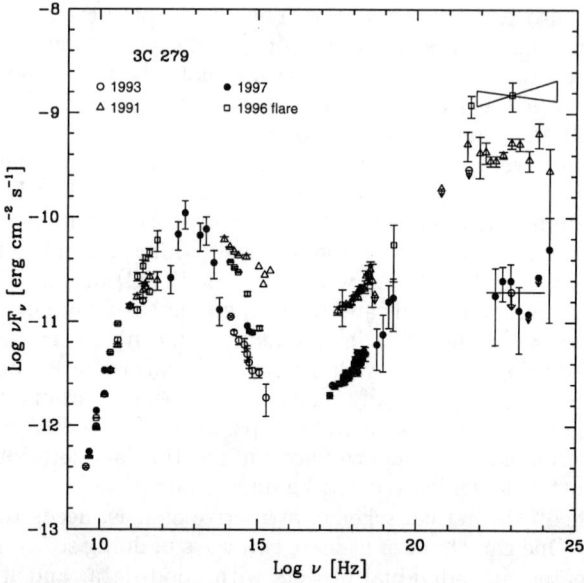

Figure 2. Quasi-simultaneous SEDs of the quasar 3C279 taken in the different epochs. The *Beppo*SAX and EGRET data taken in 1997 are almost exactly contemporaneous, while the ISO spectrum is taken one month before.

rather low state, analogous to the low state observed in 1993 (also shown). Also shown for comparison are the two highest states recorded in 1991 and 1997.

A general correlation of the optical X-ray and γ-ray fluxes is apparent. The FIR data strongly suggest an additional, highly luminous thermal dust component, presumably observed at low inclination with respect to a putative accretion disk. In any case the synchrotron peak is difficult to localize.

The situation is better for blue blazars. In several sources of this class the Synchrotron component peaks in the X-ray band, where numerous satellites can provide good data. In few bright extreme BL Lac objects the high energy γ-ray component is observable from ground with TeV telescopes (for a general account see Catanese & Weekes 1999). In these particular cases the contemporaneous X-ray/TeV monitoring demonstrated well the correlation between the Synchrotron and the IC components. A very good example is Mkn 421 for which we obtained the observation of a simultaneous TeV/X-ray flare with Whipple and *Beppo*SAX in 1998, probing for the first time the existence of correlation on short (hour) time scales (Maraschi et al. 1999; see also Takahashi et al. 1999, Catanese & Sambruna 2000). When the position of the two peaks can be well determined observationally, as is possible in this type of sources, robust estimates of the physical parameters of the jet can be obtained (e.g. Tavecchio et al. 1998). This was done for both Mkn 421 and Mkn 501; the case of Mkn 501 is shown in Fig. 3a.

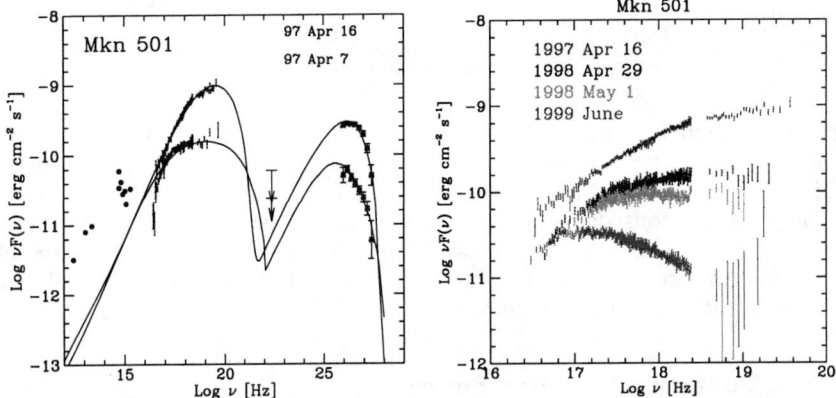

Figure 3. *Left*: Overall SED of Mkn 501 observed simultaneously by *Beppo*SAX and CAT during the major flare in April 1997 (from Tavecchio et al. 2000, in preparation). The solid line is the spectrum obtained with the SSC model. The possibility to constrain both the two peaks allows to obtain robust estimates of the physical parameters of the jet. *Right*: Spectral evolution of Mkn 501 during the period 1997–1999 (from Tavecchio et al. 2000, in preparation). It is clear the strong relation between the position of the peak and the total luminosity.

The broad band response of *Beppo*SAX allows a reliable determination of the position of the synchrotron peak, $E_{\rm peak}$, if it falls between 0.1 and 100 keV. We could verify that during the flare of Mkn 421 mentioned previously $E_{\rm peak}$, moved to higher energies with increasing intensity (Fossati et al. 2000). The same behavior was exhibited in a more dramatic way by Mkn 501. Its synchrotron peak moved to $E > 100$ keV during the extraordinary activity in April 1997. Subsequent snapshot spectra obtained with *Beppo*SAX showed a systematic decrease of $E_{\rm peak}$ down to $\simeq 0.1$ keV in June 1999 while the source was fading (Fig. 3b, Tavecchio et al., in preparation). Over this two year period the X-ray light curve as measured by the ASM aboard XTE was not monotonic with an overall decay interrupted by flares. Thus our observations show that $E_{\rm peak}$ correlates with luminosity not only along individual flares, but also on much longer timescales.

The $E_{\rm peak}$ vs. luminosity relation observed in the time dependent behavior of these two sources (higher $E_{\rm peak}$ for higher luminosity) is opposite to that found in the "spectral sequence," where the peak falls at *lower* frequencies for objects of higher luminosity.

A general scenario that could include both types of behavior is the following. Let us suppose that the Lorentz factor of particles emitting at the peak $\gamma_{\rm p}$ is determined by the equilibrium between the cooling and acceleration processes, namely $t_{\rm acc}(\gamma_{\rm p}) = t_{\rm cool}(\gamma_{\rm p})$. Given that $t_{\rm cool} = {\rm const}/U\gamma$, where $U = U_{\rm rad} + U_B$ is the total energy density, and using the general expression $t_{\rm acc}(\gamma) = \gamma t_{\rm o,acc}$, found in the theory of diffusive shock acceleration (see e.g. Kirk et al. 1998) one can write:

$$\gamma_p = \left(\frac{\text{const}}{Ut_{o,\text{acc}}}\right)^{1/2} \quad (1)$$

This expression is consistent with the correlation found by Ghisellini et al. (1998), $\gamma_p \propto U^{-1/2}$, provided that the acceleration timescale is, on average, similar in all sources. Flares in single sources can then be interpreted as due to the temporaneous decrease of $t_{o,\text{acc}}$ due to changes in the physical process of acceleration. This scenario seems to apply quite well to Mkn 501 (Tavecchio et al. 2000, in preparation), for which it is possible to reproduce the observed variability with the only change of γ_p.

4. Jet Power vs. Accretion Power

Finally, we wish to discuss some recent results on luminous blazars with emission lines. They are at the high-luminosity end of the sequence, with the Synchrotron peak in the FIR region. In these sources the X-ray emission is believed to be produced through the IC scattering between soft photons external to the jet (produced and/or scattered by the Broad Line Region) and electrons at the low energy end of their energy distribution. Thus measuring the X-ray spectra and adapting a broad band model to their SEDs yields reliable estimates of the total number of relativistic particles involved, which is dominated by those at the lowest energies. This is interesting in view of a determination of the total energy flux along the jet (e.g. Celotti et al. 1997; Sikora et al. 1997). The "kinetic" luminosity of the jet can be written as

$$L_j = \pi R^2 \beta c U \Gamma^2 \quad (2)$$

(e.g. Celotti et al. 1997), where R is the jet radius, Γ is the bulk Lorentz factor and U is the total energy density in the jet, including radiation, magnetic field, relativistic particles and finally protons. If one assumes that there is 1 (cold) proton per relativistic electron, the proton contribution is usually dominant.

In high luminosity blazars the UV bump is often directly observed and/or can be estimated from the measurable emission lines, yielding important information on the accretion process. Thus the relation between accretion power and jet power can be explored.

For three sources of this type we performed a detailed analysis of the *Beppo*SAX data and modeled the overall SED (Fig. 4, Tavecchio et al. 2000). For the other 6 sources also observed with *Beppo*SAX a similar analysis is in progress and the results reported here are still preliminary.

In all cases we estimated the kinetic luminosity of the jet and the luminosity of the disk as described above. We can then compare the two in Fig. 5b. We include also four BL Lac objects (namely BL Lac, ON231, Mkn 501 and Mkn 421) for which we have reliable information on the power carried by the jet. Unfortunately we can set only upper limits on the luminosity of their putative accretion disks except for BL Lac, where the presence of a broad H_α line allows us to estimate of the ionizing continuum (e.g. Corbett et al. 2000).

The diagram in Fig. 5a shows the comparison between the total radiative luminosity L_{rad} of the jet and the power transported by the jet including the

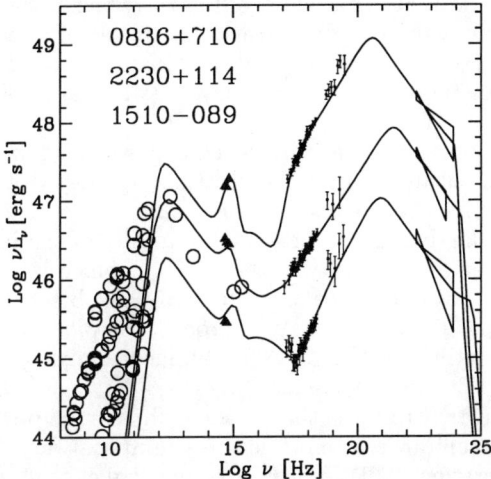

Figure 4. Overall SEDs of three powerful emission-lines blazars (from Tavecchio et al. 2000). The objects are characterized by the presence of a strong UV-bumps, allowing the determination of the luminosity of the accretion disk.

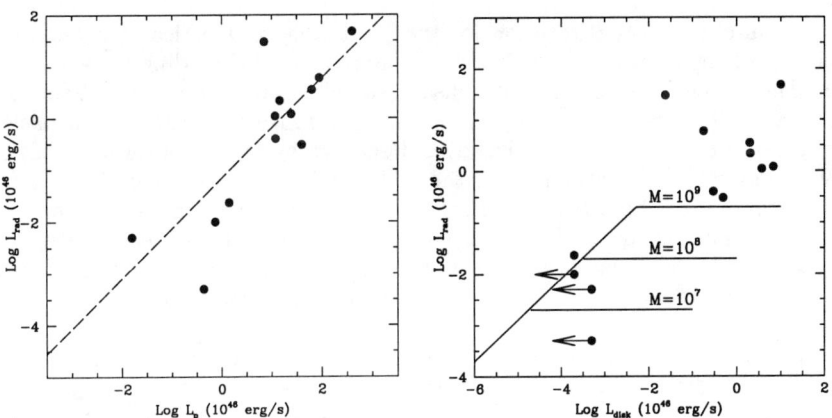

Figure 5. *Left*: Radiative luminosity vs. jet power for the sample of blazars discussed in the text. The dashed line indicates the least-squares fit to the data. *Right*: Radiative luminosity of jets vs. disk luminosity. The solid lines represent the *maximum* jet power estimated for the Blandford & Znajek model for black holes with different masses (units are in Solar masses).

proton contribution which is dominant. The ratio between these two quantities gives directly the "radiative efficiency" of the jet, which turns out to be $\eta \simeq 0.1$, though with large scatter. The line traces the result of a least-squares fit: we found a the slope ~ 1, indicating a rather constant radiative efficiency along the blazar sequence (note that the data cover a wide range of about 5 orders of magnitude).

As discussed above, we have reliable estimates of the luminosity released by the accretion disk, calculated directly with the luminosity of the blue bump or inferred by the luminosity of emission lines. Therefore we can directly compare the luminosity of the disk, $L_{\rm accr}$, and the luminosity of the jet, $L_{\rm jet}$. In Fig. 5b we report the radiative luminosity of the jet $L_{\rm rad}$, which is a *lower limit* of $L_{\rm jet}$ versus $L_{\rm accr}$. Note that in the case of the four BL Lac objects the disk luminosity is formally an upper limit. A similar approach was pioneered by Celotti et al. (1997), but their estimates of $L_{\rm jet}$ were obtained applying the SSC theory to VLBI data which refer to larger scales.

The first important result is that in some objects the power transported by the jet is much *larger* (at least an order of magnitude) than the luminosity released through accretion. This result is a strong challenge for models elaborated to explain the formation of jets.

Two main lines of approach consider either extraction of rotational energy from the black hole itself or magnetohydrodynamic winds associated with the inner regions of accretion disks. We briefly discuss here the first case, treated by Blandford & Znajek (1977). The result of their complex analysis is summarized in the well known expression:

$$L \simeq 10^{37} B^2 M_8^2 \quad {\rm erg\ s^{-1}} \tag{3}$$

Assuming maximal rotation for the black hole, the critical problem is the estimate of the intensity reached by the magnetic field threading the event horizon. Several authors have recently discussed this difficult and subtle issue: here we use the results obtained by Ghosh & Abramovicz (1997) on the basis of equipartition within an accretion disk described by the Shakura and Sunyaev (1973) model. Their estimates for the rotational power are shown in Fig. 5 for various values of the mass of the central black hole as a function of the luminosity observed from the disk. The latter is related to the accretion rate which appears in the formulae of Ghosh & Abramovicz (1997) adopting an efficiency of 10%. Clearly the model fails to explain the large power observed in the jets of bright quasars, even for BH masses ($M \sim 10^9$ $M_\odot$).

In another class of models it is assumed that the energy powering the jet is extracted directly from the accretion disk. Assuming a similar efficiency, $\eta \sim 0.1$, for the conversion of both the accretion power and the jet power into radiation, the plot in Fig. 5 suggests that the mechanism responsible for the formation of the jet is able to split the total accreted power $\dot{M}c^2$ in such a way that $L_{\rm accr} \sim L_{\rm jet}$.

5. Conclusions

The study of broad band SEDs and their variability is essential for understanding blazars. A unified approach is possible and valuable since it can be tested and possibly disproved. While the phenomenological framework is suggested to be "simple" (e.g. "red" blazars are highly luminous and emit GeV gamma-rays while "blue" blazars have low luminosity and emit TeV gamma-rays), we do not yet know what determines the emission properties of jets of different power nor what determines the jet power in a given AGN. There is however the exciting prospect that such problems can be tackled with data that can be gathered in the near future.

References

Blandford, R. D. & Znajek, R. L. 1977, MNRAS, 179, 433
Blandford, R. D. & Rees, M. J. 1978, in Pittsburgh Conference on BL Lac Objects, ed. A. N. Wolfe (Pittsburgh: University of Pittsburgh Press), 328
Blandford, R. D. 1990, in Saas-Fee Advanced Course 20. Lecture Notes, (Berlin: Springer-Verlag), 161
Catanese, M. & Weekes, T. C. 1999, PASP, 111, 1193
Catanese, M. & Sambruna, R. M. 2000, ApJ, 534, L39
Celotti, A., Padovani, P., & Ghisellini, G. 1997, MNRAS, 286, 415
Corbett, E. A., et al. 2000, MNRAS, 311, 485
Fossati, G., et al. 1998, MNRAS, 299, 433
Fossati, G., et al. 2000, ApJ, 541, 166
Ghisellini, G., et al. 1998, MNRAS, 301, 451
Ghosh, P. & Abramowicz, M. A. 1997, MNRAS, 292, 887
Haas, M., et al. 1998, ApJ, 503, L109
Impey, C. 1996, AJ, 112, 2667
Kirk, J. G., Rieger, F. M., & Mastichiadis, A. 1998, A&A, 333, 452
Kubo, H., et al. 1998, ApJ, 504, 693
Landau, R., et al. 1986, ApJ, 308, 78
Maraschi, L., et al. 1999, ApJ, 526, L81
Sambruna, R. M., Maraschi, L., & Urry, C. M. 1996, ApJ, 463, 444
Shakura, N. I. & Sunyaev, R. A. 1973, A&A, 24, 337
Sikora, M., et al. 1997, ApJ, 484, 108
Takahashi, T., et al. 1999, Astroparticle Physics, 11, 177
Tavecchio, F., Maraschi, L., & Ghisellini, G. 1998, ApJ, 509, 608
Tavecchio, F., et al. 2000, ApJ, in press (astro-ph/0006443)

The HST View of the Nuclei of Radio Galaxies: Implications for the Radio-loud AGN Unification Models

Marco Chiaberge, Annalisa Celotti

SISSA/ISAS, Via Beirut 2-4, 34014 Trieste, Italy

Alessandro Capetti

Osservatorio Astronomico di Torino, Str. Osservatorio 20, 10025 Pino Torinese (TO), Italy

Gabriele Ghisellini

Osservatorio Astronomico di Brera, Via Bianchi 46, 13807 Merate (LC), Italy

Abstract. We present results of the analysis of the nuclear regions of nearby ($z < 0.3$) 3CR radio galaxies as seen in HST images. Unresolved nuclear sources are detected by the HST in the great majority of a complete sample of FR I radio galaxies. The emission of these central compact cores is anisotropic and shows a striking linear correlation with the radio core one. We identify this emission with optical synchrotron radiation produced in the inner regions of a relativistic jet. In order to explore how the FR I/FR II dichotomy is related to the innermost nuclear structures, we also studied a complete sample of 26 FR II with $z < 0.1$. Their nuclei show a more complex behavior, which is however clearly related to their optical spectral classification. Most importantly we find a new class of narrow or weak-lined galaxies which possess optical nuclei essentially indistinguishable from those seen in FR I. In order to investigate the viability of FR I/BL Lacs unification, we compare the optical and radio properties of these two classes of objects, taking advantage of the newly discovered optical cores of FR I. In order to reconcile the observed properties with the usually adopted AGN unification model, velocity structures in the jet are suggested.

1. The HST View of FR I Radio Galaxies

1.1. The Sample

Our sample is formed by all 33 radio galaxies belonging to the 3CR catalogue and morphologically identified as FR I radio sources. The redshifts of the galaxies are in the range $z = 0.0037$–0.29 (median redshift $z = 0.03$) and the total radio luminosities at 178 MHz are in the range log L_{178} [W Hz^{-1}] = 23.7–28.1 ($H_0 = 75$ km s^{-1} Mpc^{-1} and $q_0 = 0.5$). Since the morphological classification is in many cases somewhat subjective, we also consider separately objects below and above $L_{178} = 2\times10^{26}$ W Hz^{-1}, which corresponds to the 'transition' between

FR I and FR II. Two thirds of the sources lies below this fiducial radio power separation. HST/WFPC2 observations are available in the public archive for 32 out of 33 sources. The whole sample was observed using the F702W filter as part of the HST snapshot survey of 3CR radio galaxies (e.g. Martel et al. 1999). Details on the data analysis, together with an extensive discussion of the results, are given in Chiaberge et al. (1999).

2. Results

In 22 cases, the radial brightness profiles of the nuclear regions is clearly indicative of the presence of a central compact core (CCC), unresolved at the HST resolution. We performed aperture photometry of the 22 CCC, adopting the internal WFPC2 flux calibration, which is accurate to better than 5 per cent. However, the dominant photometric error is the determination of the background in regions of varying absorption and steep brightness gradients, especially for the faintest CCC, resulting in a typical error of 10% to 20%.

In order to investigate the origin of this nuclear emission, we explore possible (cor-)relations between the CCC flux/luminosity and other observed properties. No trend is found between CCC luminosities and total radio power or absolute visual magnitude of the host galaxy. Conversely, the CCC emission bears a clear connection with the radio core. With the aim of quantifying this relation we consider the two subsamples separately. In the low luminosity subsample (formed by 21 objects), we detected 18 CCC. They show a tight correlation between optical and radio core emission, both in flux and luminosity (Fig. 1a). It extends over 4 orders of magnitude, it has an extremely high statistical significance, a small dispersion and its slope is consistent with unity. The correlation coefficient is $r = 0.88$, which gives a probability that the points are taken from a random distribution of $P = 3.1 \times 10^{-6}$. Only one point (3C 386) is well separated from the others. This object, which is classified as a fat-double source, appears to be peculiar, as it presents a broad $H\alpha$ line, atypical of FR I. A similarly strong correlation is present also between radio core and CCC luminosities. The fact that the correlation is found both in flux and luminosity gives us confidence that it is not induced by either selection effects or a common redshift dependence. Since the 3CR sample has been selected according to the total radio flux at low frequency and no trend is found between CCCs and extended (radio and optical) properties, the result does not appear to be induced by selection effects.

The behaviour of the 11 sources with higher total radio luminosity is more complex: sources either follow the core fluxes and luminosities correlation (in 4 cases) or CCC are not detected. More precisely, 2 are obscured by extended nuclear dust lanes while the last 5 only show the smooth galaxy emission and no central source.

2.1. The Origin of CCC Emission

- As the radio core emission is certainly originated by non-thermal synchrotron radiation, the presence of a tight linear correlation between radio core and CCC emission is a strong suggestion that also this optical component is produced by the same process. The radio-optical spectral indices of the CCC span the range $\alpha_{\rm ro} = 0.6$–0.9 ($F_\nu \propto \nu^{-\alpha}$), fully consistent with the $\alpha_{\rm ro}$ typical of blazars and optical jets which are indeed dominated by synchrotron radiation.

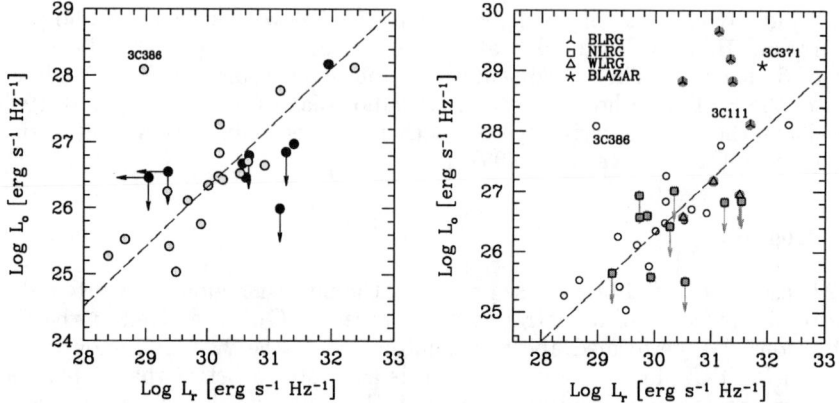

Figure 1. *Left panel*: optical versus radio (5 GHz) core luminosities. Different symbols mark sources belonging to the low (open circles) and high luminosity (filled circles) subsamples. The dashed line is the best linear fit, calculated considering only the low power galaxies. *Right panel*: Optical nuclear luminosity versus radio (5 GHz) core luminosity for both the FR I (open circles) and FR II (filled circles) samples. Different symbols are used to identify different spectral classifications. The dashed line is the correlation found for FR I galaxies.

- We also find indication that the CCC emission is beamed. In fact, in order to explain why jets with optical counterparts are smaller (they are foreshortened), brighter and one-sided (they are relativistically beamed) with respect to typical radio jets, it has been argued that jets are detected in the optical band only when pointing towards the observer. Five sources of our sample (namely 3C 66B, 3C 78, 3C 264, 3C 274 and 3C 346) show optical jets and these very same objects clearly stand out for being among the brightest CCC of their respective subsamples, implying that beaming plays an important role. This is also supported by the correlation found between CCC luminosity and the orientation of the nuclear dusty disks with respect to the line of sight (Capetti & Celotti 1998).

- The detection of CCCs indicates that we have a direct view of the innermost nuclear regions of FR I. The high rate detection in low luminosity sources ($\sim 85\%$) suggests that if obscuring material is present in these objects, it has to be distributed in a geometrically thin structure, or thick tori are present only in a minority of FR I. This argue against the presence of standard pc-scale dusty tori in FR I, which would be expected in analogy with other AGN classes. Therefore, the absence of broad emission lines in the FR I class cannot be attributed to obscuration.

- CCC fluxes also represent upper limits to any thermal/disc emission. For a 10^9 $M_\odot$ black hole, this limit translates to a fraction as small as $\lesssim 10^{-5}$ of the Eddington luminosity, suggesting that accretion might take place in a low efficiency radiative regime similar to what observed in elliptical galaxies.

The picture which emerges is that the innermost structure of FR I radio galaxies differs in many crucial aspects from that of the other classes of AGN. In fact they lack of substantial BLR, tori and thermal disc emission, which conversely are usually associated with active nuclei. The origin of this difference is of course a key issue in the understanding of the nuclear activity.

3. The HST View of the FR I/FR II Dichotomy

In order to explore how the differences in radio morphology are related to the optical nuclear properties, we analyzed HST images of a sample of 26 extended radio-galaxies morphologically classified as FR II, belonging to the 3CR catalog and with $z < 0.1$. Analogously to what we have done for FR I, we derive the brightness profiles of the nuclear regions. We detected unresolved optical cores in 13 FR II. In the L_o vs. L_r plane, they show a more complex behavior with respect to FR I CCC, which is however clearly related to their optical spectral classification (see Fig. 1b).

- Broad line FR II radio galaxies (BLRG) are located well above the FR I correlation, suggesting that a contribution from thermal (disc) emission is present. Three narrow line (NLRG) and one weak line radio galaxy (WLRG), in which no nuclear source is seen, might be interpreted as the obscured counter-parts of BLRG, in agreement with the current unification schemes.

- Most importantly, in 5 sources of the sample, all of them NLRG or WLRG, optical cores are located on the same correlation defined by FR I and with similar radio and optical luminosities. This suggests that, in analogy to FR I, their emission is dominated by synchrotron radiation and represents the optical counter-part of the non-thermal radio cores. Interestingly, all these galaxies are located in clusters, an environment typical of FR I.

Our results suggest that the radio morphology is not univocally connected with the optical properties of the innermost structure of radio galaxies. In fact, at least at low redshifts, there is not a single homogeneous population of FR II. Clearly, a classification based on the optical nuclear properties, as seen in these HST images, is more likely to reflect true similarities (or differences) on the nature of the central engine (such as, e.g. the rate of radiative dissipation in the accretion disc) than the traditional dichotomy of radio morphology.

Notice that the rate of detection of FR II optical cores implies that the covering fraction of any obscuring material is less than ~ 0.54. This can be even smaller if at least some of the upper limits are actually just below the detection threshold. Therefore, our determination is only marginally consistent with what is observed in higher redshift $(0.5 < z < 1)$ 3CR quasars and radio galaxies (Barthel 1989). Instead, our results appear to be consistent with the lower quasar fraction in low luminosity objects recently found by Willot et al. 2000.

These results have also interesting bearings from the point of view of the unified models. In fact, this picture argues against the idea that all FR II radio galaxies constitute the parent population of radio-loud quasars. We propose instead that galaxies with FR II morphology and an FR I-like core are possibly misaligned counter-parts of BL Lac objects. This can account for the observation that some radio-selected-type BL Lacs show radio morphologies more consistent with FR II than with FR I (e.g. Kollgaard et al. 1996).

4. FR I and BL Lac Unification: Clues for the Presence of Velocity Structures in Jets

In the frame of the AGN unification models FR I are believed to be the parent population of BL Lacs. In this section we directly compare the properties of the optical and radio cores of FR I with two complete samples of BL Lacs derived from the 1 Jy catalog (almost all of them being Low Energy peaked BL Lacs, LBL) and from the *Einstein* Slew Survey (almost all of them being High Energy Peaked BL Lacs, HBL).

We first consider the optical and radio core luminosities separately, comparing the nuclear emission of FR I and BL Lacs for bins of equal extended radio power (see Figs. 1 and 3 of Chiaberge et al. 2000). BL Lacs are at least 4 orders of magnitude brighter in the optical than FR I nuclei. As the core radiation of BL Lacs is enhanced by relativistic beaming, we can derive the bulk Lorentz factors Γ in order to account for the observed spread, if typical observing angles are known. If BL Lacs are observed at an angle $\theta \sim 1/\Gamma$ and $\theta \sim 60°$ is the median observing angle of FR I galaxies, bulk velocities with $\Gamma \sim 4$–6 are required to account for the optical enhancement of BL Lacs in each bin of extended luminosity. A similar result is obtained by comparing the radio core emission (e.g. Kollgaard et al. 1996). However, such values are significantly and systematically lower than those required by other independent means, such as superluminal motions and high energy spectral constraints in both LBLs and HBLs. These in fact require values of the Doppler factor $\delta = [\Gamma(1 - \beta \cos\theta)]^{-1}$ ($\delta = \Gamma$ for a viewing angle $\theta = 1/\Gamma$) in the range 15–20 for the region emitting most of the radiation in both HBLs and LBLs.

In Fig. 2a we show the optical vs. radio core luminosity for the three samples. In order to determine how beaming affects the observed luminosities and thus how objects could be connected in the $L_o - L_r$ plane, we consider the SED of BL Lacs, observationally much better determined, and calculate the observed spectrum of the misaligned objects, by taking into account the relativistic transformations. In fact, an important and previously neglected point is that these transformations depend on the spectral index in the band considered, which in itself might change as a function of the degree of beaming. We consider the two sources Mkn 421 and PKS 0735+178, which are representative of HBL and LBL, respectively. We derive both a continuous description of their SED and the bulk Lorentz factors of the two sources by adopting a homogeneous synchrotron self-Compton emission model. We then calculate the corresponding observed SEDs for different orientations. Clearly the net effect of debeaming is a "shift" of the SED towards lower luminosities and energies. As expected, for $\theta = 60°$ the resulting "debeamed" BL Lac component is about four orders of magnitude below the radio galaxy region in the optical, and two/four in the radio band (see Fig. 2b). The simplest and rather plausible hypothesis to account for this discrepancy within the unification scenario is to assume a structure in the jet velocity field, in which a fast spine is surrounded by a slow, but still relativistic, layer (see Chiaberge et al. 2000 for an extensive discussion). A slower component in the jet has the effect to "stop" the debeaming trails.

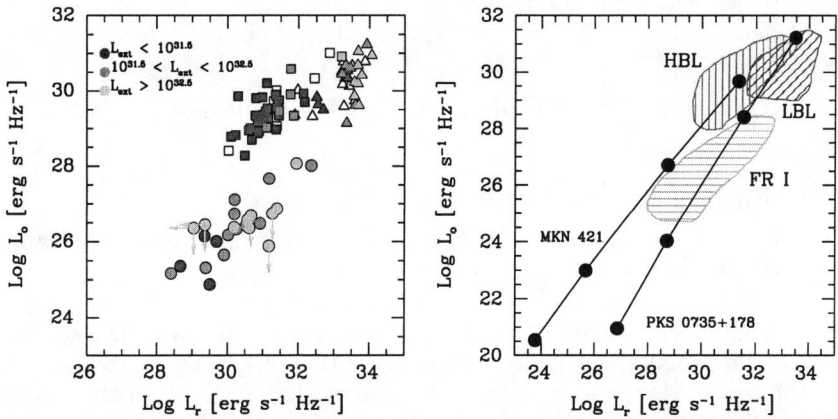

Figure 2. *Left panel (a)*: BL Lacs and FR I radio galaxies in the $L_r - L_o$ plane. Empty symbols are objects with no data on their extended radio power. *Right panel (b)*: debeaming trails in the radio-optical luminosity plane for Mkn 421 and PKS 0735+178, in the frame of a single emitting region model. The filled circles correspond to the predicted luminosities of objects at different viewing angles. Top to bottom: $\theta = 1/\Gamma$, $10°$, $30°$ and $60°$.

While equivalently incompatible with the FR I population properties, the HBL and LBL move on different trails. This effect is due to the different shape of their SED. In particular, HBL trails run almost parallel to the FR I correlation. However, they should be compared with lower luminosity radio galaxies (i.e. the B2 sample), as they do not share the same range of extended power of 3CR sources. A direct implication of the presence of velocity structures in the jet is that the observed flux should be dominated by the emission from either the spine or the slower layer, in the case of aligned and misaligned objects, respectively. Interestingly, the existence of velocity structures in the jet has been suggested by various authors (e.g. Laing 1999) in order to explain the observed radio properties of FR I jets.

References

Barthel, P. D. 1989, ApJ, 336, 606
Capetti, A. & Celotti, A., 1999, MNRAS, 304, 434
Chiaberge, M., Capetti, A., & Celotti, A. 1999, A&A, 349, 77
Chiaberge, M., et al. 2000, A&A, 358, 104, 112
Kollgaard, R. I., et al. ApJ, 465, 115
Laing, R. A., et al. 1999, MNRAS, 306, 513
Martel, A. R., et al. 1999, ApJS, 122, 81
Willott, P., et al. 2000, MNRAS, 316, 449

Blazars in Low-Luminosity and Radio-Weak AGN?

Heino Falcke, Sera Markoff, Peter L. Biermann

Max-Planck-Institut für Radioastronomie, Auf dem Hügel, D-53121 Bonn, Germany

Abstract. Typical blazars seem to be associated with FR I and FR II radio galaxies and radio-loud quasars. However, what happens at lower powers? Do blazars exist in low-luminosity AGN or do they exist in radio-quiet AGN? Our recent detection of superluminal motion in a supposedly radio-quiet Seyfert raises the question whether beaming can play an important role in some of these objects as well. Moreover, VLBI observations of nearby low-luminosity AGN reveal compact flat-spectrum radio cores very similar to those in bright radio-loud blazars. Furthermore, with the detection of X-ray emission from the least luminous AGN we can study, Sgr A* in the Galactic Center, this source seems to be dominated entirely by non-thermal emission—like in BL Lacs. The same may be true for some X-ray binaries in the Low/Hard-state. Inclusion of low-power radio jets into the overall picture provides some clues for what type of accretion is important, what the power, radiative efficiency and matter-content of jets is, and what mechanism could be responsible for making jets radio-loud. We specifically discuss whether proton-proton collisions in a hot accretion flow could provide the switch for the radio-dichotomy.

1. Introduction

Over the last decades our basic picture of what blazars are has solidified. It is not even discussed anymore that the cause of the blazar and BL Lac phenomenon are relativistic jets and their non-thermal emission. In many cases this jet emission dominates the spectrum of these sources over the entire electromagnetic spectrum—from radio through TeV. The main discussions today revolve mainly around the internal structure and parameters of the jets, and the particle processes responsible for the emission.

Within the blazar community we have become so accustomed to jets that we take them for granted, however, outside the community jets are still considered something exotic: one tries to get away with thermal, spherical or disk-only models as long as possible. One example certainly is the Gamma-Ray Burst (GRB) field. This are the most violent, non-thermal events we know in the universe. After all the experience with the history of AGN research, it should have been natural to start with the assumption that these energetic sources have a setting similar to the other violent sources we know in the universe—blazar jets. Still, only after many years and under the overwhelming weight of extremely luminous GRBs like GRB971214, the fireball model collapsed into various jet models (one of which was developed by us and which we cannot resist to reference, i.e. Pugliese, Falcke, & Biermann 1999).

Another example are X-ray binaries, which are considered micro-quasars (Mirabel & Rodríguez 1999) and where many jets have been observed (Fender & Hendry 2000; Fender & Kuulkers 2000). However, so far we have not identified anything similar to a blazar or BL Lac in this field, and most emission models seem to ignore completely the non-thermal jet component. Is this justified in all cases?

Finally, we need to mention the AGN-field itself. Clearly, blazars are just the tip of the iceberg since we usually are only looking at the most luminous AGN. Only some 10% of these can be considered radio-loud, and of these only a small fraction will point towards us to reveal the amplified non-thermal spectrum through boosting. Of course there are many more AGN around: radio-quiet quasars and low-luminosity AGN that we never really consider to be blazars or blazar-like, yet even here jets are important ingredients.

Thus, the question we want to address is: can we identify blazars or blazar-like sources in other classes of astrophysical sources? How do we find them and what are we looking for? In the end this boils down to the question what the definition of a blazar actually is. Does "blazar" in this context imply an observational classification or a description of a physical state? For example, in luminous AGN relativistic boosting is very important since it makes it possible for the non-thermal emission to dominate over any thermal emission. But what happens if for some reason the thermal emission is strongly suppressed anyway, e.g. by a hot, radiatively inefficient accretion flow? In this case we will still see the non-thermal jet emission dominating even from large angles of the line-of-sight to the jet axis and we may still have substantial variability—do we call this a blazar even without relativistic boosting being important? And if not, how are we going to tell such a source apart from a boosted jet in surveys? Determining the Doppler factor of jets is still very difficult.

Rather than answering all these question here, we will rather discuss a few recent observational results on non-traditional blazars sources, that nevertheless show how difficult such a classification can become. Since some of the distinctions among AGN are related to the question of radio-loudness we will also discuss this issue here, since it was hotly debated at this workshop.

2. Radio Dichotomy

The first quasars were actually discovered through their radio identification (Hazard, Mackey, & Shimmins 1963) and belonged to the blazar class. It was soon found that many other quasars (QSOs) exist that do not show such strong radio emission. Indeed, support for a bimodal distribution of radio luminosities (relative to optical luminosity) came from radio observations of optically (Strittmatter, Hill, & Pauliny-Toth 1980; Kellermann, Sramek & Schmidt 1989; Falcke, Sherwood, & Patnaik 1996) and X-ray selected quasar samples (Della Ceca et al. 1994). In some cases with lower statistics the evidence for an actual bimodality remained ambiguous (Hooper et al. 1996) and White (this conference, see also Helfand et al. 1999) claimed that in the (radio-selected) FIRST quasar survey a bimodality was not confirmed. Similarly, at this conference Meg Urry raised the question whether in fact there is a continuous distribution of radio-to-optical luminosities rather than a dichotomy. Are we fooled similar to the apparently artificial distinction between X-ray and radio-selected BL Lacs that is now becoming replaced by a continuum of sources just peaking at different frequencies?

The answer to this challenge is probably that both camps are not entirely wrong, depending on what the actual question is. The problem is, that any dichotomy in the physics of AGN and of jet formation can be easily washed out by various effects. A survey that contains a mixed bag of AGN—old and young, luminous and faint, obscured and face on—may in fact not show any bimodality while the underlying physics still is. Why is this?

The dichotomy is seen so far mainly in optically selected samples of quasars with rather homogeneous properties: strong and luminous UV bump and broad emission lines. Assuming this is indeed emission from radiatively efficient accretion disks, the optical luminosity should be a good measure of the accretion power onto the central black hole. Similarly, if the radio properties of quasars are homogeneous then the radio-luminosity can be a useful estimate for the jet power. A constant radio-to-optical flux density ratio (the R-parameter) then reflects nothing else but a constant fraction of the accretion power being channeled into a jet. This is one of the motivations behind the jet-disk symbiosis idea (Falcke & Biermann 1995) and explains the radio optical correlations of quasars (Baum & Heckman 1989; Miller, Rawlings, & Saunders 1993; Falcke, Malkan, & Biermann 1995). A bimodality in the R-parameter than implies a dichotomy in the jet formation—whatever that mechanism is.

On the other hand the R-parameter is a rather shaky measure for such a physical effect. First of all, the radiative efficiency of jets need not be constant and in fact depends on many external factors. Doppler boosting of the core certainly affects it, and for hotspots the pressure of the surrounding medium and their location of the jet terminus (inside or outside the galaxy) can change the radio output quite strongly for a fixed jet power. The latter is seen for example in the declining radio power for jets as they grow from GPS (Gigahertz-Peaked-Spectrum) to CSS (Compact-Steep-Spectrum) sources and finally evolve into large radio galaxies (e.g., O'Dea 1998).

On the other hand, the optical luminosity can easily be obscured or could be entirely absent so that a reliable estimate of the accretion power is not possible. The optical selection of the PG quasar sample, on which many of the papers mentioned above are based on, seemed to have avoided many of these problems. In this context "bias" can be good since it provides one with a rather clean sample that gives us some insight into the physics of jet formation. This is illustrated in Figure 1 where we show the distribution of R-parameters in the PG quasar sample, however, with all known flat-spectrum ($\alpha > -0.5$ at 5 GHz) sources taken out—this discriminates strongly against those quasars affected by boosting and also against GPS sources. The resulting distribution is rather clean and bimodal (but is also present if one does not take out the flat-spectrum quasars).

The answer to the question whether jet formation in luminous quasars is bimodal, is therefore most likely "yes." On the other hand, the question whether

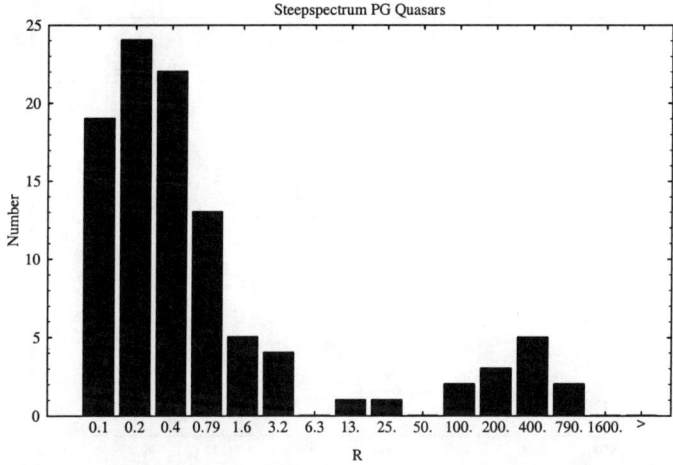

Figure 1. Distribution of the radio-to-optical flux density ratio R for steep-spectrum PG-quasars (from Falcke, Sherwood, & Patnaik 1996)

the R-parameter distribution for 'all the AGN on the sky' is bimodal, is most likely "no."

3. Radio-Quiet Quasars & Seyferts

One interesting finding from studying the R-parameter distribution in PG quasars was that the region intermediate between the radio-loud and radio-quiet distribution was not empty but populated with many flat-spectrum sources (Miller, Rawlings, & Saunders 1993; Falcke, Malkan, & Biermann 1995; Falcke, Sherwood, & Patnaik 1996), called radio-intermediate quasars (RIQs). Some of these sources had radio properties similar to those of blazars: core-dominated, flat-spectrum, and variable. The suggestion was that they constitute a population of relativistically boosted radio-quiet quasars, i.e. something one might call radio-weak blazars.

To make this case watertight one would have liked to see superluminal motion in these sources. Early VLBI observations did not reveal anything but a compact core. However, one of the galaxies, III Zw 2, then became target of a monitoring campaign during a major outburst starting in 1997. Barely resolved in the first three epochs it suddenly started to expand for a brief period of about a few months (Brunthaler et al. 2000b) and then held steady while continuing to decrease in flux (Fig. 2). The spectral evolution monitored with the VLA indicated that the expansion happened on an even shorter time scale and depending on which timescale one takes the implied expansion speed was between 1.2 and 2.7 c, i.e. superluminal even in the most conservative case.

This finding is interesting, since it shows that indeed the RIQs contain relativistic jets. Apart from the bright radio core, III Zw 2 has all the properties of a Seyfert galaxy with a spiral host and very faint, uncollimated, and short radio lobes—certainly not an FR I or FR II radio galaxy. Is this then a radio-weak

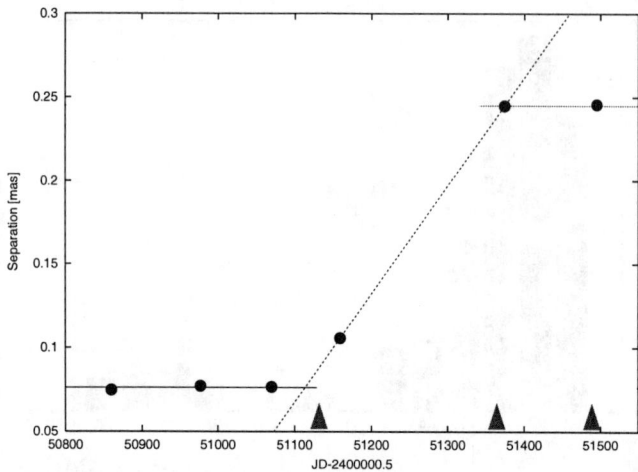

Figure 2. Component separation as a function of time for the Seyfert galaxy III Zw 2 as measured by the VLBA at 43 GHz. There is a short period of superluminal expansion indicating the presence of a relativistic jet in this Seyfert galaxy. (From Brunthaler, Falcke, & Bower et al. 2000)

blazar? It certainly has all criteria one might require: core-dominance, variability, and superluminal motion. However, there is one additional factor complicating things: the evolution of the outburst—with a stop and go behavior—suggests that the jet itself is strongly interacting with the surrounding ISM. (Brunthaler et al. 2000a) suggest that this could be explained with an 'inflating balloon' model and the formation of ultra-compact hotspots in this galaxy, similar, yet smaller in size, to those seen in GPS and CSS sources.

Consequently the large radio flux and compact size is not entirely due to boosting alone even though a relativistic jet is clearly present. Here we may be pointed also into another direction, if III Zw 2 can be regarded as a luminous and perhaps extreme case of a Seyfert galaxy—i.e., a supposedly radio quiet AGN. Like in radio-loud quasars, the jet might start out relativistically but then is disrupted already on the sub-parsec scale, possibly interacting with the torus in some cases, and then propagates outwards sub-relativistically as seen on parsec-scales with VLBI (Ulvestad et al. 1998; Roy et al. 2000) forming the characteristic uncollimated Seyfert radio lobes (Ulvestad & Wilson 1989). The denser ISM in spirals compared to ellipticals may play a role here.

Whether or not the jet-ISM interaction in III Zw 2 is important, the relativistic jet suggests that blazar-like radio-quiets should be detectable, possibly at rather high radio-frequencies. A few attempts to find them with the VLBA are underway (Blundell & Beasley 1998). An alternative way to look for radio-weak blazars is to do variability studies. Recently Barvainis et al. (in preparation, see also Falcke et al. 2001) made multi-epoch (11 epochs) observations with the VLA of a sample of 30 radio-quiet, radio-intermediate, and radio-quiet quasars at 8.5 GHz. They found up to 20% variability within one year in the cores of radio-quiet and radio-intermediate quasars exceeding that of cores in (non-blazar) radio-loud quasars. Even among a sample of ill-selected core-dominated radio-sources (used as phase-calibrators but dominated by blazars) only a few sources showed somewhat more variability on the same timescale. Clearly, the compact radio-emission in these variable radio-quiet/intermediate quasars is AGN-dominated and most likely from a jet. It would be worth investigating some of the top variability performers (after III Zw 2 which tops that list in this survey) in greater detail with the VLBA to look for jets and superluminal motion again.

4. Low-Luminosity AGN

Of course, blazars could be dim not only because they are radio-weak, but also because the jet power and the accretion rate onto the black hole are low. Blazars and BL Lacs have been associated with both FR I and FR II radio galaxies (Urry & Padovani 1995) and hence span a large range of luminosities and jet powers already. Does this continue further down to even less luminous AGN? Marcha et al. (1996) studied a sample of fainter core-dominated AGN and found a number of blazar-like sources, however, with a large spread in optical and line-emission properties that makes it once more difficult to decide what to call a BL Lac. Clearly, once the power is very low, emission from the galaxy—hot gas and stars—will dilute every blazar spectrum.

This saga continues at even lower powers. Nagar et al. (2000), Falcke et al. (2000a), and Falcke et al. (2000b) studied a sample of Low-Luminosity AGN (LLAGN), selected from the spectroscopic survey of Ho, Filippenko, & Sargent (1995) with the VLA and the VLBA. A large fraction of these AGN showed flat-spectrum compact radio cores in their nuclei. In the brightest cores the VLBA resolved the radio emission into jet-like structures. Morphology and spectral index together therefore suggest that the compact radio emission, like in radio-loud and radio-quiet quasars, is produced in an outflow rather than in an advection dominated accretion flow (ADAF, e.g. Yi & Boughn 1998). Finally, a comparison of the radio flux densities at several epochs revealed rather strong intra-year variability with peak-to-peak variations of up to 200–300% in some cases. X-ray emission is also detected in many of these LLAGN (Terashima, Ho, & Ptak 2000) as are optical/UV continuum point sources (Maoz et al. 1995). Of course, the UV-to-X-ray emission might come from an accretion disk, as commonly assumed, but how can we be sure? Could some of it also be non-thermal emission from a relativistic jet pointing towards us? How could we possibly pick out a very weak BL Lac from a 'normal' source?

This confusion is highlighted in Fig. 3 where we show the radio-optical correlation for a mixed sample of radio-loud AGN together with the predictions from the jet-disk symbiosis model from Falcke & Biermann (1996) outlining regions of constant inclination angle to guide the eye. Besides an overall scaling of the radio core power with optical emission it is interesting that the radio cores of giant elliptical LLAGNs seem to connect to the radio cores in FR I radio galaxies. They are also close to the "blazar" line (solid) and are slightly offset from the rest of the LLAGN population. It is not clear at present, whether this represents another radio-loudness dichotomy at lower luminosities or not. There

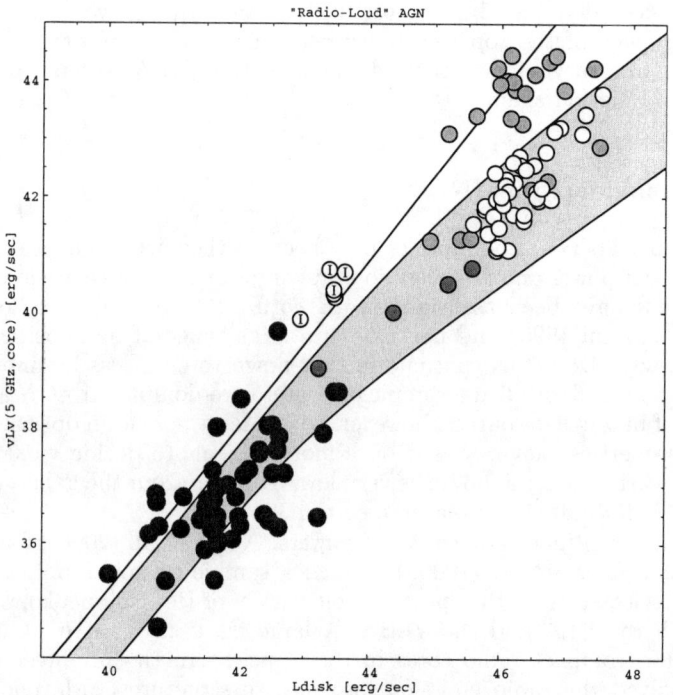

Figure 3. 5 GHz radio core vs. bolometric nuclear luminosity (derived from narrow Hα luminosities) for radio-loud AGN as given in Falcke, Malkan, & Biermann (1995) and Falcke (1996). Solid black dots are LLAGN from Falcke, Nagar, & Wilson (2000). Circles marked "I" are FR I radio galaxies from Rawlings & Saunders (1991), open circles are quasars. Gray shades indicate flat-spectrum, core-dominated quasars, dark gray shades are radio-intermediate quasars and Seyferts. The lines are the radio-loud jet-disk symbiosis model from Falcke & Biermann (1996), where the thick line represents sources with inclination angles at the boosting cone, i.e. what is expected for blazars. Interestingly the LLAGN at this line (black dots) are all giant ellipticals and seem to connect to FR I radio galaxies. These sources may be underluminous in Hα (i.e. $L_{\rm disk}$) emission and might need to be shifted to the right.

is some evidence that FR Is are simply underluminous in emission lines (Zirbel & Baum 1995), perhaps due to obscuration (Falcke, Gopal-Krishna, & Biermann 1995, but see Chiaberge, Capetti, & Celotti 2000) or due to radiatively deficient disks. This would shift the entire population to the right and thus below the "blazar" line.

Clearly the FR Is and their low-luminosity siblings cannot be all blazars in the sense that they are all pointing towards us. On the other hand (Chiaberge, Capetti, & Celotti (2000; also at this workshop) argue convincingly that if an optical point source is found in an FR I radio galaxy it is probably non-thermal synchrotron emission. Therefore these sources may be dominated by non-thermal emission but perhaps are not "blazing." In line with the BL Lac/FR I unification scheme one should still also find some sources that are well above the "blazar" line shown in Fig. 3 because of strong relativistic boosting and suppression of Hα emission. So far the sample sizes for LLAGN are probably too small to find such sources but some of them might have been included in the Marcha et al. (1996) survey.

5. Sgr A* and the Nature of the Radio-dichotomy

In our journey to lower luminosities we now want to take a brief look at the least luminous AGN we know of, and which may offer some clues for the physics of jets and blazars. Sgr A* is now generally believed to be the central black hole of the Galaxy and its emission mechanism has been strongly debated (see Melia & Falcke 2001 for a review). However, recent X-ray observations have provided some interesting new insight. Baganoff et al. (2000) detected a point source at the position of Sgr A* with Chandra, however, with a rather soft spectrum—too soft for thermal bremsstrahlung. On the other hand this X-ray spectrum fits nicely with the expectations one has from synchrotron self-Compton (SSC) emission from the submm-wave emission region ("submm-bump") in Sgr A*. The parameters of this bump are rather well-constrained (Falcke 1996b; Beckert & Duschl 1997) and hence calculating SSC emission is straight forward. In fact the entire radio-through-X-ray spectrum of Sgr A* is now very well fit by pure non-thermal emission from a radio jet (Falcke & Markoff 2000), the main SSC contribution coming from the nozzle.

As one can see from the model fit to Sgr A* in Fig. 4 that the spectrum up to X-rays can be explained by two humps (radio and SSC), very similar to what is seen in BL Lacs (Fossati et al. 1998). The absence of any thermal emission component provides another similarity. A major and interesting difference, however, is the fact that Sgr A* does not show an optically thin power-law at high frequencies: the strong IR limits require an almost exponential cut-off of the synchrotron spectrum. The absence of such an electron power-law also explains the compactness of the jet (as seen also in M81; Bietenholz, Bartel, & Rupen 2000), since, in contrast to the flat-spectrum core which is the compact $\tau = 1$ surface of the jet, the extended jet emission in AGN is always optically thin, steep-spectrum emission from a power-law.

In Sgr A* we are very confident that the highest synchrotron frequencies come from the smallest region, just a few Schwarzschild radii from the black hole (e.g., Krichbaum et al. 1998). Hence, the electron population we see in Sgr A*

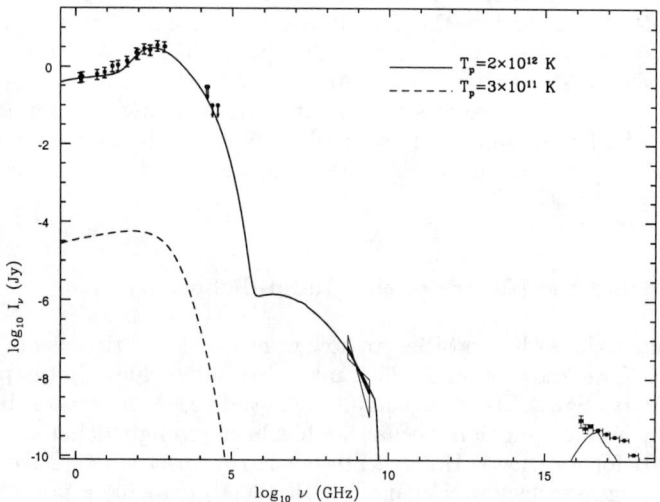

Figure 4. Radio through γ-ray spectrum for Sgr A* resulting from proton-induced $e^{\pm}$s in a hot ($T_p = 2 \times 10^{12}$ K) accretion flow fed into a plasma jet. This includes radio emission from jet and nozzle, X-ray emission via synchrotron self-Compton in the jet, and γ-rays from π^0-decay in the accretion flow. The γ-ray data are considered upper limits because of the large beam of the observations. The fitted parameters for the jet are: nozzle width $2r_0 \sim 5 \times 10^{12}$ cm, height $z_0/r_0 = 8$, inclination angle of jet with respect to line-of-sight $\theta_i = 23°$, and $B_0 = 20$ G (implying an equipartition factor of $k = 0.2$, i.e., magnetically dominated). The fit is very tightly constrained and the remaining free parameters given here have an inherent scale, i.e. gravitational radius or equipartition value. The dashed line shows the spectrum for the same parameters but with $T_p = 3 \times 10^{11}$ K.

is probably the freshly injected particle population at the base of the jet and the inner region of the accretion flow *before* the particles are redistributed by shock acceleration. Therefore Sgr A* may offer a unique perspective into jet formation and particle acceleration. From the frequency of the SSC peak relative to the synchrotron peak and the shape of the submm-bum one can then directly derive that the characteristic electron Lorentz factor has to be around $\gamma_e \sim 10^2$. This fits nicely expectations for the minimum Lorentz factor of electrons in radio-loud quasar jets and hence also in blazars (Celotti & Fabian 1993; Falcke & Biermann 1995).

One can then ask what creates these particles at the base of the jet? The best explanation for the absence of a thermal bump in the Sgr A* spectrum is a hot, optically thin accretion flow (Rees 1982; Melia 1992; Narayan et al. 1998) and some of the synchrotron emission in the jet could come from the hot electrons near the inner edge of the accretion disk being advected in a jet. An alternative proposal (Markoff, Falcke, & Biermann 2000) is that proton-proton collisions could be responsible. As soon as the protons in the flow reach a threshold temperature around 10^{12} K, *pp*-collisions become inelastic and inevitably produce pairs, neutrinos, and γ-rays in hadronic cascades. The resulting electron/positron pair-spectrum peaks around 30 MeV, essentially what is needed for Sgr A*. The pair-production rate is a function of the accretion rate and the viscosity parameter α of the accretion flow. For the parameters of Sgr A*, the right number of pairs is produced for accretion rates of order $\alpha^{-1}\dot{M} \sim 10^{-4}$ M$_\odot$ yr^{-1}, just what is expected from Bondi-Hoyle accretion of stellar winds.

The spectrum from this process in conjunction with a jet model is what is actually shown in Fig. 4. It also includes the expected γ-emission from the pion-decay. What makes the *pp*-process so interesting is that it drops drastically when the temperature falls below 10^{12} K. This is the case at the inner edge of hot accretion flows when one reduces the spin of the black hole (e.g., Manmoto 2000). As a consequence the jet quickly switches to a "radio-quiet" state when the spin drops below maximal (dashed line in Figure 4). Hence, *pp*-collisions are an interesting mechanism for particle injection into jets, naturally providing a fundamental switch between radio-loud and radio-quiet jets, and establishing a link between black hole spin and radio-loudness.

In this picture radio-loud jets would be a mixture of a pair plasma and a normal plasma and require hot accretion flows around maximally spinning black holes. The difference in accretion disk structure in radio-loud/radio-quiet AGN could be reflected in their X-ray spectra as reported by Eracleous, Sambruna, & Mushotzky (2000).

6. Summary

Let us now summarize the findings of our exploration into the non-blazar space. First we need to define what we actually mean by BL Lac or blazar-like when applying these terms to weaker sources. The main feature of BL Lacs is their dominant non-thermal broadband spectrum. Domination here needs to be taken relative to other emission components from the *black hole system*, such as the thermal emission from the accretion flow, and not, e.g., relative to the host galaxy. This means that classifications based on ground-based spectroscopy (break contrast, equivalent width) will loose their usefulness for low AGN luminosity levels (Marcha et al. 1996).

Relativistic boosting is another important feature that characterizes a luminous blazar. It mainly serves to enhance the non-thermal over the more isotropic thermal emission. If, however, the thermal emission is suppressed by other means, boosting may not be so important anymore. The physics inside the jet, particle acceleration and non-thermal emission mechanisms, will remain the same with or without boosting. In fact from viewing BL Lacs or blazars at different angles and studying them at different power levels and in different environments we may have yet a lot to learn about them.

To account for these two different definitions, that may not necessarily always go together, we have tentatively started to use the word low-luminosity BL Lac in cases where we mean "non-thermal dominance" and the word low-luminosity or radio-weak "blazar" when talking about boosting.

So, where are the low-luminosity BL Lacs and blazars? In radio-quiet quasars we have now some good evidence for relativistic jets—the variability and the superluminal motion. This clearly speaks for the presence of radio-weak blazars in some of them, such as III Zw 2. They will, however, hardly ever appear as BL Lacs, since even if boosted the non-thermal emission cannot overwhelm the accretion disk emission that is always there.

On the contrary, we find a number of BL Lac-like sources in the cores of low-luminosity AGN, such as LINERS, with core dominance and strong variability. In many cases thermal X-rays are very weak, e.g. in some ellipticals (Mushotzky, priv. comm.) and it could well be that the non-thermal emission is the most important. So, they can look like BL Lacs without necessarily being boosted. This is probably what is seen in Sgr A* in the Galactic Center and hence we might call it the least luminous BL Lac.

Finally, we want to mention that another interesting direction to look in, for future studies, are X-ray binaries. Especially in the Low/Hard-state, X-ray binaries seem to have very little thermal emission and show a flat spectrum emission that extends from radio to optical and then continues in an X-ray power-law. In trying to fit the broadband spectrum of the newly discovered X-ray binary XTE J1118+480 we were able to account for the entire spectrum from radio to X-rays by emission from a mildly relativistic jet (Markoff, Falcke, & Fender 2000) alone. The question to be asked therefore is whether this actually is the BL Lac analogue for stellar mass black holes.

References

Baganoff, F., Bautz, M., Brandt, N., Cui, W., Doty, J., Feigelson, E., Garmire, G., Maeda, Y., Morris, M., Pravdo, S., Ricker, G., & Townsley, L. 2000, ApJ, submitted

Baum, S. A. & Heckman, T. 1989, ApJ, 336, 702

Beckert, T. & Duschl, W. J. 1997, A&A, 328, 95

Bietenholz, M. F., Bartel, N., & Rupen, M. P. 2000, ApJ, 532, 895

Blundell, K. M. & Beasley, A. J. 1998, MNRAS, 299, 165

Brunthaler, A., Falcke, H., Bower, G. C., Aller, M. F., Aller, H. D., & Teräsranta, H. 2000a, in Proceedings of the 5th EVN Symposium (Gothenburg, Sweden: Onsala Space Observatory, Chalmers Technical University), in press

Brunthaler, A., Falcke, H., Bower, G. C., Aller, M. F., Aller, H. D., Teräsranta, H., Lobanov, A. P., Krichbaum, T. P., & Patnaik, A. R. 2000b, A&A, 357, L45

Celotti, A. & Fabian, A. C. 1993, MNRAS, 264, 228

Chiaberge, M., Capetti, A., & Celotti, A. 2000, A&A, 355, 873

Della Ceca, R., Zamorani, G., Maccacaro, T., Wolter, A., Griffiths, R., Stocke, J. T., & Setti, G. 1994, ApJ, 430, 533

Eracleous, M., Sambruna, R., & Mushotzky, R. F. 2000, ApJ, 537, 654

Falcke, H. 1996a, in ASP Conf. Ser., Vol. 102, The Galactic Center (San Francisco: Astronomical Society of the Pacific), 453

Falcke, H. 1996b, in IAU Symp. 169: Unsolved Problems of the Milky Way, Vol. 169, 169

Falcke, H. & Biermann, P. L. 1995, A&A, 293, 665

Falcke, H. & Biermann, P. L. 1996, A&A, 308, 321

Falcke, H., Gopal-Krishna, & Biermann, P. L. 1995, A&A, 298, 395

Falcke, H., Lehár, J., Barvainis, R., Nagar, N., & A. S. W. 2001, in Probing the Physics of Active Galactic Nuclei by Multiwavelength Monitoring, ASP Conf. Ser. (San Francisco: Astronomical Society of the Pacific), in press

Falcke, H., Malkan, M. A., & Biermann, P. L. 1995, A&A, 298, 375

Falcke, H. & Markoff, S. 2000, A&A, 362, 113

Falcke, H., Nagar, N. M., Wilson, A. S., & Ulvestad, J. S. 2000a, in Black Holes in Binaries and Galactic Nuclei, ESO Workshop, eds. P. W. L. Kaper, E. P. J. van den Heuvel (Springer Verlag), in press

Falcke, H., Nagar, N. M., Wilson, A. S., & Ulvestad, J. S. 2000b, ApJ, 542, 197

Falcke, H., Sherwood, W., & Patnaik, A. R. 1996, ApJ, 471, 106

Fender, R. P. & Hendry, M. A. 2000, MNRAS, 317, 1

Fender, R. P. & Kuulkers, M. A. 2000, MNRAS, submitted

Fossati, G., Maraschi, L., Celotti, A., Comastri, A., & Ghisellini, G. 1998, MNRAS, 299, 433

Hazard, C., Mackey, M. B., & Shimmins, A. J. 1963, Nature, 197, 1037

Helfand, D. J., Becker, R. H., Gregg, M. D., Laurent-Muehleisen, S., Brotherton, M., & White, R. L. 1999, BAAS, 195, 1701

Ho, L. C., Filippenko, A. V., & Sargent, W. L. 1995, ApJS, 98, 477

Hooper, E. J., Impey, C. D., Foltz, C. B., & Hewett, P. C. 1996, ApJ, 473, 746

Kellermann, K. I., Sramek, R., Schmidt, M., Shaffer, D. B., & Green, R. 1989, AJ, 98, 1195

Krichbaum, T. P., Graham, D. A., Witzel, A., Greve, A., Wink, J. E., Grewing, M., Colomer, F., de Vicente, P., Gomez-Gonzalez, J., Baudry, A., & Zensus, J. A. 1998, A&A, 335, L106

Manmoto, T. 2000, ApJ, 534, 734

Maoz, D., Filippenko, A. V., Ho, L. C., Rix, H., Bahcall, J. N., Schneider, D. P., & Macchetto, F. D. 1995, ApJ, 440, 91

Marcha, M. J. M., Browne, I. W. A., Impey, C. D., & Smith, P. S. 1996, MNRAS, 281, 425

Markoff, S., Falcke, H., & Biermann, P. L. 2000, in preparation

Markoff, S., Falcke, H., & Fender, R. 2000, ApJ, submitted

Melia, F. 1992, ApJ, 387, L25

Melia, F. & Falcke, H. 2001, ARA&A, 39, submitted

Miller, P., Rawlings, S., & Saunders, R. 1993, MNRAS, 263, 425

Mirabel, I. F. & Rodríguez, L. F. 1999, ARA&A, 37, 409

Nagar, N. M., Falcke, H., Wilson, A. S., & Ho, L. C. 2000, ApJ, 542, 186

Narayan, R., Mahadevan, R., Grindlay, J. E., Popham, R. G., & Gammie, C. 1998, ApJ, 492, 554

O'Dea, C. P. 1998, PASP, 110, 493

Pugliese, G., Falcke, H., & Biermann, P. L. 1999, A&A, 344, L37

Rawlings, S. & Saunders, R. 1991, Nature, 349, 138

Rees, M. J. 1982, in AIP Conf. Proc. 83: The Galactic Center, 166

Roy, A., Wilson, A., Ulvestad, J., & Colbert, E. 2000, in Proceedings of the 5th EVN Symposium (Gothenburg, Sweden: Onsala Space Observatory, Chalmers Technical University), in press

Strittmatter, P. A., Hill, P., Pauliny-Toth, I. I. K., Steppe, H., & Witzel, A. 1980, A&A, 88, L12

Terashima, Y., Ho, L. C., & Ptak, A. F. 2000, ApJ, 539, 161

Ulvestad, J. S. & Wilson, A. S. 1989, ApJ, 343, 659

Ulvestad, J. S., Wrobel, J. M., Roy, A. L., Wilson, A. S., Falcke, H., Krichbaum, T., Bower, G., & Zensus, A. 1998, BAAS, 193, 9003

Urry, C. M. & Padovani, P. 1995, PASP, 107, 803

Yi, I. & Boughn, S. P. 1998, ApJ, 499, 198

Zirbel, E. L. & Baum, S. A. 1995, ApJ, 448, 521

Connection between Superluminal Ejections and γ-Ray Flares in Blazars

Svetlana Jorstad, Alan Marscher

Institute for Astrophysical Research, Boston University, 725 Commonwealth Ave., Boston, MA 02215, USA

Margo Aller, Hugh Aller

Astronomy Department, University of Michigan, Ann Arbor, MI 48109 USA

Abstract. We examine the coincidence of times of high γ-ray flux and epochs of zero separation of superluminal components from the core in EGRET blazars based on a 1993.9–1997.6 VLBA monitoring program at 22 and 43 GHz. In 19 cases of γ-ray flares for which sufficient VLBA data exist, 10 of the flares fall within 2σ uncertainty of the extrapolated epoch of zero separation from the core of a superluminal radio component. The number expected by random chance ≤ 3 at 95% confidence and ≤ 5 at 99.9% confidence. As additional evidence for a connection between γ-ray flares and ejection of superluminal knots we find a higher level of polarization after ejection of VLBI components and during a number of γ-ray flares.

1. Introduction

We have completed a program of monitoring of the milliarcsecond-scale structure of γ-ray blazars (42 sources) with the VLBA at 22 and 43 GHz during the period from November 1993 to July 1997. We have determined velocities of jet components in 33 sources (Jorstad et al. 2001) and compared the epoch of zero separation from the (presumed stationary) core with the γ-ray light curves obtained from the 3rd EGRET catalog (Hartman et al. 1999) in order to determine whether γ-ray flares are associated with major energetic disturbances that propagate down the jet.

To identify γ-ray flares we have determined the mean γ-ray flux of every source as a weighted average value of all measurements, including the upper limits of the flux, with weight equal to $1/\sigma$, where σ is the measurement error; in the case of an upper limit σ is equal to the upper limit itself. We assume that a γ-ray flare is detected if the flux measurement exceeds the average flux value by a factor of 1.9 or more and if the uncertainty of the measurement is less than the deviation of the measurement from the average value. In 29 cases the VLBA data are contemporaneous with γ-ray observations; in 19 of these 29 cases a gamma-ray flare is detected. In 10 of these 19 cases the epoch of zero separation coincides to within 1σ (7 cases) and within 2σ (3 cases) uncertainties, with the time of the γ-ray flare. On the other hand, a number of superluminal ejections do not correspond to γ-ray flares and there are γ-ray flares which are not accompanied by new superluminal knots. Nevertheless, numerical simulations involving 10^6 samples of random epochs of zero core-knot separation shows that in a sample of 19 γ-ray events the number of chance coincidences ≤ 3 at the 95% confidence level. If the number of coincidences ≥ 5 the γ-ray flares and superluminal ejections are associated with each other at the 99.9% confidence level (see Marchenko-Jorstad et al. 2000).

2. Comparison of γ-ray and Radio Polarized Flux Light Curves

The University of Michigan Radio Astronomy Observatory data base contains polarization data for 8 sources with γ-ray flare/superluminal ejection associations, however, in the case of 0440−003, 1222+216, 1622−253, 1622−297, and 1730−130 there are insufficient radio data during the γ-ray outbursts. Fig. 1 displays the γ-ray (circles), total radio flux density (triangles), and polarized radio flux density (squares) light curves on a logarithmic scale for the 5 remaining sources. To these we add the quasar 1156+295, in which there is a prominent γ-ray flare with no associated moving VLBI component detected, and in which there are ejections of superluminal knots at times of which there are no γ-ray data. The solid vertical lines show the times of ejection of superluminal components with 1σ uncertainties (solid horizontal line). The dotted line corresponds to the observed maximum of the γ-ray flare, which is usually imprecise owing to sparse time coverage. The radio light curves are taken at the frequency with the best coverage near the time of the γ-ray flare. The plots show that there is an increase in the γ-ray emission and the polarized radio flux density shortly after the ejection of a superluminal component. A summary of the data is given in Table 1, which presents: (1) the source name; (2) the average γ-ray flux ($\bar{S}_\gamma$) in units of 10^{-8} phot cm^{-2} s^{-1} and its standard deviation; (3) the factor of increase of the γ-ray flux during the flare compared with the average γ-ray flux (f_γ); (4) the time of the maximum of the γ-ray flux (T_γ); (5) the average level of polarization ($\bar{p}$) and its standard deviation; (6) the level of polarization of the local maximum of the polarized flux nearest to the time of the γ-ray flare or extrapolated zero core-knot separation ($p_{\max}$) and its measurement error; (7) the time of the local maximum of the polarized radio flux (T_p); (8) the time of zero separation from the core of the VLBI component (T_o).

We have determined the average delay between the time of ejection of VLBI components and time of γ-ray peaks as a weighted average with weight equal to $1/\sigma$, where σ corresponds to the uncertainty of epoch of zero separation. This indicates that the ejection of the superluminal component leads the γ-ray flare by a time $(T_o - T_\gamma)=0.12\pm0.15$ yr, while the γ-ray flare precedes the local maximum of the radio polarized flux by 0.08 ± 0.06 yr. Both averages differ from zero lag insignificantly, with a probability less than 60%. In the case of quasar 1156+295, a significant increase of the polarized radio flux is observed shortly after a prominent γ-ray flare; in addition, a marginal rise of the level of the radio polarization is found after ejection of superluminal knots.

3. Discussion

The detailed connection between γ-ray flares and ejection of superluminal knots is difficult to specify owing to the sparse character of the γ-ray light curves and

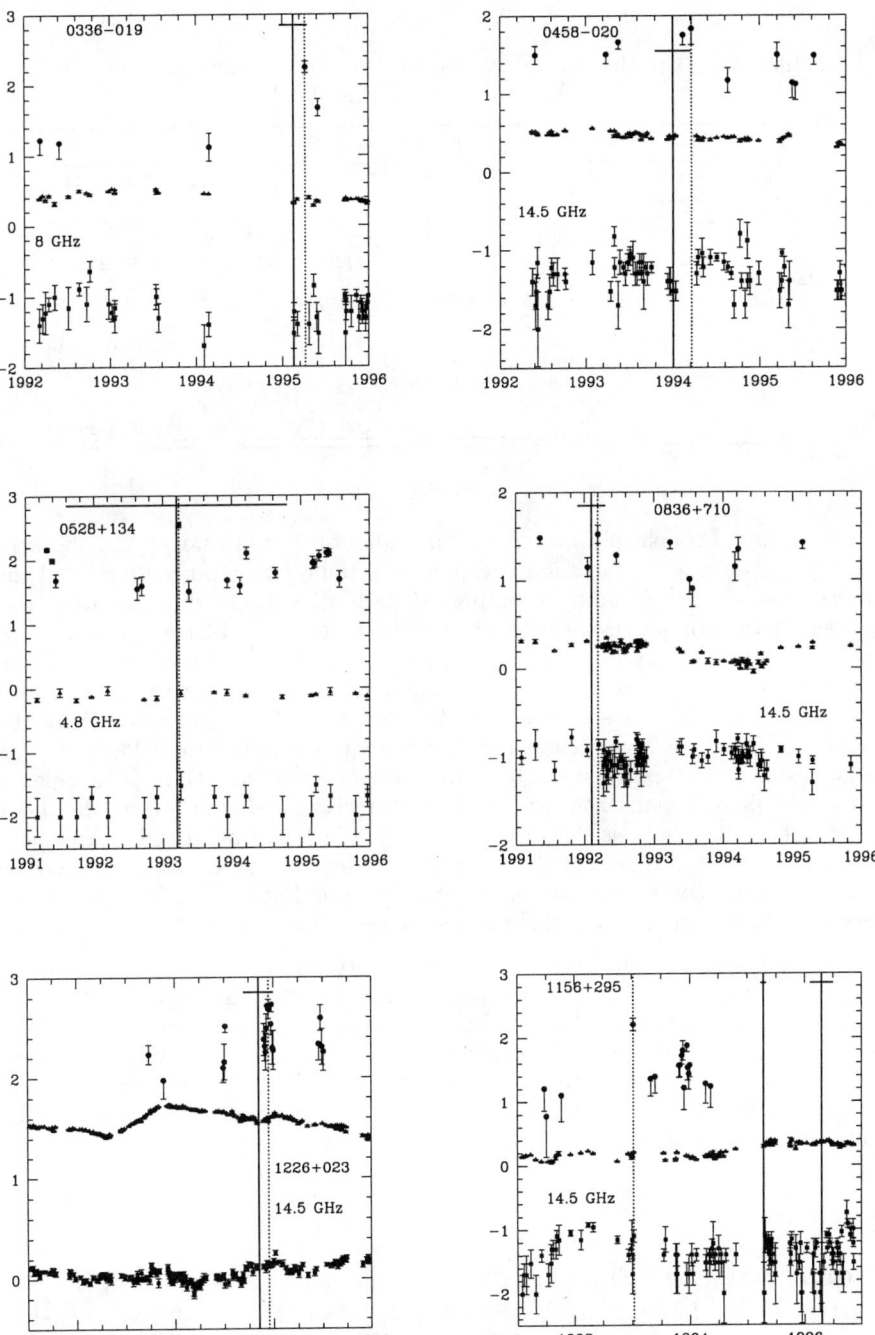

Figure 1. γ-ray flux(circles), total radio flux density (triangles), and polarized radio flux density (squares) light curves; solid lines indicate times of zero core-knot separations, and dotted lines correspond to observed maxima of the γ-ray flares.

Table 1. Epochs of Zero Separation and Gamma-Ray and Radio Polarized Flux Peaks

Name	$\bar{S}_\gamma$	f_γ	T_γ	$\bar{p}$ [%]	p_{max} [%]	T_p	T_o
0336−019	34±10	5.3	1995.266	2.7±1.9	7.19±2.87	1995.361	1995.13±0.16
0440−003	28±8	3.0	1994.632				1994.21±0.19
0458−020	21±10	3.2	1994.212	2.1±1.2	3.43±1.70	1994.333	1994.02±0.08
0528+134	87±19	4.0	1993.233	2.0±1.6	3.41±1.51	1993.257	1993.2 ±1.6
0836+710	18±7	1.9	1992.197	5.3±2.6	7.71±1.10	1992.203	1992.11±0.25
1222+216	21±8	2.3	1992.984				~1993.1
1226+023	22±10	2.2	1993.899	3.3±1.1	4.09±0.17	1994.031	1993.7 ±0.3
1622−253	41±15	2.0	1995.729				1995.95±0.26
1622−297	68±20	4.7	1995.460				1995.07±0.13
1730−130	48±15	2.2	1995.510				1995.54±0.05
1156+295	27±20	6.1	1993.024	2.5±2.3	5.12±0.82	1993.026	
					3.95±3.33	1995.299	1995.28±0.03
					3.22±0.54	1996.389	1996.3 ±0.2

uncertainties of epochs of zero separation of about 0.2 yr. However, the observed relation between γ-ray emission and polarized radio flux reported here and the time delays we derive lead us to propose a speculative model that explains these characteristics. In our scenario, in which the ejection of a VLBI component leads the γ-ray flare, followed by a rise in the radio polarized flux, the γ-ray emission is caused by the synchrotron self-Compton mechanism in the very thin forward layer of a shock. This shock is first detected near the radio core, where the density of the jet is high. The shock front orders the magnetic field, hence the polarization increases when the front first becomes optically thin. If turbulence causes the polarization of the knot to decrease with the distance from the shock front (whose flux falls as it proceeds down the jet), the polarized flux density will decline after this, although the total flux density can continue to increase until the entire component becomes optically thin. Future multi-frequency observations are required to confirm this scenario.

Acknowledgments. This work was supported in part by NSF grant AST-9802941 and by NASA grants NAG5-2508, NAG5-3829, and NAG5-7323. The VLBA is an instrument of the National Radio Astronomy Observatory (NRAO), which is a facility of the National Science Foundation operated under cooperative agreement by Associated Universties, Inc. The UMRAO observations were partially supported by NSF grant AST-9421979 and preceeding grants and by the University of Michigan Department of Astronomy.

References

Hartman, R. C., Bertsch, D. L., Bloom, S. D., et al. 1999, ApJS, 123, 79
Jorstad, S. G., Marscher, A. P., Mattox, J. R., et al. 2001, ApJS, submitted
Marchenko-Jorstad, S. G., Marscher, A. P., Mattox, J. R., et al. 2000, in Astrophysical Phenomena Revealed by Space VLBI, eds. H. Hirabayashi, P. G. Edwards, & D. W. Murphy (Sagamihara, Japan: ISAS), 305.

The Emission Line Properties of DXRBS Blazars

Hermine Landt

Space Telescope Science Institute, 3700 San Martin Drive, Baltimore, MD 21218, USA

Hamburger Sternwarte, Gojenbergsweg 112, D-21029 Hamburg, Germany

Paolo Padovani

Space Telescope Science Institute, 3700 San Martin Drive, Baltimore, MD 21218, USA

Affiliated to the Astrophysics Division, Space Science Department, European Space Agency

On leave from Dipartimento di Fisica, II Università di Roma "Tor Vergata," Via della Ricerca Scientifica 1, I-00133 Roma, Italy

Eric S. Perlman

Joint Center for Astrophysics, University of Maryland, 1000 Hilltop Circle, Baltimore, MD 21250, USA

Paolo Giommi

BeppoSAX Science Data Center, ASI, Via Corcolle 19, I-00131 Roma, Italy

Abstract. We present preliminary results on the emission line measurements of flat spectrum radio quasars and BL Lacs from DXRBS, a blazar sample reaching relatively faint radio and X-ray fluxes. We find that the two FSRQ subclasses, high-energy and low-energy peaked FSRQ, do not differ significantly in the luminosity of their emission line regions. The optical continuum luminosity, on the other hand, is about half an order of magnitude higher for high-energy peaked FSRQ. Furthermore, we observe a continuity in emission line luminosity between FSRQ and BL Lacs, which indicates that the separation of these two blazar subclasses by emission line strength might not necessarily be a strictly physical one.

1. The Sample

The Deep X-ray Radio Blazar Survey (DXRBS) (Perlman et al. 1998; Landt et al. 2001) is currently the largest existing sample of blazars reaching relatively faint radio ($\sim$ 50 mJy) and X-ray (a few $\times 10^{-14}$ erg cm^{-2} s^{-1}) fluxes. It comprises $\sim$ 370 candidates and its identification is almost ($\sim$ 90%) complete.

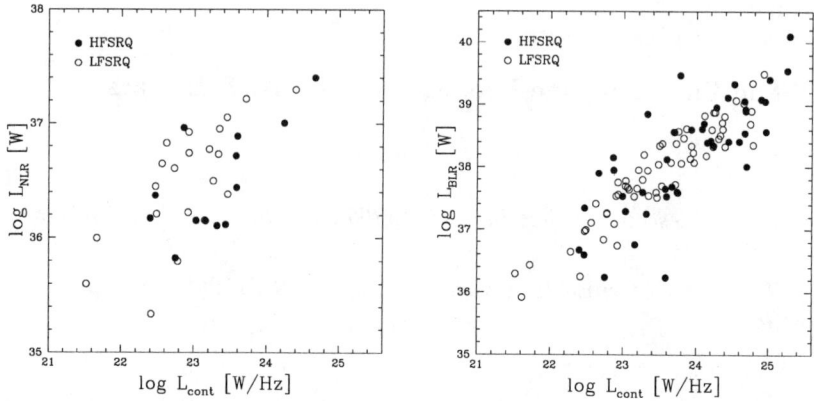

Figure 1. The luminosity of the NLR and BLR vs. the optical continuum luminosity for HFSRQ (filled circles) and LFSRQ (open circles).

The DXRBS selected its candidates based on the fact that blazars are strong radio and X-ray emitters and have a flat radio spectrum. The X-ray database *ROSAT* WGACAT was correlated with several radio catalogs (GB6 and NORTH20 in the north and PMN in the south), restricting the candidate list to sources with a radio spectral index $\alpha_r \leq 0.70$, where $S_\nu \propto \nu^{-\alpha}$.

The DXRBS contains, as of October 2000, 330 optically identified sources: 252 radio-loud quasars (193 flat-spectrum radio quasars: FSRQ [$\alpha_r \leq 0.50$] and 59 steep-spectrum quasars: SSRQ), 43 BL Lacs, and 35 narrow-line radio galaxies (NLRG). Of these sources, 221 have been spectroscopically identified by us and 109 are previously known objects.

With the help of DXRBS, which is a sample containing for the first time a large number of blazars selected in a uniform and unbiased way, we hope, amongst other things, to clarify in detail the properties of the FSRQ class and its relation to the less luminous BL Lacs.

In the following sections we present preliminary results on the emission line measurements of 135 DXRBS blazars (109 FSRQ and 26 BL Lacs).

2. The Flat Spectrum Radio Quasars

We have found in the DXRBS (and similarly in the RASS Green Bank (RGB) Survey; Padovani et al., in preparation) a large population of FSRQ whose synchrotron emission peak seems to be located in the UV/soft X-ray band. These objects were named, in analogy to the subdivision of the BL Lac class, high energy-peaked FSRQ (HFSRQ). A dividing value of $\alpha_{rx} = 0.78$ is chosen to separate them from their siblings, the low energy-peaked FSRQ (LFSRQ).

As a first step in order determine the differences between the two FSRQ subclasses, we have computed the power of the line emitting regions, both narrow- and broad-line (NLR and BLR), from the measured emission line fluxes. The NLR luminosity was determined from the luminosities of the narrow lines [OII]

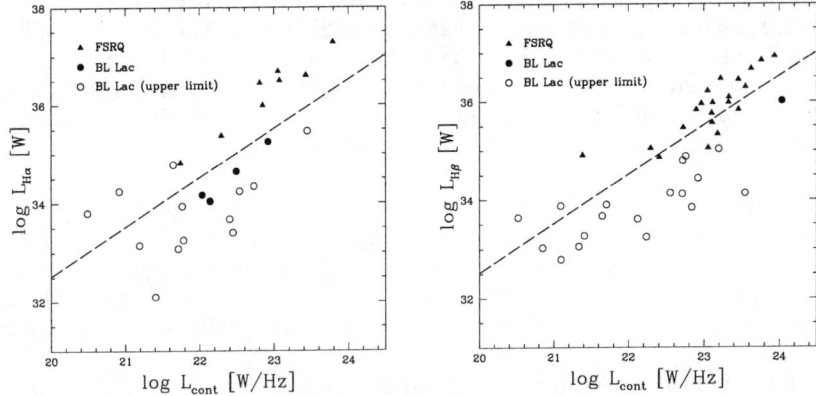

Figure 2. The Hα and Hβ line luminosity vs. the local optical continuum luminosity for FSRQ (triangles) and BL Lacs (circles). Open circles denote upper limits. The dashed line indicates the locus of constant EW = 48 Å.

3727 and [OIII] 5007, as described in Rawlings & Saunders (1991), while the luminosity of the BLR was computed following Celotti, Padovani & Ghisellini (1997).

Fig. 1 shows the NLR luminosity, L_{NLR}, for 34 objects, and the BLR luminosity, L_{BLR}, for 109 objects, vs. the optical continuum luminosity, L_{cont}. First, we find that LFSRQ reach about an order of magnitude lower optical continuum luminosities than HFSRQ. In addition, according to a Student's t-test, the mean optical continuum luminosity of LFSRQ is about half an order of magnitude lower than that of HFSRQ ($P = 98.0\%$). The test was performed for the L_{cont},L_{BLR} distribution, where the statistical uncertainties are lower given the much higher number of objects. This result confirms an interpretation that the jet emission responsible for the continuum in FSRQ peaks at different frequencies for LFSRQ and HFSRQ (IR/optical band and UV/soft X-ray band respectively). Therefore, in HFSRQ we believe to observe a rise of the continuum power in the optical band (towards its peak), whereas this is declining in the spectra of LFSRQ.

Regarding the power of the emitting line regions, we find that LFSRQ and HFSRQ have NLR and BLR similar in luminosity. In summary, the two FSRQ subclasses differ in their non-thermal (jet) energy distributions, but seem to be indistinguishable considering the power of their emitting line regions, suggesting that the ionizing continuum is not related to the jet.

3. Flat Spectrum Radio Quasars and BL Lacertae Objects

The blazar class is split into two subclasses depending on the appearance of the optical spectrum: FSRQ have strong, broad and narrow emission lines, while BL Lacs show no or very weak emission features.

Current blazar surveys separate FSRQ from BL Lacs by the strength of the emission lines observed in the optical spectrum. Following Marchã et al. (1996), the rest frame equivalent width of the strongest of these lines is required to be below 48 Å for an object to be called a BL Lac. This limit is set for a Ca H&K break (an absorption feature typically seen in the spectra of elliptical galaxies) of 0.4, and is lower if this break is decreased by the non-thermal emission coming from the jet.

The studies of Marchã et al. (1996) were based on Hα 6563. This emission line, which is one of the strongest, appears in the optical band ($\sim$ 3000–9000 Å) only if an object has a redshift of $z < 0.4$. For BL Lacs with redshifts $0.4 < z < 1.4$ any separation from FSRQ will need to be based on the strength of the Hβ 4861 or [OIII] 5007 emission line (above $z \sim 1.4$, MgII 2798 lies in the optical band).

We have investigated how FSRQ and BL Lacs behave in terms of their Hα and Hβ emission line luminosities. Fig. 2 shows the Hα and Hβ luminosities vs. the local optical continuum luminosity for the FSRQ and BL Lacs from DXRBS. The dashed line indicates the locus of constant EW = 48 Å.

For both emission lines we observe a continuity in luminosity between FSRQ and BL Lacs. This result is consistent with the studies of Scarpa & Falomo (1997) based on MgII 2798. Therefore, a division by emission line strength between the two blazar classes has probably not a strict physical meaning.

4. Conclusions

We have measured the emission line properties of the flat spectrum radio quasars from the DXRBS. We find that the two FSRQ subclasses, high-energy peaked FSRQ (HFSRQ) and low-energy peaked FSRQ (LFSRQ), do not differ significantly in the luminosity of their narrow line region or broad line region. On the other hand, HFSRQ have on average about half an order of magnitude higher optical continuum luminosities. Therefore, the only difference between LFSRQ and HFSRQ seems to lie in the energy distribution of their synchrotron jets.

Current surveys separate the two blazar subclasses, FSRQ and BL Lacs, by the strength of their emission lines. Our preliminary results, based on measurements of the Hα 6562 and Hβ 4861 emission lines, seem to indicate that this division might not be necessarily a physical one. We observe rather a continuity in emission line luminosity between the two blazar subclasses.

References

Celotti, A., Padovani, P., & Ghisellini, G. 1997, MNRAS, 286, 415
Landt, H., Padovani, P., Perlman, E. S., et al. 2001, MNRAS, in press
Marchã, M. J. M., Browne, I. W. A., Impey, C. D., & Smith, P. S. 1996, MNRAS, 281, 425
Perlman, E. S., Padovani, P., Giommi, P., et al. 1998, AJ, 115, 1253
Rawlings, S. & Saunders, R. 1991, Nature, 349, 138
Scarpa, R. & Falomo, R. 1997, A&A, 325, 109

On the Relation between Radio and Non-Radio Elliptical Galaxies

Riccardo Scarpa

European Southern Observatory, Alonso de Cordova, Casilla 19001, Santiago, Chile

C. Megan Urry

Space Telescope Science Institute, 3700 San Martin Drive, Baltimore, MD, 21218, USA

Abstract. Empirical evidence suggests that elliptical galaxies hosting a radio source may not be different from normal non-radio ellipticals. To test this possibility, we use Monte Carlo simulations to reproduce the distribution of radio galaxies in the radio-optical luminosity plane. The input parameters of the simulation are the optical luminosity function (LF) of ellipticals and the radio LF of radio galaxies, linked by a function giving the probability for an elliptical to be a radio galaxy.

Simulations reproduce the observations well, supporting unification of radio and non-radio ellipticals, provided that the probability of an elliptical hosting a radio source is proportional to the square of its optical luminosity.

The difference of ~ 0.5 mag in average optical luminosity between FRI and FRII radio galaxies is also explained in this framework.

1. Introduction

Regardless of nuclear activity, all ellipticals lie in the same fundamental plane, have similar ellipticity distributions, isophotal twists, and colors (Homabe & Kormendy 1987; Ledlow & Owen 1995; Govoni et al. 2000; Urry et al. 2000). Furthermore, the presence of massive black holes at the centers of elliptical galaxies is the rule rather than the exception (Ho 1998; Magorrian et al. 1998; Richstone et al. 1998; van der Marel 1999). This suggests all ellipticals have the potential of experiencing a phase of intense nuclear activity.

To test this attractive possibility, we used Monte Carlo simulations to check whether observed samples of radio galaxies can be random selections of elliptical galaxies.

We start from the optical luminosity function (LF) of ellipticals, which in this scenario gives the number of potential radio galaxies as a function of optical magnitude. Then we introduce a probability function for the fraction of galaxy, of a given magnitude, to host a radio source.

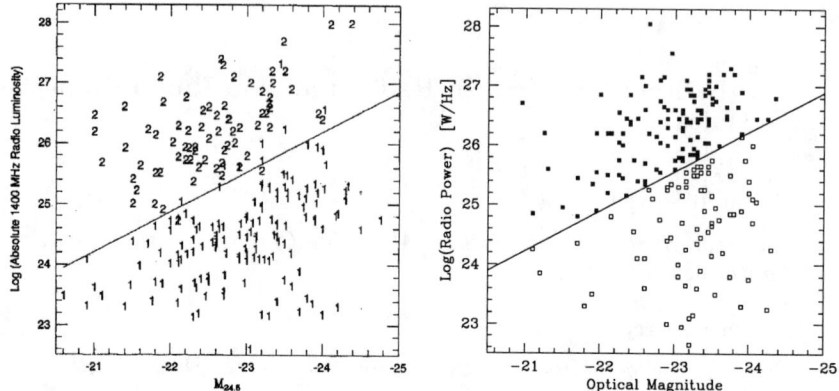

Figure 1. **Left Panel:** The distribution of radio galaxies (1=FRI and 2=FRII) as derived by Ledlow & Owen (1996; their Fig. 1; for consistency $H_0 = 75$ km/s/Mpc was used), plotting data from a complete flux-limited survey, to 0.1 Jy at 1.4 GHz, with no redshift limit. The solid line separating FR I from FR II is the one originally used by Ledlow & Owen. **Right Panel:** Representative Monte Carlo simulation matched to the Ledlow & Owen (1996) sample. The simulation nicely reproduces the almost uniform coverage of the plane in the region $-25 < M_R < -21$ mag and $23 < Log(P) < 28$ W/Hz, with maximum concentration around the center of this region. Solid squares represent FR II, open squares FR I.

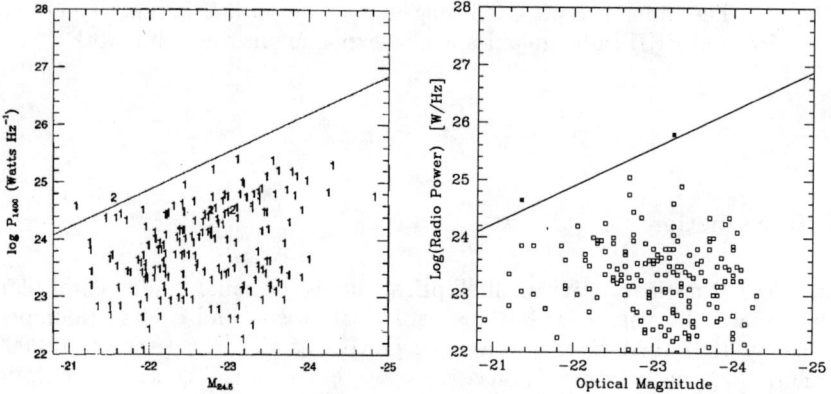

Figure 2. **Left Panel:** Radio and optical luminosity for a sample of 188 radio sources, complete out to $z = 0.09$ down to a radio flux of 0.01 Jy at 1.4 GHz, from Ledlow & Owen (1996; their Fig. 3; for consistency $H_0 = 75$ km/s/Mpc was used). **Right Panel:** Distribution derived from our Monte Carlo simulations. As in the real data, there are no very bright radio sources because of the small volume surveyed. In particular, basically all sources should be FR I (open squares), as observed.

2. Calculation

Based on available empirical results for radio galaxies, the following general assumptions are made:

1. The distribution of ellipticals in optical luminosity L is given by a Schechter function, with $M_R^* = -22.8$ mags and $\alpha = +0.2$, as found in the Stromlo-APM experiment (Loveday et al. 1992).

2. All elliptical galaxies of all optical luminosities have the potential of being radio sources, with probability $S(L) = S^*(\frac{L}{L^*})^h$, where S^* sets the overall normalization. From the bivariate LF it is known that $S(L) \propto L^2$ (Ledlow & Owen 1996), so we set $h = 2$ (the results are not very sensitive to the exact value of h).

3. Regardless of their optical luminosity, all active ellipticals produce radio sources with total power P distributed following the known radio luminosity function $\frac{dN}{dP} \propto P^\beta$, with $\beta = -2$ (Auriemma et al. 1977; Toffolatti et al. 1987; Urry & Padovani 1995; Ledlow & Owen 1996).

4. In the radio-optical luminosity plane, FR I and FR II are separated by a transition line roughly proportional to L^2 (Bicknell 1995).

The number of radio sources per unit volume, having optical luminosity L is the product of the optical LF times the probability S (points 1 and 2), which is a Schechter function with exponent $(\alpha + h) = 2.2$. The final distribution of radio sources in the radio-optical luminosity plane, as derived in a radio-flux-limited survey, is given by the product of $N(L)$ times the function $\frac{dN}{dP}$, assumed to be the same for all optical luminosities (point 3), times the volume $V(P)$ over which sources of power P can be observed above the flux limit of the survey. Once a random set of galaxies has been generated, sources are divided into FR I and FR II according to point 4.

Given the assumptions, there are essentially no free parameters, but we note the exponent h is not well determined. Also, different groups have found significantly different values of M^* and α (Muriel et al. 1995; Lin et al. 1996, Loveday et al. 1992; Zucca et al. 1997), and there is some freedom in the position of the FR I–II transition power.

3. Results and Discussion

Result shows that under quite general assumption the observed distribution of radio galaxies in the radio-optical luminosity plane is nicely reproduced, as is the observed difference in optical luminosity between FR I and FR II.

The physical basis for our result is that all ellipticals should have a central black hole (van der Marel 1999; Macchetto et al. 1999). Active and non-active galaxies are linked by a probability function, found to be $\propto L^2$. We do not attempt to explain why radio sources are preferentially observed in giant ellipticals, however, the L^2 dependence of the probability of radio activity comes from the observed shape of the bivariate radio LF, and is similar to the dependence

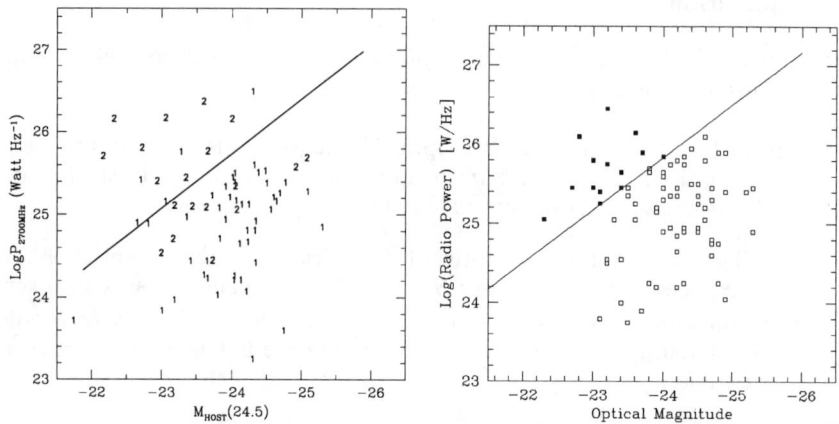

Figure 3. **Left Panel:** Distribution of radio galaxies studied by Govoni et al. (2000; their Fig. 7; for consistency $H_0 = 50$ km/s/Mpc was used). The sample includes all radio galaxies in the redshift range $0.01 < z < 0.12$, down to a flux limit, at 2.7 GHz, of 2 Jy for part of the sample and 0.25 Jy for the rest. **Right Panel:** Monte Carlo simulation matched to the Govoni et al. sample. The agreement is excellent. We are able to explain nicely the distribution of the sources in the radio-optical luminosity plane, as well as the relative population of FR I (open squares) and FR II (solid squares).

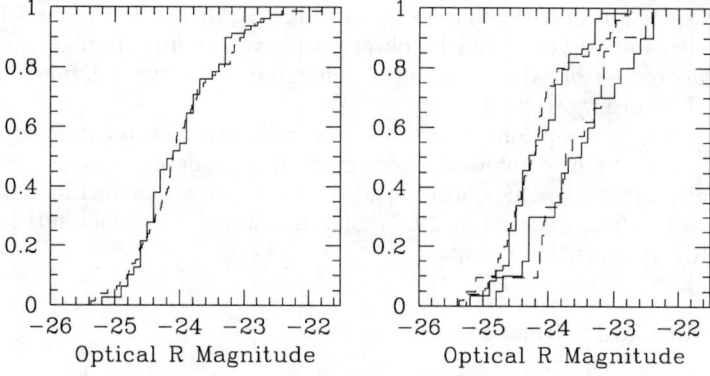

Figure 4. Distributions of optical magnitudes from Govoni et al. (2000) **Left Panel:** Cumulative distribution of absolute magnitudes for the real (dashed line) and simulated (solid line) data set shown in Fig. 3. **Right Panel:** Cumulative distribution of absolute magnitudes for FR I (Left) and FR II (Right) radio galaxies separately. The agreement is excellent, explaining the difference in average optical luminosity between FR I and FR II as a subtle selection effect.

of the transition power from FR I to FR II. There must be a deeper physical meaning for this.

Once accretion onto the black hole has began, the strength of the radio emission should depend mostly on the accretion rate, which is independent from the galaxy size, justifying assumption 3.

The $\sim L^2$ dependence of the transition power from FR I to FR II imposes that to be an FR II, a bright galaxy must be associated with a very powerful radio source, a very improbable combination given the steepness of both radio and optical LF. Thus, from probability alone, the association FR II—faint galaxies is favored, causing the difference in observed optical luminosity between the two classes. No deeper physical difference between FR I and FR II host galaxies is required.

References

Auriemma, C., Perola, G. C., Ekers, R., et al. 1977, A&A, 57, 41
Bicknell, G. V. 1995, ApJS, 101, 29
Govoni, F., Falomo, R., Fasano, G., & Scarpa, R. 2000, A&A, 353, 507
Ho, L. C. 1998, in Observational Evidence for Black Holes in the Universe, ed. S. K. Chakraberti, (Kluwer), 157
Homabe, M. & Kormendy, J. 1987, in Structure and Dynamics of Galaxies, IAU symp. No. 127, ed. T. de Zeeuw, (Dordrecht: Reidel), 379
Ledlow, M. J. & Owen, F. N. 1995, AJ, 109, 853
Ledlow, M. J. & Owen, F. N. 1996, AJ, 112, 9
Lin, H., Kirshner, R. P., Shectman, S. A., et al. 1996, ApJ, 464, 60
Loveday, J., Peterson, B. A., Efstathious, G., & Maddox, S. J. 1992, ApJ, 390, 338
Macchetto, F. D. 1999, in press (astro-ph/9910089)
Magorrian, J., Tremaine, S., Richstone, D., et al. 1998, AJ, 115, 2285
Muriel, H., Nicotra, M. A., & Lambas, D. G. 1995, AJ, 110, 1032
Richstone, D., Ajhar, E. A., Bender, R., et al. 1998, Nature 395, 14
Toffolatti, L., Franceschini, A., Danese, L., & de Zotti, G. 1987, A&A, 184, 7
Urry, C. M. & Padovani, P. 1995, PASP, 107, 803
van der Marel R. 1999, AJ, 117, 744
Zucca, E., Zamorani, G., Vettolani, G., et al. 1997, A&A, 326, 477

Part 2
Jet Physics

Blazar Jets: The Spectra

Gabriele Ghisellini

Osservatorio Astron. di Brera, V. Bianchi 46, I-23807 Merate, Italy

Abstract. The radiation observed by blazars is believed to originate from the transformation of bulk kinetic energy of relativistic jets into random energy. A simple way to achieve this is to have an intermittent central power source, producing shells of plasma with different bulk Lorentz factors. These shells will collide at some distance from the center, producing shocks and then radiation. This scenario, called *internal shock model*, is thought to be at the origin of the γ-rays observed in gamma-ray bursts and can work even better in blazars. It accounts for the observed key characteristics of these objects, including the fact that radiation must be preferentially produced at a few hundreds of Schwarzschild radii from the center, but continues to be produced all along the jet. At the kpc scale and beyond, the slowly moving parts of a (straight) jet can be illuminated by the beamed radiation of the core, while the fast parts of the jet will see enhanced cosmic microwave radiation. In both cases the Inverse Compton process can be the dominant radiation process, leading to a copious production of high energy (X-rays and beyond) radiation in both radio loud quasars and radio-galaxies.

1. Introduction

We believe that the radiation we see from blazars comes from the transformation of bulk kinetic into random energy of particles, which then produce beamed emission. How to produce the large velocities of the plasma in the jet and which is the dissipation mechanism are still a matter of debate, but there is no doubt that nature succeds in producing collimated outflows with bulk Lorentz factors $\Gamma \sim 5$–20 for blazars, and even higher for gamma-ray bursts. Only in recent years we began to estimate the power of jets, through the radiation they emit (e.g. Celotti & Fabian 1993) and especially through the energy required to be transported to the lobes (Rawling & Saunders 1991). It has been found that the observed jet radiation must be a small fraction of the total energetics, even of the last decade witnessed a factor 10 increase in the power observed to be emitted by blazars as a class, thanks to the high energy γ-ray observations of EGRET, onboard the Compton Gamma Ray Observatory satellite. The EGRET observations, and the detection of a few sources (Mkn 421, Mkn 501, 2344+514 and PKS 2155−304) in the TeV band by ground based Cherenkov telescopes, renewed the interest about blazars, allowing the discovery that their Spectral Energy Distribution (SED) is characterized by two broad peaks, whose location is a function of the observed bolometric luminosity (Fossati et al. 1998; Ghisellini

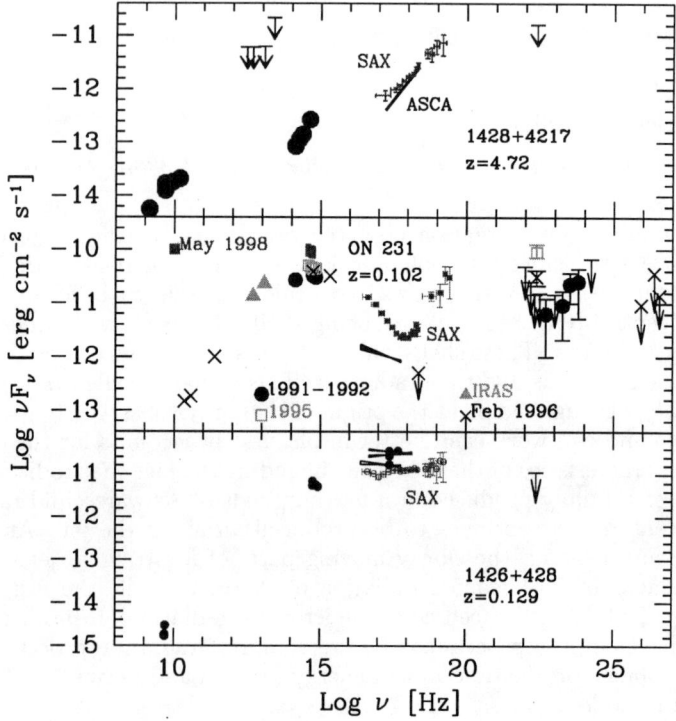

Figure 1. Three examples of SED of blazars to illustrate the blazar sequence. The top panel shows the most distant radio-loud AGN known, at $z = 4.72$ (from Fabian et al. 2000). Its luminosity in hard X-rays exceeds, if isotropic, 10^{49} erg s^{-1}. Note that the peak of the inverse Compton emission is in the MeV band, and that the hard X-ray emission dominates the power output. In the middle panel we show the intermediate BL Lac object ON 231, in which the synchrotron emission dominates the steep soft X-ray flux, and a very flat inverse Compton component dominates above a few keV (from Tagliaferri et al. 2000). In the bottom panel we show the SED of the extreme BL Lac 1426+428, in which the synchrotron component peaks above 100 keV (from Costamante et al. 2000). This is the third example of BL Lac object with a synchrotron peak located above 100 keV, besides Mkn 501 and 1ES 2344+514. In these low luminosity class of sources, the emitting electrons can attain the highest energies, making these objects the best candidates to be detected in the TeV band. Note the broad band X-ray range of *Beppo*SAX and how useful it is to characterize the SED in all three classes of blazars.

et al. 1998). These peaks have been interpreted as due to synchrotron and inverse Compton radiation, respectively. Blazars form a well defined sequence, with low powerful objects having both peaks at a similar level of luminosity, and located at higher frequencies than in more powerful objects, in which the inverse Compton peak dominates the emission. In Fig. 1 we show three examples of SED of blazars with different power, to illustrate the overall behavior and what can be the contribution of X-ray observations in these three classes of objects. Recent observations of high redshift ($z > 4$) blazars (Celotti, these proceedings), and of low power BL Lacs (Costamante et al., these proceedings) have extended the blazar sequence at both ends, confirming the original trend.

At the high luminosity end of the sequence we find interesting lower limits on the bulk kinetic power that the jet can carry, requiring it to be larger than the power dissipated in radiation, derived dividing the apparent luminosity (assuming isotropy and no beaming) by the square of the bulk Lorentz factor. Results indicate that jet of FSRQ (flat spectrum radio quasars) must have a large kinetic power (Celotti, these proceedings). As an example, PKS 0836+710 has an apparent luminosity of 10^{49} erg s^{-1}, which requires a jet power of at least $10^{47}/\Gamma_1^2$ erg s^{-1} (Tavecchio et al. 2000a). Note that in these sources most of the jet power must not be dissipated through radiation, but must feed the extended radio structures.

At the low luminosity end of the blazar sequence we find objects whose synchrotron spectrum peaks in the X-ray band, indicating very large energies of the emitting electrons. Here we can learn about the acceleration mechanism, and find good candidates to be detected in the TeV band (Costamante et al., these proceedings).

Here I will focus on two main topics, namely how the *internal shock scenario* can explain the main characteristics of blazars, and how the large (and very large) scale X-ray jets recently observed by *Chandra* can be interpreted.

2. Internal Shocks

The key idea of the internal shock scenario is to assume a central engine working intermittently, i.e. producing discrete blobs or shells of plasma moving at slightly different velocities. In this case there will always be a later faster shell catching up a slower earlier one. If the initial separation of the two shells is R_0 and the Lorent factors Γ differ by a factor 2, the collision will take place at $R \sim R_0 \Gamma^2$.

This idea is not new: Rees (1978) proposed it to explain some features of the M87 jet, by it was almost forgotten in the AGNs field, even if it became the leading scenario to explain the γ-ray radiation of gamma-ray bursts.

2.1. Points in Favor

"Low" Efficiency. Consider two shells with bulk Lorentz factors Γ_1 and Γ_2 and mass m_1 and m_2. Conservation of energy and momentum implies that a fraction η of the total bulk kinetic energy must be dissipated:

$$\eta = 1 - \Gamma_f \frac{m_1 + m_2}{\Gamma_1 m_1 + \Gamma_2 m_2} \qquad (1)$$

where $\Gamma_f = (1 - \beta_f^2)^{-1/2}$ is the bulk Lorentz factor after the interaction and is given through (see e.g. Lazzati, Ghisellini, & Celotti 1999)

$$\beta_f = \frac{\beta_1 \Gamma_1 m_1 + \beta_2 \Gamma_2 m_2}{\Gamma_1 m_1 + \Gamma_2 m_2} \qquad (2)$$

The above relations imply, for shells of equal masses and $\Gamma_2 = 2\Gamma_1 = 20$, $\Gamma_f = 14.15$ and $\eta = 5.7\%$. The fraction η is not entirely available to produce radiation, since part of it is in the form of hot protons and magnetic field. This is the efficiency for a single collision. Merged shells (that have already collided) can however collide again with other shells (or merged shells), increasing the total fraction of kinetic energy transformed into radiation to 5–10%. The rest is transported to the outer radio structures of the jet. The small efficiency in producing radiation is a major problem in the field of gamma-ray bursts, but is indeed a positive feature for blazars, since we need to transport most of the power to the outer radio lobes.

Right Location. One of the most important implications of the EGRET observations of blazars is the realization that most of the luminosity of these sources must be emitted in a well localized region of the jet (Ghisellini & Madau 1996). This region cannot be too close to the jet apex, to avoid absorption of γ-rays by X-ray radiation produced by the jet itself or by the accretion disk and its corona. On the other hand the rapid γ-ray variability suggests that the γ-ray emitting zone is not too far from the jet apex. Hundreds of Schwarzschild radii are indicated. In powerful blazars, this distance is conveniently close to the distance of the Broad Line Region (BLR), which can produce seed photons for the formation of the γ-ray flux.

On the other hand, the entire jet must emit some radiation, particularly at radio frequencies, where synchrotron self-absorption limits the emission in the inner part of the jet. In the internal shock scenario the emission at large scales is due to collisions between shells that have already collided once (or more times). The efficiency in this case is lower, since the bulk Lorentz factors have already averaged out somewhat.

Variability. Internal shocks are a very simple way to produce variability. In this scenario there is a typical variability timescale (at least for the first collisions) connected to the initial separation between two colliding shells and their width. If the initial separation of two consecutive shells is R_0, they will collide at the distance $R = R_0 \Gamma^2$, but the corresponding time will be observed Doppler contracted by the factor $(1 - \beta \cos \theta) \sim 1/\Gamma^2$ and will be of the same order of the initial separation R_0/c. The duration of each flare is linked to the duration of the collision, which will be of the order of the shock crossing time. Inhomogeneities within the shells and small scale instabilities, if present, can produce variability on shorter timescales.

Correlated Variability. High frequency emission is mainly produced by shells colliding for the first time at $R \sim 10^{17}$ cm from the jet apex. Lower frequency (radio and far IR) flux is produced further out in the jet, when merged shells collide with other merged shells. Therefore there should be some correlations between the light curves at different frequencies, especially between the γ-ray and optical fluxes and the mm-radio flux.

2.2. Internal Shocks: A Powerful Blazar

We (Spada et al. 2000) simulated the case of a powerful blazar jet, of average bulk kinetic power of 10^{48} erg s^{-1}, carried by shells or blobs injected in the jet, on average, every few hours, with a bulk Lorentz factor chosen at random in the range [10–30]. The shell width is initially of the same order of the initial shell-shell separation. Material in the shell (both protons and electrons) are assumed to be initially cold, and consequently the shell is assumed to have a constant width until the first collision takes place. After that, the shell width is assumed to expand with the sound velocity. The first collisions happen at a few$\times 10^{16}$ cm, well within the Broad Line Region (BLR), assumed to be located at 5×10^{17} cm and to reprocess 10% of the disk luminosity, assumed to be equal to 10^{46} erg s^{-1}. For simplicity, a fixed and constant fraction of the energy dissipated during each collision is assumed to go into the electron and to the magnetic field components. The emitting relativistic particles are assumed to have a broken power-law energy distribution throughout the entire emitting zone. This energy distribution is derived by assuming to inject in the source electrons with a single power law distribution with minimum electron energy γ_{min} whose value is found by energetic considerations. Limits in computing time do not allow us to consider details of spectral changes on a timescale faster than the light crossing time of a single shell (few hours when $R \approx 10^{17}$ cm and a month on the parsec scale).

Particles emit by synchrotron, synchrotron self-Compton (SSC), and Compton scattering off the external radiation (EC) produced by the BLR.

We simulate the evolution of the total spectrum summing the locally produced spectra of those regions of the jet which are simultaneously active in the frame of the observer. In Fig. 2 we show some spectra, each corresponding to one particular shell-shell collisions at a different distance, as sketched in the bottom panel. The entire time dependent evolution can be seen in the form of a movie at the URL: http://www.merate.mi.astro.it/~lazzati/3C279/index.html.

As can be seen, the predicted spectra are extremely variable (more so at the higher frequencies). First collisions are the most efficient in converting bulk energy into radiation, since in this case the "Γ-contrast": (i.e. Γ_2/Γ_1) is the largest. These collisions, taking place inside the BLR, make the inverse Compton process the most important cooling agent. The corresponding spectrum therefore peaks in the γ-ray band. Collisions taking place outside the BLR (preferentially between shells that have already collided once) have a smaller Γ–contrast, and see relatively less seed photons. This makes the synchrotron component to dominate (dashed and dotted lines in Fig. 2).

In Fig. 2 we also show the observational data of 3C 279 during three observational campaigns. While the agreement is gratifying, we stress that we have not yet tried to obtain "a best fit" for this source. We have rather used fiducial numbers for the initial time separation of the shells, their initial width and the overall average bulk kinetic energy of the jet.

We have performed a cross-correlation analysis of the light curves at different frequencies. As expected, there is no lag between the γ-ray and the X-ray

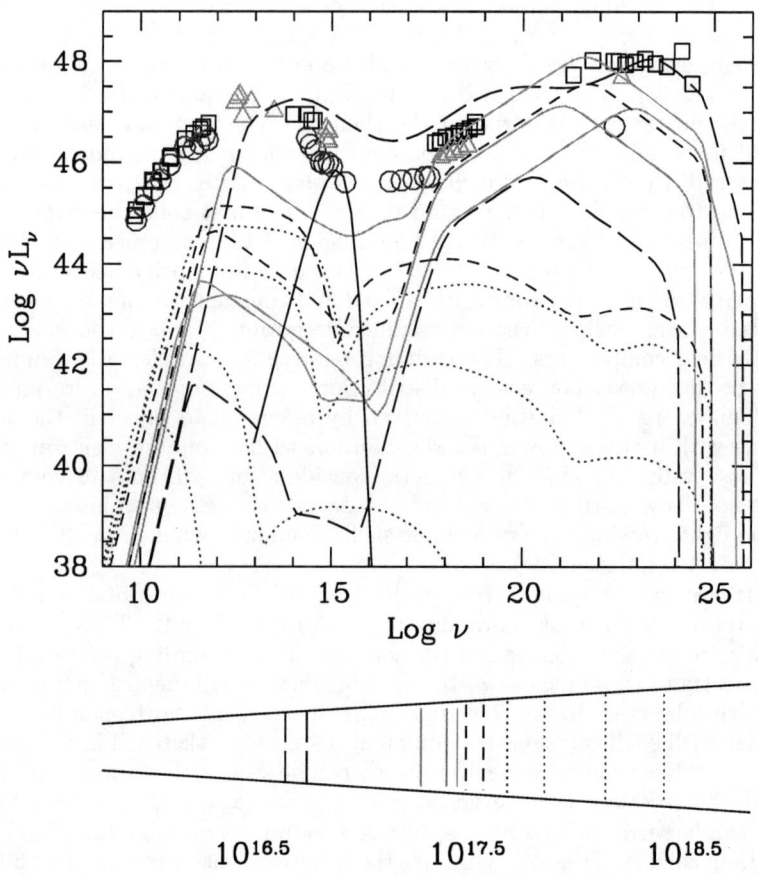

Figure 2. Some spectra calculated in the internal shock scenario, produced at different jet locations, as labelled in the lower panel. For illustration, we have superimposed the SED of 3C 279 during three simultaneous observing campaigns (i.e. in 1991, 1993 and 1996, see Maraschi et al. 1994 and Wehrle et al. 1998). The blackbody peaking at 10^{15} Hz is the assumed spectrum of the accretion disk. 10% of this luminosity is reprocessed by the BLR located at 5×10^{17} cm.

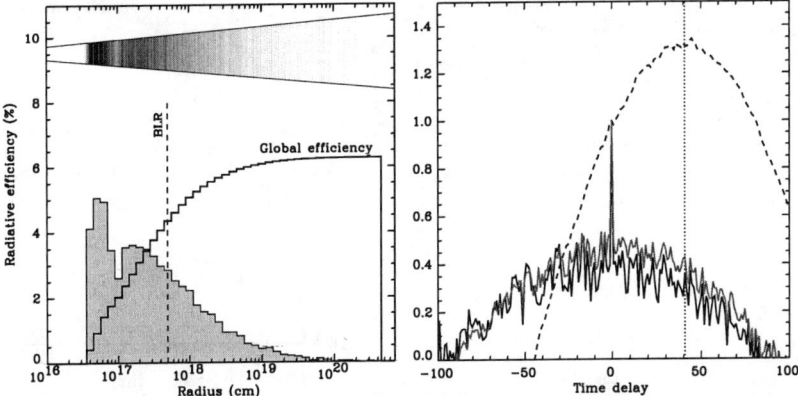

Figure 3. *Left*: The radiative efficiency versus the collision radius for the particular simulation of Fig. 2. The solid line refers to the global efficiency, i.e. the fraction of the total kinetic energy of the wind radiated on scales smaller than a given radius; the shaded histogram shows instead the differential efficiency, i.e. the fraction of bulk kinetic energy radiated at a given radius interval (multiplied by a factor of 10 for clarity). The cone at the top shows a grey-tone representation of the differential efficiency of the jet. The darker the color the higher the efficiency. *Right*: Cross correlation of the simulated light curves, between γ- and X-rays, γ-rays and optical, and γ-rays and the mm band (short dashed line).

and the optical fluxes, which are mainly produced in the same (inner) zones of the jet. Instead, there is a well defined delay of ~ 40 days between the γ-ray and the far infrared (1 mm) fluxes. This is easily explained by the fact that the mm radiation is preferentially produced at some pc from the center, yielding a time delay of

$$\Delta t = \frac{\Delta R}{c\Gamma^2} \sim 38.5 \, \Delta R_{19} \, \Gamma_1^{-2} \text{ days}, \qquad (3)$$

One of the main assumptions of our simulations has been to calculate the value of the magnetic field considering only the energy dissipated in each collisions, and neglecting any seed magnetic field which, surviving from previous collisions, can be amplified by shock compression. As a consequence, the magnetic field B scales with distance R (from the jet apex) as $B \propto R^{-1.5}$, resulting in very small magnetic field values on the outer zones. In turn this implies long cooling times and small variability in the radio band.

While we hope to "cure" this in future work, we would like to stress that there are other possibilities for enhanced dissipation at large distances. The relativistic plasma could in fact be shocked by "obstacles" in the jet, or it can interact with the (steady) walls of the jet. This case resembles what in the gamma-ray burst field is called "external shock scenario," in which the collisions are much more efficient in converting bulk into random energy than internal shocks (i.e. the shell decelerates much more). In this case it is natural to ex-

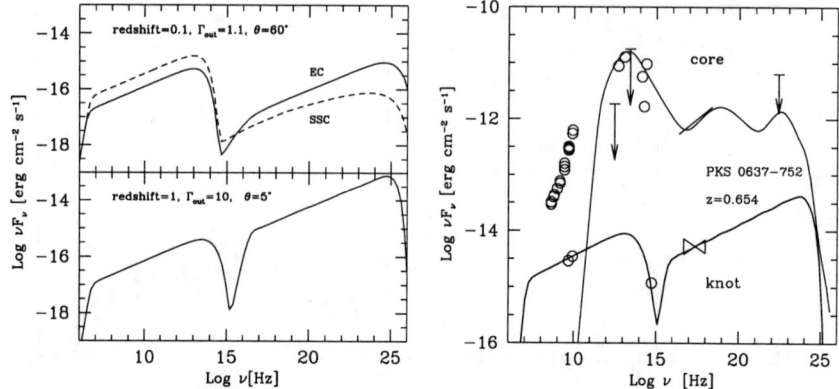

Figure 4. *Left*: SED calculated assuming that at distances of 10 kpc from the center a region of 1 kpc of size embedded in magnetic field of 10^{-5} G radiates an intrinsic power of 3×10^{41} erg s^{-1}. The nuclear (blazar) component emits an intrinsic power of 10^{43} erg s^{-1}. The *upper panel* shows the emission from a layer with $\Gamma_{\rm layer} = 1.1$ and viewing angle $\theta = 60°$. The dashed line corresponds to the SED assuming that electrons emit SSC radiation only. The solid line takes into account the radiation field coming from the core of the jet and illuminating the region. The *bottom panel* shows the emission from a spine moving with $\Gamma_{\rm spine}$ at a viewing angle $\theta = 5°$. *Right*: The SED of the core and the large scale knot of PKS 0637−752, together with the models for both components (solid lines). From Celotti, Ghisellini, & Chiaberge, (2000).

pect some velocity structure across the jet, with a fast "spine" along the axis and slower "layers" along the borders. Indeed, Chiaberge et al. (2000) found evidences for this structure, for explaining the spectra of radiogalaxies.

3. Chandra jets

We have shown (Celotti, Ghisellini, & Chiaberge 2000) that if some dissipation takes place in the jet at distances greater than 1–100 kpc, then the Inverse Compton scattering process with external radiation is favored with respect to the synchrotron self Compton process, leading to an X-ray flux larger than what expected by a pure SSC model. This study was motivated by the recent detection by Chandra of large scale X-ray jets both in radio-galaxies and in quasars, especially in PKS 0637−752 (Chartas et al. 2000; Schwartz et al. 2000). This source is particularly interesting because radio VSOP observations detected superluminal motion in the (pc scale) jet (Lovell 2000), which implies $\Gamma > 17.5$ and a viewing angle $\theta < 6.4$ degrees. This in turn implies a de-projected X-ray jet length of almost a Mpc. We have proposed that, at these scales, the jet is still relativistic $\Gamma = 10$–15) and then its randomly accelerated particles can efficiently interact with the cosmic microwave background (CMB), producing beamed X-ray rays through the inverse Compton process.

It may seem unusual to invoke relativistic bulk motion at these extremely large scales. But:

i) If the jet decelerates, it has to dissipate, and a sizable fraction of the initial energy must be radiated, contrary to the requirement that most of the jet power survives up to the radio structures. More so if the jet becomes only mildy relativistic, through, e.g. entrainment.

ii) Suppose that the 100 kpc jet is only mildly relativistic. The produced radiation is then only marginally beamed, *enhancing* the requirements on the total energetics with respect to a relativistic jet (Ghisellini & Celotti 2000, in preparation).

Celotti, Ghisellini, & Chiaberge (2000) were able to fit both the core and the large scale jet emission of PKS 0637−752 (see Fig. 4), with a bulk Lorentz factor of 17 for the core and 14 for the jet, *conserving the bulk kinetic power of the jet* between the two emission sites, and with nearly equipartition between magnetic and electron energy densities in the large scale jet emission site (see Tavecchio et al. 2000b for another solution).

In the frame comoving with the jet, the energy density of the cosmic background radiation is $\propto \Gamma^2(1+z)^4$. Therefore distant blazars with fast large scale jets should be even brighter than PKS 0637−752 in X-rays with respect to their synchrotron radio-optical components, making the efforts of Chandra to detect them easier.

As mentioned, besides fast "spines" we can have slowly moving "layers," and also these components can emit more inverse Compton radiation with respect to a pure SSC model, because the layers can be illuminated by the beamed radiation produced by the core, if the core and the large scale jets are aligned. Fig. 4 (top panel) shows one example, with the comparison with a pure SSC spectrum. Being slow, and possibly only mildly relativistic, the layers produce radiation which is much more isotropic than the spine: at small viewing angles (blazar case) this component is outshined by the spine component, but it can become visible as the viewing angle increases (i.e. in radio-galaxies).

4. Discussion

Internal shocks can dissipate the right amount of jet power at the right location, originating in a natural way the violent variability observed in blazars. This scenario has the virtue to be at the same time simple and quantitative, offering a coherent view of almost all the radiative jet, from milliparsecs to kiloparsecs. In the γ-ray band we expect the most pronounced variability, in agreement with observations, and we are now working to find how the γ-ray duty cycle (i.e. the fraction of the time spent in high γ-ray states) scales with the initial separation of the shells and their initial bulk Lorentz factor to compare it with data coming from the foreseen γ-ray satellites AGILE and GLAST.

We hope that this scenario will also explain the observed "blazar sequence," linking the overall blazar spectrum with the jet power (i.e. the bulk motion power). We also hope, with longer simulations, to be able to see if there is a relation between the strongest γ-ray flares and the birth of radio superluminal blobs at the pc scale.

The emission predicted by the internal shock scenario (as it is now) beyond the kpc scale likely underestimates the X-ray flux (both soft and hard). At these distances the energy densities in magnetic field and locally produced synchrotron radiation are very small, and seed photons of the cosmic microwave background on one hand and of the core of the jet on the other hand are important contributors to the inverse scattering process for fast spines and slow layers, respectively.

References

Celotti, A. & Fabian, A. C. 1993, MNRAS, 264, 228

Celotti, A., Ghisellini, G., & Chiaberge, M., 2000, MNRAS, submitted (astro-ph/0008021)

Chartas, G., et al. 2000, ApJ, in press (astro-ph/0005227)

Chiaberge, M., Celotti, A., Capetti, S. & Ghisellini, G. 2000, A&A, 358, 104

Costamante, L., et al. 2000, in preparation

Fabian, A. C., Celotti, A., Iwasawa, K., & Ghisellini, G. 2000, MNRAS, submitted

Fossati, G., Celotti, A., Comastri, A., Maraschi, L., & Ghisellini, G. 1998, MNRAS, 299, 433

Ghisellini, G. & Madau, P. 1996, MNRAS, 280, 67

Ghisellini, G., Celotti, A., Fossati, G., Maraschi, L., & Comastri, A. 1998, MNRAS, 301, 451

Ghisellini, G. 2000, in Stellar Endpoints, AGN and the Diffuse Background, Bologna, Sept. 1999, in press

Lazzati, D., Ghisellini, G., & Celotti, A. 1999, MNRAS, 309, L13

Maraschi, L., et al. 1994, ApJ, 435, L91

Rawlings, S. G. & Saunders, R. D. E. 1991, Nature, 349, 138

Rees, M. J. 1978, MNRAS, 184, P61

Schwartz, D. A., et al. 2000, in press (astro-ph/0005255)

Spada, M., Ghisellini, G., Lazzati, D., & Celotti, A. 2000, MNRAS, submitted

Tagliaferri, G., et al. 2000, A&A, 354, 431

Tavecchio, F., et al. 2000a, ApJ, in press

Tavecchio, F., Maraschi, L., Sambruna, R. M., & Urry, C. M. 2000b, ApJL, submitted (astro-ph/0007441)

Wehrle, A. E., et al. 1998, ApJ, 497, 178

Jets in Quasars

Marek Sikora

N. Copernicus Astronomical Center, 00716 Warsaw, Bartycka 18, Poland

Abstract. In my review of jet phenomena in quasars, I focus on the following questions: How powerful are jets in radio-loud quasars? What is their composition? How are they launched? And why, in most quasars, are they so weak? I demonstrate the exceptional role that blazar studies can play in exploring the physics and structure of the innermost parts of quasar jets.

1. Introduction

The jet activity in quasars is common, but very diverse. As radio observations indicate, jet powers can differ by several orders of the magnitude within the same optical luminosity range. The most powerful jets produce hundred kiloparsec-scale double radio structures. They can be characterized as composed from a pair of edge-brightened radio lobes, with the hotspots matching their luminosity peaks. Additionally, one sided jets are often observed, connecting one of two hotspots with the radio core in the center of the host galaxy. The above structure has a good interpretation in terms of a dynamical model which involves propagation of light, relativistic jets in the IGM (Scheuer 1974; Begelman & Coffi 1989). The hotspots are located at the ends of channels drilled by the jets through the IGM. They mark the regions where material of the jet is shocked and spreads sideways, forming the radio lobe. Relativistic speeds of jets explain their one-sided appearance, while lightness (jet density lower than IGM density) is necessary to explain formation of extended radio lobe structures and the non-relativistic speeds of hotspots.

Highly polarized and relatively steep radio spectra of extended radio structures are uniquely interpreted in terms of optically thin synchrotron radiation. The synchrotron spectra extend from $\sim$ 10 MHz, up to IR/optical. I discuss briefly, in §2, how these data can be used to estimate the jet power.

Quasar jets can be traced in radio down to parsec-scale central regions. There the jets are more relativistic (bulk Lorentz factor $\Gamma \sim 10$, Padovani & Urry 1992; Ghisellini et al. 1993; Homan et al. 2000) than on kpc-scales ($\Gamma \sim 3$, Wardle & Aaron 1997), and only those which are oriented close to the line of sight are bright enough to be observed in detail. Parsec-scale jets viewed "pole-on" can be decomposed into radio-cores and one-sided linear structure. The linear structure is very inhomogeneous, with some bright regions propagating with relativistic speeds and appearing to us as "superluminal" sources. Parsec-scale radio sources show spectra with a low-energy break due to synchrotron-self-absorption. The frequency of this break is larger the closer to the center one measures, and superposition of spectra from all radio components give the characteristic flat radio spectrum, with a energy spectral index $\alpha < 0.5$. Quasars with such radio spectra are called FSRQ (flat-spectrum-radio-quasars). I discuss the energetics and composition of quasar parsec-scale jets in §3.

As high frequency VLBI observations of nearby radio galaxies show, extragalactic jets are launched much deeper than the angular-resolution and synchrotron-self-absorption limited observations can follow in quasars (Lobanov 1998; Junor, Biretta, & Livio 1999). Fortunately, sub-parsec scale jets radiate a lot at higher, non-radio frequencies: up to optical/UV by the synchrotron mechanism, and in the X-ray and γ-ray bands via Comptonization of synchrotron and external diffuse radiation fields (Sikora, Begelman, & Rees 1994; Błażejowski et al. 2000). This radiation, Doppler boosted into our direction, often dominates over thermal quasar components, such as: UV/optical radiation of the accretion disk, X-ray radiation of a disk-corona, and IR radiation of dust located in a molecular torus and heated by a disk. FSRQ which have spectra dominated by the nonthermal radiation from a jet are called blazars. Historically, this category also includes BL Lac objects, those sources with thermal signatures too weak present at all to place them in the quasar category. Hence, in order to avoid confusion while talking about "quasar-hosted" blazars, I will call them "Q-blazars."

The broad-band spectra of Q-blazars can in general be superposed from the radiation produced over a large distance range. However, at least during short-term high amplitude flares the spectra are dominated by radiation produced co-spatially, very likely in short-lived shocks, somewhere at 0.1–1.0 pc from the center. Thus, as discussed in §4, multiwavelength studies of flares in blazars provide exceptional tools for exploring the structure and physics of jets on sub-parsec scales.

Jets are predicted to produce radiation not only by relativistic electrons (here and after the term 'electrons' is used for both electrons and positrons), but also by cold electrons. Streaming with a bulk Lorentz factor $\Gamma \sim 10$, the cold electrons Compton scatter the external optical/UV photons and boost them up to the soft X-ray range (Begelman & Sikora 1987; Sikora et al. 1997). It should be emphasized that cold electrons are an unavoidable constituent of jets near their base, where even mildly relativistic electrons cool faster than they propagate. And they can be present up to distances where non-thermal flares are produced, dragged by as-yet-unshocked portions of the flow. As yet, no soft X-ray excesses have been confirmed. The upper limits imposed on the number of cold electrons by observed soft X-ray fluxes exclude pure e^+e^--jets and provide strong constraints on the minimum distance of jet acceleration and collimation (Sikora & Madejski 2000). These constraints, together with possible jet production scenarios, are discussed in §5.

Of course, any model of jet production should be able to explain the huge range of jet powers. Recent discoveries that many luminous radio-quiet quasars reside—like the radio-loud quasars—in giant ellipticals (Taylor et al. 1996; Kukula et al. 2000), and that the galaxy environments of the same luminosity radio-quiet and radio-loud quasars is similar (McLure & Dunlop 2000), challenge the previous claims that radio-loudness can be related to the morphology of the host galaxy or its clustering richness. Furthermore, optical/UV spectral similarities

(Francis, Hooper, & Impey 1993; Zheng et al. 1997) and recent discoveries that BAL (broad absorption line) systems exist also in radio-loud quasars (Brotherton et al. 1998; Becker et al. 2000) suggest that radio-loudness is not very dependent on the parsec-scale environment, as well. All the above strongly supports the so-called spin paradigm, according to which powerful jets, giving rise to radio-loud quasars, can be produced only with the help of rapidly rotating black holes. How the spin paradigm relates to central engine models and the evolution of quasars is discussed toward the end of §5.

2. Radio Lobes

What can we learn about jets from radio lobes? First of all, they provide very useful information about the energetics of jets, which, contrary to that derived from radio-core scales and smaller, is not biased by such uncertainties as jet bulk Lorentz factor, dissipation efficiency and variability. The procedure for deriving the jet power is simple, but not free of assumptions and approximations. The first step is to recover from the observed electromagnetic spectrum the energy distribution of electrons. This can be done using the following approximate formula:

$$L_{\nu,syn} d\nu \simeq (N_\gamma d\gamma) m_e c^2 |\dot\gamma| \qquad (1)$$

where

$$|\dot\gamma| \simeq \frac{4}{3} \frac{c \sigma_T u_B \gamma^2}{m_e c^2} \qquad (2)$$

is the rate of electron synchrotron energy losses; $u_B = B^2/8\pi$ is the magnetic energy density; and $\nu \propto \gamma^2 B$. For power-law synchrotron spectrum $L_\nu \propto \nu^{-\alpha}$, formula (1) gives $N_\gamma \propto \gamma^{-s}$, where $s = 2\alpha + 1$. The above procedure is not sufficient to determine the normalization of the electron energy distribution, however. Additionally, one needs to know the intensity of the magnetic field, which can be estimated by assuming energy equipartition between electrons and magnetic fields. With this assumption, the electron (and magnetic) energy content of the quasar radio lobes is found to be in the range 10^{59}–10^{61} ergs which, when divided by spectrally or dynamically determined ages of the radio lobes, $t_{\text{lobe}} \sim 3 \times 10^7$ years, gives jet powers 10^{45}–10^{47} ergs s^{-1} (Rawlings & Saunders 1991).

There are several reasons why the above estimates should be considered as lower limits. First, the equipartition condition corresponds almost exactly with the minimum total energy of electrons and magnetic fields. Second, from the observed radio spectra one can deduce that most of the energy carried by electrons is contained in the low energy part of their distribution. Assuming that the electron distribution has a break at an energy corresponding with the lowest observable frequency ~ 10 MHz (limited by reflection of radio waves in ionosphere), one finds that $\gamma_{\min} \sim 500/\sqrt{B/10\mu G}$. Since there is no proof that radiation has an intrinsic cutoff at 10 MHz, the adopted value of $\gamma_{\min}$ can be greatly overestimated and the total electron energy content underestimated. Third, Rawlings & Saunders assumed no energy contribution from protons.

How much might the jet powers be underestimated due to the above assumptions? Detection of X-rays from hotspots of several nearby radio galaxies and their interpretation in terms of the SSC (synchrotron-self-Compton) process allowed one to derive the magnetic field and electron energy densities without assuming equipartition (see Wilson, Young, & Shopbell 2000 and references therein). Departure from the equipartition condition have been found to be very small. However, recent observations of X-rays produced around quasar radio lobes show that the pressure of the external gas is several times larger than the pressure in the radio lobes obtained assuming equipartition between electrons and magnetic fields and no proton contribution (Hardcastle & Worral 2000). This inconsistency cannot be resolved, even assuming that the electron distribution extends down to $\gamma_{\min} \sim 1$, if the equipartition condition is kept. There must be significant departure from equipartition between electrons and magnetic fields and/or the lobe pressure is dominated by protons. There are some observations suggesting that, indeed, the equipartition conditions can be violated in radio lobes, with particle pressure dominant over magnetic pressure (see Blundell & Rawlings 2000 and references therein). Yet another possibility is that extra pressure in radio lobes is provided by cosmic rays accelerated via the Fermi process operating in the boundary layer between a jet and the surrounding medium (Ostrowski 2000).

The rate at which energy is delivered to radio lobes can be estimated more directly, just from the bolometric luminosities of hotspots. The rate is

$$L_j = \frac{L_{HS}}{\eta_e \eta_{\mathrm{rad}}} \qquad (3)$$

where η_e is the fraction of kinetic energy of a jet converted in the shock to relativistic electrons and η_{rad} is the fraction of electron energy lost by radiation. If, in the radiative regime, the electromagnetic spectrum has a slope $\alpha \simeq 1$, then $\eta_{\mathrm{rad}} \sim \ln(\nu_{\max}/\nu_c)/\ln(\nu_{\max}/\nu_{\min})$, where ν_c is the "cooling" break, and $\nu_{\min} \leq 10$ MHz. Applying this for hotspot D in Cyg A, where $\nu_c \sim 10^{10}$ Hz and $\nu_{\max} \sim 10^{12}$ Hz, one can find that for $\nu_{\min} = 10$ MHz, $\eta_{\mathrm{rad}} \sim 0.4$. Now, assuming that the energy dissipated in the hotspots is equally shared by electrons and magnetic fields, i.e. $\eta_e = 1/2$, and adopting from Meisenheimer et al. (1997) $L_{HS} \simeq 4 \times 10^{44}$ ergs s^{-1}, we obtain $L_j \sim 2 \times 10^{45}$ ergs s^{-1}. This estimate of L_j is consistent with that deduced from the radio-lobe energetics (Carilli & Barthel 1996). In both cases the energy is underestimated only by a factor 3/2, if there are protons and they equally share energy with electrons and magnetic fields. Unfortunately, hotspot spectra up to the highest synchrotron frequencies are currently available only for nearby radio galaxies and, therefore, the bolometric-luminosity method cannot be applied to distant quasars.

What about the pair content of radio-lobe plasmas? Noting that jets are approaching the hotspots with relativistic speeds, the average energy of shocked protons is expected to be of the order of the jet Lorentz factor. Thus, if the dissipated kinetic energy of a jet is initially shared equally by electrons, protons, and magnetic fields, the average electron energy should be $\gamma \sim (1/3)\Gamma(m_p/m_e)(n_p/n_e)$ $\sim 600\Gamma(n_p/n_e)$. Since radio observations cannot follow electrons with energies lower than $\gamma \sim 500/\sqrt{(B/10\mu G)}$, the pair content cannot be verified by radio lobe observations.

3. Parsec-scale Jets

In order to recover the physical parameters of radiating plasma in compact radio sources, one needs: to use the emissivity formula (like that in equation [1], but written in the source comoving frame); to transform the comoving luminosity and frequency to the observed ones; and to take advantage from two of the following:

(a) the value of the bulk Lorentz factor Γ (if available from VLBI observations);

(b) X-ray flux, provided the X-rays are produced by the SSC process;

(c) synchrotron-self-absorption break;

(d) equipartition condition.

Such analyses have been performed for a large sample of compact radio sources by Ghisellini et al. (1992), who used (b) and (c), and for the series of compact radio components in 3C 345 and 3C 279 by Hirotani et al. (1999; 2000), who used (c) and (d). Among other aspects, they calculated electron energy distributions, $n_\gamma = C\gamma^{-s}$, assuming $\gamma_{\min} = 1$, where $s = 2\alpha + 1$. This allows one to obtain electron energy densities, $u'_e \equiv n'_\gamma \langle\gamma\rangle m_e c^2$, and then energy fluxes of relativistic electrons

$$L_e \sim u'_e \Gamma^2 \pi a^2 c \tag{4}$$

where a is the cross-sectional radius of the source. For the studied sources, values of L_e are in the range 10^{45}–10^{47} ergs s^{-1}, provided $\gamma_{\min} = 1$, and 2–3 times smaller, if $\gamma_{\min} \gg 1$.

Hence, the energy fluxes of relativistic electrons alone come close to satisfying the energy requirements of radio lobes. However, one should note that the total energy flux also includes other forms of energy and, in general, we have:

$$L_j = \frac{L_e}{\eta_{\mathrm{diss}}\eta_e(1-\eta_{\mathrm{rad}})} \tag{5}$$

where η_{diss} is the fraction of the total jet energy which is dissipated and used to accelerate electrons, to heat protons and to amplify magnetic fields; η_e is the fraction of the dissipated energy which is used to accelerate electrons; and η_{rad} is the fraction of electron energy lost by radiation during the source lifetime. Thus, with the derived values of L_e, the total energy fluxes, L_j, become dangerously high, particularly if $\eta_{\mathrm{diss}} < 0.1$ as intrinsic shock theories predict. Therefore, provided that the derived densities of relativistic electrons are not affected by systematic errors, one needs to postulate external shock models, with dissipation efficiencies $\eta_{\mathrm{diss}} \geq 0.5$ (Dermer & Chiang 1998).

What is the pair content of the compact radio sources? Assuming that the dominant energy carriers are protons, we have

$$L_j \simeq L_{p,0} + L_{\mathrm{diss}} = L_{p,0} + \delta L_p + L_B + L_e, \tag{6}$$

where $L_{p,0} = n'_p m_p c^3 \pi a^2 \Gamma^2$ and $\delta L_p \simeq n'_p m_p (\langle \gamma_p \rangle - 1) c^3 \Gamma^2 \pi a^2$. Since

$$\frac{L_e}{L_{p,0}} = \frac{n'_e \langle \gamma \rangle m_e}{n'_p m_p} = \frac{\eta_e \eta_{\rm diss}(1 - \eta_{\rm rad})}{1 - \eta_{\rm diss}} \quad (7)$$

where $\eta_e = 1/(1 + L_B/L_e + \delta L_p/L_e)$, we obtain

$$\frac{n'_+}{n'_p} \simeq \frac{n_e}{2 n_p} \simeq \frac{50}{\gamma_{\min}} (3\eta_e) \frac{\eta_{\rm diss}}{1 - \eta_{\rm diss}} (1 - \eta_{\rm rad}) \quad (8)$$

where I used $\langle \gamma \rangle \sim 6 \gamma_{\min}$, which corresponds with $\alpha \simeq 0.6$.

Unfortunately, due to the synchrotron-self-absorption it is impossible to follow electrons with energies $\gamma < 50 \sqrt{(\nu_{\rm abs}/1{\rm GHz})/(B/0.1{\rm G})}$ and determine $\gamma_{\min}$ directly from the observed spectra. One can eventually try to determine the upper limits for the minimum electron energies from measurements of circular polarization provided such polarization results from the Faraday conversion mechanism. Measurements of circular polarization have been completed for several extragalactic sources (Homan & Wardle 1999) and at least in 3C 279 there are indications that circular polarization is produced by this mechanism and that $\gamma_{\min} < 20$ (Wardle et al. 1998). For such $\gamma_{\min}$s the number of pairs per proton can range from a few up to tens. There are several indirect arguments in favor of rather low pair contents: (i) if pairs are produced in the central engine or its vicinity, then their flux is very limited by the annihilation process (Ghisellini et al. 1992); (ii) if pairs are created by nonthermal pair cascades operating in the jet shocks, they would produce much softer X-ray spectra than observed in Q-blazars (Ghisellini & Madau 1996); (iii) jets with a large number of cold electrons would produce soft X-ray bumps by Comptonization of UV disk radiation and BELs and such bumps have not been confirmed (Begelman & Sikora 1987; Sikora et al. 1997).

4. Q-blazars

Strong and fast variability in blazars is commonly interpreted in terms of the shock-in-jet model. Producing a flare of the observed time scale t_{fl}, the shock passes a distance range

$$\Delta r_{\rm fl} \sim c t_{fl} \Gamma^2 \sim 100 (t_{fl}/1 {\rm day})(\Gamma/10)^2 \; {\rm lt-days}. \quad (9)$$

Sharp profiles of flares and comparable time scales of their rise and decay suggest that shocks are launched at distances $r_{\rm fl} \sim \Delta r_{\rm fl}$ (Sikora et al. 2000). The distance of flare production can be also estimated from the spectral location of the γ-ray luminosity peak, provided the peak is related to the break in the electron energy distribution caused by the cooling effect, i.e. that above the peak electrons radiate on time scales shorter than the lifetime of the shock. If γ-ray production is dominated by the ERC (external-radiation-Compton) process, and the luminosity of the ERC component is larger than the luminosity of the synchrotron component, then the spectral distance is

$$r_{\rm sp} \sim c t'_{\rm sp} \Gamma \sim \frac{m_e c^2}{\sigma_T} \sqrt{\frac{\nu_{\rm diff}}{\nu_c}} \frac{1}{u_{\rm diff}} \quad (10)$$

where the following relations were used: $t'_{sp} = |\gamma_c/\dot\gamma_c|$; $|\dot\gamma| \sim \Gamma^2 \gamma_c^2 \sigma_T u_{\text{diff}}/m_e c$; and $\gamma_c \simeq \sqrt{\nu_c/\nu_{\text{diff}}}/\Gamma$. It can be checked that both the variability distance, r_{fl}, and the spectral distance, r_{sp}, are of the same order if $u_{\text{diff}} \sim 0.005$ ergs cm^{-3}, i.e. if $L_{\text{BEL}}(r_{\text{fl}}) \sim 10^{44}$ ergs s^{-1}, and/or covering of the central source by dust at $T = 1000$ K is ~ 0.1. Both quantities are consistent with our knowledge about BELR and near-IR radiation in quasars (Błażejowski et al. 2000).

Equations (9) and (10), combined with emissivity formulae for the production of X-rays via the ERC process (Błażejowski et al. 2000) and with given value of $L_{\text{diff}}(r_{\text{fl}})$, can be used to calculate the number of relativistic electrons involved in flare production and their energy flux, L_e. For $L_{\text{BEL}} \sim 10^{45}$ ergs s^{-1} and Γ calculated from the model, it can be found that $L_e \sim 10^{45} L_{\text{SX},46}$ ergs s^{-1}, where L_{SX} is the soft X-ray luminosity (Sikora et al., in preparation). Since soft X-rays are very likely dominated by the SSC process (Inoue & Takahara 1996; Kubo et al. 1998; Błażejowski et al. 2000), the above estimate should be considered only as the upper limit.

The electron energy flux, L_e, can be estimated also using the bolometric luminosity procedure. We have

$$L_e \sim \frac{\Omega}{4\pi} \frac{L_{\text{QB}}}{\eta_{\text{rad}}} \qquad (11)$$

where $\Omega \sim \pi/\Gamma^2$ and L_{QB} is the total apparent luminosity of a blazar. During high states L_{QB} is dominated by luminosity in γ-ray bands and is of the order 10^{48-49} ergs s^{-1} (von Montigny et al. 1995). From typical high energy spectra of Q-blazars ($\alpha \sim 1$ for $h\nu > 30$ MeV and location of ν_c in the 1-30 MeV range), one can conclude that $\eta_{\text{rad}} \sim 1/2$ if $\gamma_{\text{min}} \sim 1$, and $\eta_{\text{rad}} \sim 1$ if $\gamma_{\text{min}} \sim 100$. With these numbers, and Γ taken from the model calculations, we obtain $L_e \sim 1-2 \times 10^{45}(\Gamma/15)^{-2} L_{\text{QB},48}$ ergs s^{-1}, which is of the same order as calculated using the X-ray flux. Noting that in Q-blazars L_e is about 1-2 orders lower than L_j estimated from radio lobes, and that $L_e = \eta_{\text{diss}} \eta_e L_j (1 - \eta_{\text{rad}})$, one can conclude that activity of the sub-parsec jets is governed by low dissipation efficiencies, very likely by intrinsic shocks. This conclusion can actually be reinforced by the fact that the high (flaring and radio active) states are not permanent, and after averaging over longer periods of time the values of L_e are likely to be lower by at least a factor two.

If the main carriers of energy are cold protons, then from equation (8) the pair content is $n'_+/n'_p \sim 5/\gamma_{\text{min}}$ for $\gamma_{\text{min}} < 5$, and is negligible for $\gamma_{\text{min}} > 5$, assuming $\alpha = 0.6$, $\eta_{\text{diss}} = 0.05$ and $\eta = 1/3$. Here, like in the radio lobes and compact radio sources, we have a sort of conspiracy regarding the value of γ_{min}. Since X-rays from blazars are presumably dominated by the SSC process, it is very difficult to follow the low energy portions of the ERC spectra to check whether there are any signatures of the low-energy cutoff in energy distribution of electrons. But at least we can say that a large pair content $n_+/n_p \gg few$ can be excluded. One could speculate that there is a large number of cold pairs. However, this is excluded because such pairs, Comptonizing external UV photons, would produce a huge soft X-ray bump, which is not confirmed observationally.

5. Central Engine and Spin Paradigm

The absence of soft X-ray bumps in Q-blazar spectra provides strong constraints on the jet's structure near its base. There, Comptonization of the disk radiation by cold electrons, dragged by the relativistic and well collimated jet, is predicted to produce a prominent X-ray bump even if number of pairs is zero. This suggests that powerful jets are wider and/or slower at their bases (Sikora & Madejski 2000). They can be launched by the innermost parts of the accretion disk, with the matter pulled from the disk surface and accelerated along the open magnetic field lines by centrifugal forces (Blandford & Payne 1982). Such proto-jets are probably collimated further away, by a disk corona or winds predicted by some models to be formed at distances $\geq 100 R_g$ (Rozanska & Czerny 2000; Murray & Chiang 1997; Proga, Stone, & Kallman 2000). Yet in the acceleration zone, the initially proton-electron outflows can be loaded by e^-–e^+ pairs. This is due to boosting of the coronal hard X-rays by cold electrons in the outflow up to MeV energies and the subsequent absorption of MeV photons in the $\gamma\gamma$ pair production process. Depending on the geometry and kinematics of the proto-jet, the number of pairs per proton can reach a value ranging from a few up to a few tens.

The fact that only 10% of quasars are radio-loud implies that powerful jets are rare. It is very likely that the leading parameter which decides about the jet power is the spin of the black hole, defined as $A = J/J_{\max}$ where $J_{\max} = GM^2/c$ (Wilson & Colbert 1995; Moderski, Sikora & Lasota 1998). Perhaps the fast rotation of the black hole is necessary to heat the accretion disk. Due to this heating the innermost parts of the disk can be inflated and this can help to generate large scale magnetic fields, which are required both to accelerate MHD outflows and to link a disk with the rotating black hole and/or the gas plunging into the ergosphere (Meier 2000; Krolik 2000). Furthermore, extra heating of surface layers of the disk by a rotating black hole can help to put gas on magnetic field lines, if the latter are not bent enough to allow centrifugal forces to pull the gas directly from the photosphere.

Now, provided that the above scenario(s) gives a strong enough dependence of the jet power on the spin of the black hole to explain the huge range of radio-loudness, the next question which should be answered is how Nature managed to have only a small fraction of rapidly rotating black holes. Let us recall that accretion disks, acting enough long to double the black hole mass, spin up the holes to $A \sim 1$ (Bardeen 1970; Thorne 1974). On the other hand, once the black hole is spun up, it cannot be easily slowed down, certainly not during the low accretion phases during which the rate of extraction of black hole energy is very small because of its proportionality to $B^2 \propto \dot{M}$. Hence, if the spin paradigm is right and the growth of supermassive black holes is governed by the large accretion events, then the number of radio loud quasars should be larger than of radio quiet quasars, oppositely to what is observed. A solution of the problem can be that growth of the black hole in most objects is dominated by low-mass accretion events with random angular momentum orientations. This picture is supported by observations of Seyfert galaxies, which show that AGNs are randomly oriented relative to the host galaxy planes (Wilson & Tsvetanov 1994; Schmitt et al. 1996). Low-mass accretion events are also supported by analyses of the dynamics of Seyfert jet interactions with the host galactic gas which show that the lifetime of these objects is of the order of 10^5 years only (Capetti et al. 1999). Thus, it is tempting to speculate that only major mergers which involve at least a one gas-rich galaxy can lead to the accretion disk operating long enough to double the black hole mass and spin up the black hole to $A > 0.5$. During early phases of the process, when A is still small, the quasar is predicted to show up as radio-quiet, and then slowly transform to a radio-loud one. After the fuel is off, the accretion drops, but spin of the black hole remains high and such an object can eventually be represented by FR I radio galaxies.

Acknowledgments. I am grateful to Mitch Begelman for his valuable comments which helped improve the paper. This work has been supported in part by the Polish KBN grant 2P03D 00415.

References

Bardeen, J. M. 1970, Nature, 226, 64
Becker, R. H., et al. 2000, ApJ, 538, 72
Begelman, M. C. & Cioffi, D. F. 1989, ApJ, 345, L21
Begelman, M. C. & Sikora, M. 1987, ApJ, 322, 650
Blandford, R. D. & Payne, D. G. 1982, MNRAS, 199, 883
Blundell, K. M. & Rawlings, S. 2000, AJ, 119, 1111
Błażejowski, M., Sikora, M., Moderski, R., Madejski, G. M. 2000, ApJ, in press (astro-ph/0008154)
Brotherton, M. S., et al. 1998, ApJ, 505, L7
Capetti, A., Axon, D. J., Macchetto, F. D., Marconi, A., & Winge, C. 1999, ApJ, 516, 187
Carilli, C. L. & Barthel, P. D. 1996, A&AR, 7, 1
Dermer, C. D. & Chiang, J. 1998, NewA, 3, 157
Francis, P. J., Hooper, E. J., & Impey, C. D. 1993, AJ, 106, 417
Ghisellini, G., Celotti, A., George, I. M., & Fabian, A. C. 1992, MNRAS, 258, 776
Ghisellini, G. & Madau, P. 1996, MNRAS, 280, 67
Ghisellini, G., Padovani, P., Celotti, A., & Maraschi, L. 1993, ApJ, 407, 65
Hardcastle, M. J. & Worrall, D. M. 2000, astro-ph/0007260
Hirotani, K., Iguchi, S., Kimura, M., & Wajima, K. 1999, PASJ, 51, 263
Hirotani, K., Iguchi, S., Kimura, M., & Wajima, K. 2000, astro-ph/0005394
Homan, D. C. & Wardle, J. F. C. 1999, AJ, 118, 1942
Homan, D. C., et al. 2000, astro-ph/0009301
Inoue, S. & Takahara, F. 1996, ApJ, 463, 555
Junor, W., Biretta, J. A., & Livio, M. 1999, Nature, 401, 891
Krolik, J. 2000, astro-ph/0008372
Kubo, H., et al. 1998, ApJ, 504, 693
Kukula, M. J., et al. 2000, astro-ph/0010007
Lobanov, A. P. 1998, A&A, 330, 79

McLure, R. J. & Dunlop, J. S. 2000, astro-ph/0007219
Meier, D. L. 2000, astro-ph/0010231
Meisenheimer, K., Yates, M. G., & Röser, H.-J. 1997, A&A, 325, 57
Moderski, R., Sikora, M., & Lasota, J.-P. 1998, MNRAS, 301, 142
Murray, N. & Chiang, J. 1997, ApJ, 474, 91
Ostrowski, M. 2000, MNRAS, 312, 579
Padovani, P. & Urry, C. M. 1992, ApJ, 387, 449
Proga, D., Stone, J. M., Kallman, T. R. 2000, astro-ph/0005315
Rawlings, S. & Saunders, R. 1991, Nature, 349, 138
Różańska, A., & Czerny, B. 2000, A&A, 360, 1170
Scheuer, P. A. G. 1974, MNRAS, 166, 513
Schmitt, H. R., Kinney, A. L., Storchi-Bergmann, T., & Antonucci, R. 1997, ApJ, 477, 623
Sikora, M., Begelman, M. C., & Rees, M. J. 1994, ApJ, 421, 153
Sikora, M. & Madejski, G. 2000, ApJ, 534, 109
Sikora, M., Madejski, G., Moderski, R., & Poutanen, J. 1997, ApJ, 484, 108
Taylor, G. L., Dunlop, J. S., Hughes, D. H., & Robson, E. I. 1996, MNRAS, 283, 930
Thorne, K. S. 1974, ApJ, 191, 507
von Montigny, C., et al. 1995, ApJ, 440, 525
Wardle, J. F. C. & Aaron, S. E. 1997, MNRAS, 286, 425
Wardle, J. F. C., Homan, D. C., Ojha, R., & Roberts, D. 1998, Nature, 395, 457
Wilson, A. S. & Colbert, E. J. M. 1995, ApJ, 438, 62
Wilson, A. S. & Tsvetanov, Z. I. 1994, AJ, 107, 1227
Wilson, A. S., Young, A. J., & Shopbell, P. L. 2000, astro-ph/0009308
Zheng, W., Kriss, G. A., Telfer, R. C., Grimes, J. P., & Davidsen, A. F. 1997, ApJ, 475, 496

On the Energy Content of Blazar Jets

Annalisa Celotti

S.I.S.S.A., Via Beirut 2-4, I-34014 Trieste, Italy

Abstract. The spectral properties of blazars can be used to significantly constrain—on the corresponding physical scales—both the energy carrier and the content of matter in their jets and the characteristics of the particle distribution(s).

1. Gamma-ray Blazars: Powers

Blazars detected in the gamma-ray band constitute the most interesting sources in this respect, as for them it is possible to estimate the total power radiated by the source, which indeed is often dominated by the contribution at these frequencies. Note however that the corresponding estimates strictly speaking refer to flaring states in the majority of cases.

The emission observed in these objects has been simply modeled as synchrotron and inverse Compton scattering (both on the synchrotron and on external photons) of a non-thermal distribution of particles in a homogeneous region relativistically moving in the jet. Typical distances of such region along the jet have been set by the requirement of transparency for the gamma–rays to the process of photon–photon pair creation in the nuclear soft photon field and by the variability timescales (see Ghisellini et al. 1998 for a detailed description of the model).

Under these assumptions it has been possible to estimate the particle density and the intensity of the magnetic field in the emitting plasma, and therefore the corresponding powers transported by these components in the jet. More precisely, we estimated: the kinetic power carried by the emitting particles themself and by a flow of associated protons (if there was one proton per emitting particle), the total luminosity dissipated by the plasma (integrated over the solid angle of emission), the luminosity in the synchrotron component only, and finally the power associated to the inferred magnetic field. The corresponding values are reported in Fig. 1 for the flat spectrum radio quasars and the BL Lac objects in the sample.

1.1. Results

The clearest finding of this work (Celotti & Ghisellini, in preparation) derives from the comparison of the dissipated with the kinetic and magnetic powers: the kinetic power associated to a population of (assumed cold) protons is the only one sufficient to produce the observed luminosity. About 10 per cent of such bulk energy would be radiated—and indeed the emission has to be (relatively)

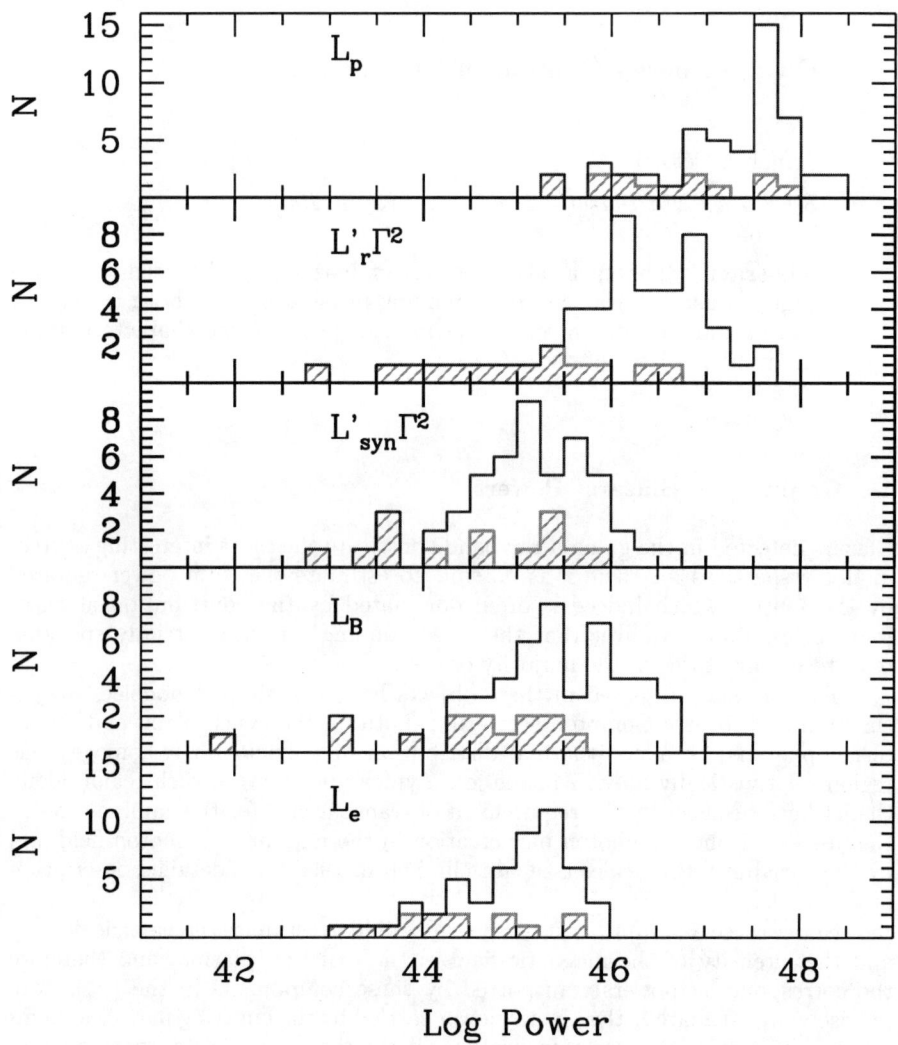

Figure 1. Histograms representing—from top to bottom—the kinetic power carried in the jet by a number of protons equal to those of non-thermal emitting particles, the total luminosity dissipated by the plasma (integrated over the solid angle of emission), the luminosity in the synchrotron component only, the power associated to the inferred magnetic field, the kinetic power of the relativistic particles them-self (Celotti & Ghisellini, in preparation). The blazars are the gamma-ray detected sources studied in Ghisellini et al. (1998) and the dashed areas refer to BL Lac objects, while the clear ones represent flat spectrum radio quasars.

inefficient if most of the power has to be carried to the outer jets and deposited in the radio lobes.

The requirement on the efficiency also implies that any component in relativistic electron-positron pairs can over-number the electron-proton population only by a factor of order unity, and thus cannot provide a significant contribution to the global energetics. In fact if pairs were a large population of emitting leptons, the corresponding proton component would be decreased with consequent decrease of the bulk power at levels comparable with the radiated luminosity.

1.2. Particle Distribution

A further interesting consequence is on the extension of the emitting particle distribution. As this is typically steep its lower limit (Lorentz factor γ_{min}) crucially determines the total particle density. As the above estimates assumed the 'maximum' particle number, i.e. a distribution extending down to non-relativistic energies ($\gamma_{min} \sim 1$), it appears that this quantity too is constrained to be of the order unity if enough power has to be transported.

Interestingly, all the observational information obtained so far on blazars supports the view that γ_{min} is of this order. Even the most recent pieces of evidence on the presence of a spectral flattening in the soft X-ray band of a few powerful blazars (e.g. Fabian et al. 2000a, 2000b; Sikora & Madejski 2000; Tavecchio et al. 2000) imply that—if this is due to a low energy cutoff in the emitting particle distribution—this has to be at such low energies.

Acknowledgments. The Italian MURST is acknowledged for financial support.

References

Fabian, A. C., Celotti, A., Iwasawa, K., Carilli, C. L., McMahon, R. G., Brandt, W. N., Ghisellini, G., & Hook, I. M. 2000a, MNRAS, submitted

Fabian, A. C., Celotti, A., Iwasawa, K. & Ghisellini, G. 2000b, MNRAS, submitted

Ghisellini, G., Celotti, A., Fossati, G., Maraschi, L. & Comastri, A. 1998, MNRAS, 301, 451

Sikora, M. & Madejski, G. 2000, ApJ, in press

Tavecchio, F., Maraschi, L., Ghisellini, G., Celotti, A., Chiappetti, L., Comastri, A., Fossati, G., Grandi, P., Pian, E., Tagliaferri, G., Treves, A., Raiteri, C. M., Sambruna, R. M. & Villata, M. 2000, ApJ, in press

Beaming in Jets with Evolving Relativistic Features

Z. Abraham

Instituto Astronômico e Geofísico, Universidade de São Paulo, Brazil

Abstract. Unified models predict that core-dominated radio sources are the beamed counterparts of radio galaxies. This hypothesis was questioned lately, in the context that the beaming factors necessary to unify the models are much smaller than those inferred from other properties, such as superluminal motions or transparency to gamma-ray emission. In this work we show that the beaming factor in sources with evolving superluminal features is strongly overestimated, changing completely the interpretation of the properties of the core dominated sources and eliminating the discrepancies between the predictions of the unified models.

1. Introduction

Unified models are successful in explaining the properties of the different types of Seyfert active nuclei. They mainly assume that the their properties are based only on orientation with respect to the line of sight. These models were also extended to radio galaxies and quasars (see Urry & Padovani 1995). It was assumed that a relativistic jet is created in the center of the radio object near the energy source, probably a black hole. The radiation emitted by this jet would be boosted when its direction is very close to the line of sight while the extended outer radio source would not be affected. In this scenario, the ratio of the power in the compact core and in the extended radio source can be used as a parameter proportional to the angle between the central jet and the line of sight, obtaining a sequence formed by the flat spectrum radio quasars (FSQ), steep spectrum quasars (SSQ), broad line radio galaxies (BLRG) and narrow line radio galaxies (NLRG) (see Morganti et al. 1997). However, although the general properties of radio loud AGNs seem to support the unified models, the beaming factors inferred from superluminal motions, transparency to gamma-ray emission, slope of spectral energy distribution and time-lags among variability at different frequencies are systematically larger than those necessary to unify BL Lacs, quasars and radio galaxies (Dondi & Gisellini 1995; Ghisellini et al. 1998; Tavecchio et al. 1998; Chiaberge et al. 2000). In this paper we show that the dependence of beaming on the Doppler factor is strongly overestimated when the main emission originates in evolving features, and as a consequence, the discrepancy between the values inferred from unified models and other processes no longer exists.

2. Beaming in Evolving Features

Let us consider the flux density $S'(\nu')$ at the observing frequency ν', emitted by a source moving with bulk velocity β in a direction which forms an angle ϕ with the line of sight. $S'(\nu')$ is related to the isotropic flux density $S(\nu)$, in the source reference frame by (Blandford & Königl 1979):

$$S'(\nu') = S(\nu)\delta^p \tag{1}$$

with $\quad \nu = \nu'/\delta, \quad \delta = \dfrac{1}{\gamma(1-\beta\cos\phi)} \quad$ and $\quad \gamma = \dfrac{1}{\sqrt{(1-\beta^2)}} \tag{2}$

If the spectrum $S(\nu)$ can be represented by a power law, with spectral index α and a maximum at frequency ν_m, where $\tau \sim 1$, then equation (1) becomes, in the case of non-evolving components:

$$S'(\nu') = \frac{S_m(\nu_m)}{(\nu_m)^{-\alpha}} \delta^{p+\alpha} (\nu')^{-\alpha} \tag{3}$$

with $p = 3$ for individual features and $p = 2$ for a continuum jet.

To calculate the dependence of the beamed flux density of an evolving feature on the Doppler factor δ, we will assumed that shock waves are formed and propagate along the relativistic jet (e.g. Marscher 1987). However, the conclusions obtained in this study can be applied to any evolving feature, as for example, individual expanding bullets ejected by the compact core. The basic assumption is that the unperturbed jet is adiabatic, implying that the electron density $N(E, z)$, energy $E(z)$ and magnetic field $B(z)$ decrease as power laws of the distance z along the jet. In the adiabatic phase the maximum in the spectrum moves to lower frequencies and its flux density decreases as the shock propagates:

$$S_m \propto \nu_m^{\alpha_{th}} \tag{4}$$

For a source with a Doppler factor δ, the flux density at the observer frequency ν' will reflect the spectrum at the smaller frequency ν, where the maximum flux density is also smaller. Quantitatively we can write:

$$S'_m(\nu') \propto \delta^{p-\alpha_{th}} \tag{5}$$

Since $p = 2$ for the continuous jet and $\alpha_{th} \sim 1$, the flux density of the individual superluminal features will be proportional to δ instead of δ^3.

3. Jet to Counterjet Intensity Ratio

The jet to counterjet intensity ratio will be also affected by shock evolution. Its calculation, however, it is not straightforward, because the maximum intensity of a given feature occurs at different projected distances $d'_\mp$ in the jet (index $-$) and counterjet (index $+$). These distances are:

$$d'_\mp(\nu') = z_m(\nu'/\delta_\mp)\sin\phi = z_m(\nu')\delta_\mp^{1/b}\sin\phi \tag{6}$$

and:
$$\frac{d'_-(\nu')}{d'_+(\nu')} = \left[\frac{1+\beta\cos\phi}{1-\beta\cos\phi}\right]^{1/b} \tag{7}$$

where $b \sim 2$. Since in most of the observed superluminal jet features the shocked region becomes optically thin at small distances from the core, the same feature in the counterjet will not be resolved even with VLBI, and its flux density will be added to the core intensity. However, when we compare the flux densities in the jet and counterjet at the same distance form the core we find:

$$\frac{S'_-(\nu')}{S'_+(\nu')} = \left(\frac{\delta_-}{\delta_+}\right)^{p+\alpha} \tag{8}$$

This is exactly the equation we obtain for a continuous non-evolving jet. Therefore, the jet to counterjet intensity ratio continues to be a measure of relativistic motion, even when the emission is dominated by shocks.

4. Flux Density of a Superposition of Evolving Features

To calculate the flux density of a superposition of features (as would be seen by a single radio telescope) we have to consider the dependence on the Doppler factor δ of the individual components formation rate and their decaying time. The decaying time $\Delta t_{1/2}$ can be defined as the time elapsed between the occurrence of the maximum in the spectrum $t_m(\delta)$ and half its value $t_{1/2}(\delta)$. From the previous equations it is possible to show that:

$$\Delta t'_{1/2} \propto \delta^{1/b-1} \tag{9}$$

This expression shows that the dependence of the decaying time with the Doppler factor is weaker than what would be expected from purely relativistic arguments.

Let us consider now that shocks are formed at a rate r in the source reference frame. The total flux density will be the combination of the flux densities of all the optically thick and thin components. If $t_m(\delta)$ is the time after shock formation at which the flux density reaches its maximum value, at the frequency $\nu_m = \nu'/\delta$, an observer will see $r t_m(\delta)$ and $r[t_m(\delta) - t_0(\delta)]$ optically thick and thin components, respectively, where $t_0(\delta)$ is the time elapsed since the first feature was formed. Due to evolution, most of the optically thin components will be very weak, and we can use the decaying time $\Delta t_{1/2}(\delta)$ as a reference for $[t_m(\delta) - t_0(\delta)]$. The contribution to the flux density at the observing frequency of all components formed at times $t_n(\delta) = t_m(\delta) + n/r$, with n an integer number, will be:

$$S_{\text{tot}}(\nu'/\delta) = S_m(\nu'/\delta) \sum_n \left[\frac{t_m(\delta)}{t_m(\delta) + n/r}\right]^{b(\alpha+\alpha_{\text{th}})} \tag{10}$$

Using equation (9) we obtain:

$$S_{\text{tot}}(\nu'/\delta) = S_m(\nu'/\delta) \left[\frac{1}{1 + n/r t_m(1)\delta^{1/b}}\right]^{b(\alpha+\alpha_{\text{th}})} \tag{11}$$

The total observed flux density will be:

$$S'_{\text{tot}}(\nu') < S_{\text{tot}}(\nu')\delta^{p-\alpha_{\text{th}}+1/b} \qquad (12)$$

We conclude that the inclusion of all the optically thin features increases the observed flux density at most by a factor $\delta^{1/b}$. The same argument applies to the optically thick components, although their contribution is even smaller due to the larger positive spectral index and therefore,the beaming factor is also smaller.

5. Discussion

The main conclusions of this work are: a) the effects of beaming are strongly overestimated when the evolution of shocked emitting regions (or expanding bullets) are not taken into account, b) the decaying time of evolving features is longer than what would be expected from only relativistic arguments, because we are looking at a source in a more advanced state of evolution, and c) the jet to counter-jet flux density, when measured at the same distance from the core is the same in evolving and non-evolving features.

How large will be the effect of beaming on the observed properties of AGNs will depend on how much the non-evolving jet and the shocked regions contribute to the total flux density at each wavelength. At radio frequencies, shocks dominate the strong, flat spectrum radio sources and the non-evolving jet is probably optically thick at these frequencies. At optical wavelengths we must consider the contribution of at least three sources: the accretion disk, the relativistic jet and the shocks. The latter can appear as short duration flares associated with radio flares, as can be seen, for example in OVV objects, but the separation of the contribution of the jet and accretion disk is more difficult.

Acknowledgments. This work was partially supported by the Brazilian financing agencies FAPESP and FINEP.

References

Blandford, R. D. & Königl, A. 1979, ApJ, 232, 34

Chiaberge M., Celotti, A., Capetti, A., & Ghisellini, G. 2000, A&A, 358, 104

Dondi, L. & Ghisellini, G. 1995, MNRAS, 273, 583

Ghisellini, G., Celotii, A., Fossati, G., et al. 1998, MNRAS, 301, 451

Marscher, A. P. 1987, in Superluminal Radio Sources, eds. J. A. Zensus & T. J. Pearson (Cambridge: Cambridge Univ. Press), 280

Morganti R., Osterloo, T. A., Reynolds, J. E., Tadhunter, C. N., & Migenes V. 1997, MNRAS, 284, 541

Tavecchio, F., Maraschi, L., Ghisellini, G. 1998, ApJ, 509, 608

Urry, C. M. & Padovani, P. 1995, PASP, 107, 803

Time Scales of Blazar Variability

Stefan J. Wagner

LSW Heidelberg, Königstuhl, 69117 Heidelberg, Germany

Abstract. Variability is one of the defining characteristics of blazars. They also exhibit very broad spectral energy distributions and their luminosity varies by a significant fraction of the long-term average on a wide range of time scales. A few common properties of variability patterns at different energies are reviewed. Structure functions in different bands are of similar power-law form and exhibit a break at a few 10^5 sec with significant steepening to shorter time scales. The characteristic time scale of this break is similar in different energy bands and sets an upper limit to the energy-dependent acceleration- and cooling-times.

1. Introduction

BL Lac objects and flat-spectrum radio quasars are unique in having been detected in all wavelength ranges accessible for astrophysical studies. Their spectral energy distributions are truly broad-band extending from the 100 MHz regime up to tens of TeV. In all of those bands blazar fluxes were found to be variable. Long-term trends lasting for decades have been found as well as changes occuring within a few minutes. The power-density spectra are broad-band as well, and variability has been detected on all time scales sampled sufficiently well. The power density spectra are not completely featureless. Characteristic time scales have been detected in the optical and radio regimes for many sources. It is now possible to extend these studies to higher frequencies and compare the statistical properties of blazar variability over a broad range of photon energies.

2. Parameter Space of Variability studies

Blazars have been studied in all frequency ranges accessible for astronomical observations. They have been observed throughout the radio-window up to the sub-mm regime, FIR to NIR observations have been carried out with IRAS and ISO, NIR and optical observations span up to a century covered in a few sources, IUE and EUVE permitted studies in the UV and EUV regions. Soft and hard X-ray emission has been detected with the WFC on ROSAT and many X-ray missions with highest X-ray energies being covered by XTE and SAX. Gamma-rays have been detected with all of the four instruments on-board the Compton GRO, and Solar-Farms as well as classical Cherenkov telescopes studied nearby Blazars in the energy range between tens of GeV and tens of TeV.

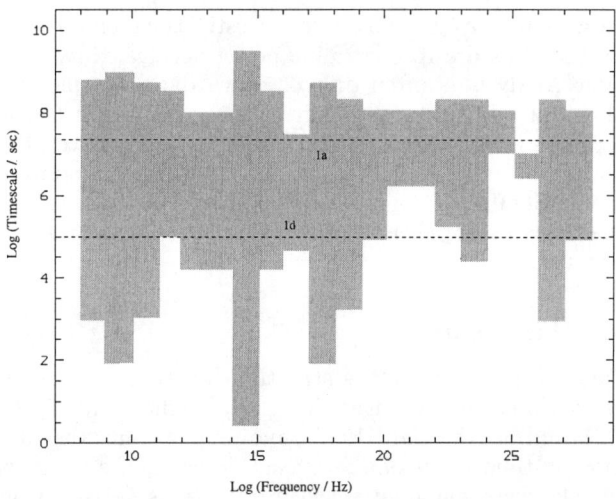

Figure 1. Parameter space of variability studies. The shaded area indicates the range of time scales probed for variability in the different energy regimes. Variability has been detected in all parts of parameter space that have been explored. Amplitudes and duty cycles are very different in the different parts of this projection of parameter space. A large fraction of the shaded area has been explored only in few dedicated studies. Especially the high-energy side of the parameter space is filled by investigations of a very small number of sources, and all of the gamma-ray range is accessible only if sources are flaring.

Apart from having been detected across the entire spectrum, Blazars were found to be variable by a significant fraction (at least a few ten percent) in all cases whenever repeated measurements have been obtained. The number of sources detected in these bands, photometric accuracies, temporal coverage, and sampling rates used, differ by several orders of magnitude and statistical samples probing variability properties are rare. It has been possible, nevertheless, to probe variability on different time scales.

3. Statistical Characteristics

The broad range of variability time scales illustrates that it would be naive to associate doubling times directly with crossing times, acceleration time scales, or cooling-times scales. Well-sampled campaigns in radio, optical, or X-ray ranges clearly indicate that many flares exhibit substructure. This may be caused by superposition of different flares (originating from physically different sub-volumes) or by the interplay between the individual time scales mentioned above. Observational biases will complicate the situation further: In any fixed energy band some flares will be monitored below the high-energy cutoff, while others will evolve such that the peak frequency shifts through the specific band and rare (but more spectacular) flares will be detected above the cutoff. Some of these effects can be disentangled by statistical investigations of the variability pattern. Some of the aims are to determine power spectra, duty cycles and flare shapes. Fourier Analysis is often difficult since data sets mostly suffer from uneven sampling, introducing window effects that are difficult to control. Structure functions, measuring average derivatives of the light-curves, do not contain phase information and are hence less affected by window functions. While the structure function $S(\tau) = \langle [I(t+\tau) - I(t)]^2 \rangle$ (Rutman 1978) is straightforward to compute, care has to be taken in error propagation, especially in the regime of low photon fluxes.

3.1. Intraday Variability

In the radio and optical regimes the structure functions on time scales longer than tens of days have been investigated, e.g., by Hughes et al. (1992), Lainela & Valtaoja (1993), and Smith et al. (1993), finding power-law distributions which reach saturation on time scales of a few years. Towards the high frequency end, pronounced breaks were found in structure functions derived from studies of faster variations in the GHz and optical regimes (Heeschen et al. 1987; Quirrenbach et al. 2000; Heidt & Wagner 1996). These breaks are seen on scales of about 10^5 seconds and led to the term 'intra-day variability.' They are a persistent feature, suggesting a preferred time scale.

Recently, long and well-sampled investigations have become possible in the X-ray regime as well. They allow meaningful statistical investigations of the variability patterns and exhibit structure functions very similar to that seen at lower frequencies. In particular, breaks can be found on time scales of about one day in several sources which exhibit synchrotron emission up to 10 keV or more (Figure 2b, Takahashi et al., in press; Kataoka et al., submitted; Tanihata et al., these proceedings).

It is interesting to note that the break, indicating a characteristic time scale, occurs at about 10^5 sec in optical and X-ray observations of high-frequency peaking Blazars as well as in radio and optical observations of low-frequency peaking objects. Irrespective of the peak-frequency of the synchrotron emission, the characteristic time scales do not seem to vary with photon energy. This illustrates that the characteristic time scale is neither set by acceleration-times nor cooling-times, since both of them would introduce a significant energy-dependence of the characteristic time scales. The similarity in time scales had been noted in a few earlier campaigns with dense sampling and sufficiently long duration, specifically on PKS 2155−304 in November 1992 (Urry et al. 1993; Brinkmann et al. 1994; Edelson et al. 1995).

Despite the power-law fall-off in the structure function towards shorter time scales, isolated, very fast flares have been detected in the optical, X-ray and VHE energy ranges. These fast flares have amplitudes comparable to the flares lasting for about one day. An X-ray flare in Mrk 501 by 60% in about 200 sec was described by Catanese & Sambruna (2000). The power-law decline of the structure function indicates the low duty cycle of such rapid events. It is obvious, nevertheless, that acceleration- and cooling-times have to be significantly faster than the characteristic time scale illustrated by the break in the structure function in order to generate the faster changes. It is speculated that t_{Break} should be associated with a geometric crossing-time scale. The similarity in different

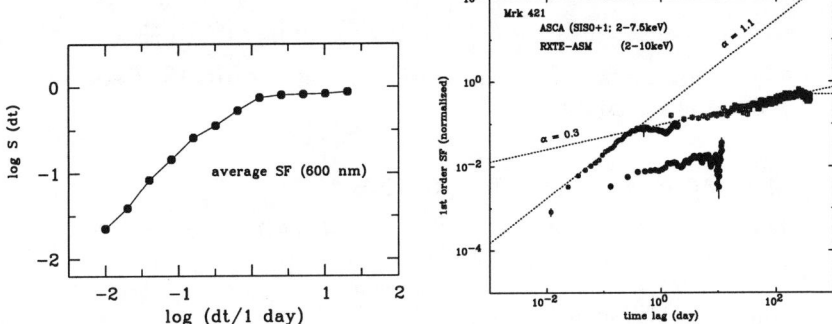

Figure 2. Comparisons of the high (temporal) frequency end of the structure functions of variability at optical wavelengths and X-ray energies. The left figure gives an average of structure functions of BL Lac objects in the sample studied by Heidt & Wagner (1996). Even though different sources are averaged without adjusting the time scales (except for cosmological dilation according to the redshifts of the sources) a pronounced break on a time scale of about 1 day is obvious. The right figures gives the structure function of Mrk 421 as determined from a long ASCA observation in 1998 and RXTE (ASM) observations (Takahashi et al., in press). Again a pronounced break ($\Delta\alpha = 0.8$) is visible on a time scale of 8×10^4 sec (~ 1 day).

energy bands would be an obvious consequence. One would expect crossing-time scales to be associated with bolometric luminosities or jet power of the sources.

Acknowledgments. I would like to thank my collaborators, specifically G. Bicknell, J. Kataoka, T. Takahashi, and A. Witzel for sharing ideas, Meg Urry and Paolo Padovani for organizing this very stimulating meeting, and the DFG for supporting this research through SFB 438.

References

Brinkmann, W., et al. 1994, A&A, 288, 433
Catanese, M. & Sambruna, R. 2000, ApJ, 534, L39
Edelson, R., et al. 1995, ApJ, 438, 120
Heeschen, D. S., et al. 1987, AJ, 94, 1493
Heidt, J. & Wagner, S. J. 1996, A&A, 305, 42
Hughes, P. A., Aller, H. D., & Aller, M. F. 1992, ApJ, 396, 469
Lainela, M. & Valtaoja, E. 1993, ApJ, 416, 485
Quirrenbach, A., et al. 2000, A&AS, 141, 221
Rutman, J. 1978, in Proc. IEEE, 66, No. 9, 1048
Smith, A. G., et al. 1993, AJ, 105, 437
Urry, C. M., et al. 1993, ApJ, 411, 614
Wagner, S. J. & Witzel, A. 1995, ARA&A, 33, 163

Size-Luminosity Scaling and Inverse Compton Seed Photons in Blazars

Markos Georganopoulos, John G. Kirk

Max Planck Institut für Kernphysik, Postfach 10 39 80, D-69029, Heidelberg, Germany

Apostolos Mastichiadis

Department of Astronomy, University of Athens, GR 15784, Athens, Greece

Abstract. We present preliminary results of our work on blazar unification. We assume that all blazars have a broad line region (BLR) and that the size of the BLR scales with the power of the source in a manner similar to that derived through reverberation mapping in radio quiet active galactic nuclei (AGNs). Using a self-consistent emission model that includes particle acceleration we show that according to this scaling, in weak sources like MKN 421, the inverse Compton (IC) scattering losses are dominated by synchrotron-self Compton scattering (SSC), while in powerful sources, like 3C 279, they are dominated by external Compton (EC) scattering of BLR photons. In agreement with other workers, we show that even in the powerful sources that are dominated by EC scattering, the hard X-ray emission is due to SSC. Finally, we show that this scaling reproduces well the observed sequence of blazar properties with luminosity.

1. Introduction

Blazars have been shown to exhibit a sequence of properties as a function of source power (Fossati et al. 1998). As the source power increases, the emission line luminosity and the ratio of Compton to synchrotron luminosity increase, while the synchrotron peak frequency ν_s and the IC peak frequency decrease. Recent multiwavelength studies support this scheme (e.g. Kubo et al. 1998), although the result that the ratio of the Compton to synchrotron luminosity increases with source luminosity suffers from limited statistics, and should be only considered tentative.

The initial division of blazars into flat spectrum radio quasars (FSRQs) and BL Lacertae objects (BLs) was based on the equivalent width (EW) of the broad emission lines. Sources with EW > 5 Å were classified as FSRQs and sources with EW < 5 Å as BLs. This difference in the EW of the emission lines has been interpreted as absence of a substantial BLR in BLs, and has been used to advance the idea that in BLs the GeV-TeV emission is due to SSC, while in FSRQs the GeV emission is due to EC. However, the EW criterion does not correspond to a dichotomy (Scarpa & Falomo 1997), because a weak BLR is present in BLs as well. Additionally, it is not the BLR luminosity L_B that is relevant for the EC luminosity, but the BLR photon energy density U_E that is measured in the comoving frame of the non-thermal source. One needs to know not only L_B, but also the BLR radius R_B and the location of the non-thermal emitter relative to the BLR in order to quantify the relative importance of SSC versus EC emission.

The size of the BLR has been measured only for radio quiet AGNs through reverberation mapping (Kaspi et al. 2000), and it has been found to scale with the ionizing luminosity $L_{\rm acc}$ of the accretion disk as $R_B \propto L_{\rm acc}^{0.7}$. It was additionally found that, assuming Keplerian motions for the BLR clouds, the mass of the central object scales with the ionizing luminosity as $M \propto L_{\rm acc}^{0.5}$.

Here we assume the scalings apply also to radio loud AGN and examine the implications for blazars. Following the formalism of internal shocks propagating in a conical jet (Rees 1978), we scale the location $D_{\rm blob}$ and the radius $R_{\rm blob}$ of the non-thermal spherical emitter with the mass M of the central object. We assume that L_B is a fraction of $L_{\rm acc}$ and scale the kinetic luminosity and the Poynting flux of the blob with $L_{\rm acc}$.

2. The Scaling and its Application

Under this scheme, the radius of the BLR scales as $R_B \propto L_{\rm acc}^{0.7}$, and the distance of the blob from the center of the system scales as $D_{\rm blob} \propto M \propto L_{\rm acc}^{0.5}$. Therefore, we assume that for powerful sources the blob radiates from inside the BLR, $D_{\rm blob} < R_B$. However, if we consider increasingly weaker sources, R_B falls faster than $D_{\rm blob}$ and, below some critical luminosity, the blob radiates from outside the BLR. For a blob inside the BLR, $U_E \propto (L_B/R_B^2)\Gamma^2$, where Γ is the bulk motion Lorentz factor of the blob. For a blob outside the BLR, the solid angle subtended by the BLR, as seen by the blob, is reduced. This affects U_E in two ways. The first is the usual geometric $1/r^2$ factor, and the second is a relativistic de-boosting that, for $D_{blob} \gg R_B$, gives $U_E \propto (L_B/D_{\rm blob}^2)/\Gamma^2$ (Dermer & Schlickeiser 1994). The second effect dominates if the blob is located less than a few R_B away from the BLR. For example, for $\Gamma = 10$, U_E at a distance of 10 R_B is smaller than U_E inside the BLR by a factor of 10^2 due to the usual $1/r^2$ and by a factor of $\approx 10^4$ due to relativistic de-boosting.

We apply this scaling using a model (Georganopoulos & Kirk 2000) for the blob which includes both particle acceleration and radiative losses with IC scattering losses treated in the Thomson regime. Particles are accelerated in an acceleration zone and escape into a radiation zone, before they eventually escape out of the system.

This simplified picture is intended to represent particles which are accelerated by a shock front, escape it, and radiate downstream before the compressed plasma re-expands. Whilst undergoing acceleration, particles simultaneously suffer synchrotron losses and losses by Compton scattering of both photons of external origin (EC) and synchrotron photons (SSC) from the radiation zone. In the radiation zone, EC, SSC and synchrotron losses occur until the particle finally escapes after a time $t_{\rm esc} = 3R_{\rm blob}/c$. The electron energy distribution (EED) is computed self-consistently, taking account of these processes.

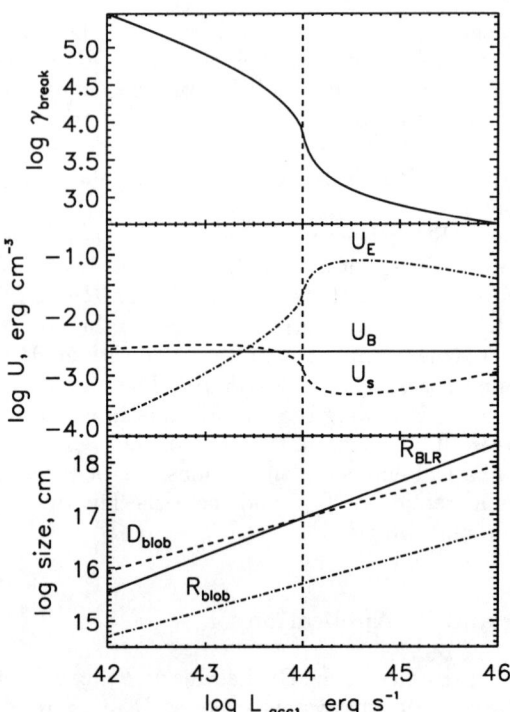

Figure 1. Sizes (lower panel), energy densities (middle panel), and the break of the EED as a function of the accretion disk luminosity.

Our assumption about the scaling of the Poynting flux implies a constant magnetic field, which we take to be $B = 0.2$ G. The blob is assumed to move with a Lorentz factor $\Gamma = 15$ at an angle $\theta = 1/\Gamma$ with respect to the line of sight. Low energy particles ($\gamma_0 = 10$) are injected into the blob at a rate of $Q = 0.1$ cm^{-3} s^{-1}. Also $L_B = 0.01\ L_{\rm acc}$. For $L_{\rm acc} = 10^{44}$ erg s^{-1}, reverberation mapping gives $R_B \approx 8.6\ 10^{16}$ cm. We assume that at this luminosity $R_B = D_{\rm blob}$ and $R_{\rm blob} = 2\ 10^{15}$ cm.

In Fig. 1, we let $L_{\rm acc}$ vary from 10^{42} erg s^{-1} to 10^{46} erg s^{-1}, and follow the behavior of the system. In the lower panel R_B, $D_{\rm blob}$, and $R_{\rm blob}$ are plotted as a function of $L_{\rm acc}$. For weak sources, the blob is located outside the BLR ($D_{\rm blob} > R_B$). As the source power increases, the blob gradually approaches and eventually enters the BLR. In the middle panel of Fig. 1 we see how this affects U_E (the magnetic field energy density U_B is constant under the adopted scaling). For weak sources, U_E is much smaller than both U_B and the self-consistently derived synchrotron photon density U_S. Gradually U_E increases, and eventually, for bright sources, dominates over both U_B and U_S. In the upper panel of Fig. 1 we plot the self-consistently derived break $\gamma_{\rm break}$ in the EED. Note that $\gamma_{\rm break}$ decreases as the power of the source decreases.

In Fig. 2 we plot basic observable quantities as a function of $L_{\rm acc}$. In the lower panel we plot the synchrotron luminosity L_S, the SSC luminosity $L_{\rm SSC}$,

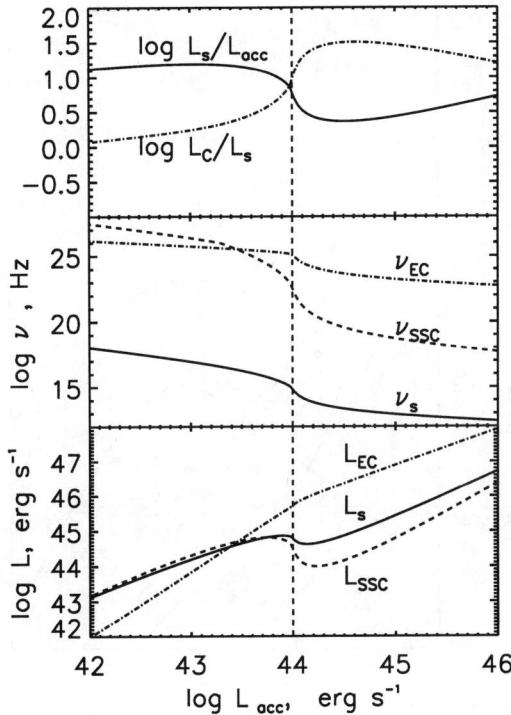

Figure 2. Luminosities (lower panel), peak frequencies (middle panel) and the ratio or synchrotron to accretion disk luminosity and IC to synchrotron luminosity (upper panel).

and the EC luminosity E_{EC}. At low powers, L_{EC} is much weaker than L_{SSC} and L_S that are roughly equal. As the source power increases, L_{EC} gradually dominates over L_{SSC} and L_S, and we end up with an EC dominated source. In the middle panel we plot the peak frequencies of the three emission components. Note how the synchrotron peak frequency ν_S decreases as the source power increases. Similar behavior is also seen for the SSC peak frequency ν_{SSC} and the EC peak frequency ν_{EC}. Finally in the upper panel of Fig. 1 we plot the ratio L_S/L_{acc} (solid line) of the synchrotron to accretion disk luminosity and the ratio L_C/L_S of the IC luminosity to synchrotron luminosity. L_S dominates over L_{acc} in weak sources in agreement with the lack of a thermal component and weak emission lines in weak sources like BLs. L_{acc} becomes more significant for powerful sources, again in agreement with the strong emission lines of FSRQs and the accretion disk signature observed in some FSRQs (e.g. in 3C 279, Pian et al. 1999). The dominance of the Compton over the synchrotron emission increases as the source power increases, in agreement with observations (e.g. Kubo et al. 1998). Note, however, that the observational case for an increasing Compton dominance with source power is based on poor statistics due to the limited sensitivity of the *CGRO*.

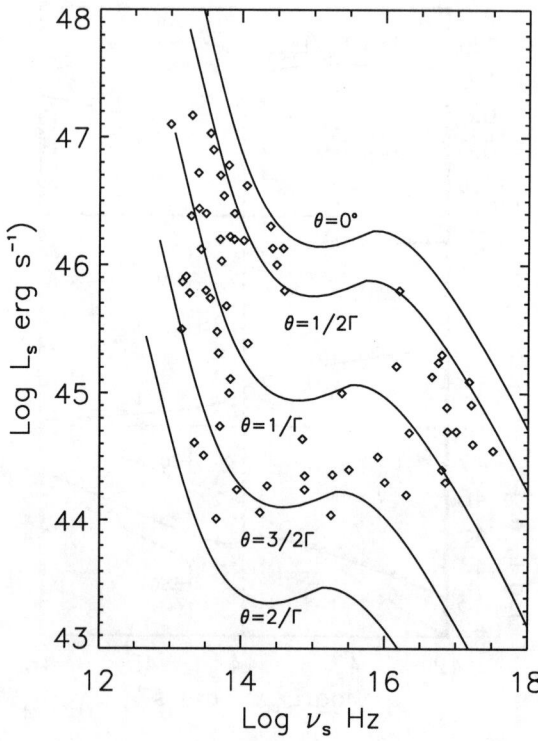

Figure 3. Synchrotron luminosity versus synchrotron peak frequency for a set of blazars (Sambruna et al. 1996; Kubo et al. 1998). Overlayed are the model tracks as a function of $L_{\rm acc}$ for a range of observing angles.

In Fig. 3 we plot the synchrotron luminosity L_s versus the synchrotron peak frequency ν_s for the blazars studied by Sambruna, Maraschi, & Urry (1996) and Kubo et al. (1998). On top of the data points we plot the model tracks as a function of $L_{\rm acc}$ for a range of observing angles. Note that these tracks represent two model derived quantities, L_S and ν_S as $L_{\rm acc}$ varies. The model covers rather well the observed parameter space. Given the track of the model under an angle $\theta = 0°$, we do not expect that any powerful ($L_S \approx 10^{47}$ erg s^{-1}) sources with high peak frequencies $\nu_S \approx 10^{17}$ Hz exist. The discovery of such sources would pose a serious problem for this model.

3. The Luminosity Scaling

In Fig. 4 we plot the peak luminosities of the three emission mechanisms for five model sources with $L_{\rm acc}$ ranging from 10^{42} up to 10^{46} erg s^{-1}. The behavior of the model is in good agreement with the general characteristics of the observed luminosity sequence (compare with Fig. 12 of Fossati et al. 1998). The synchrotron peak frequency decreases from 10^{18} Hz down to 10^{12} Hz, as the syn-

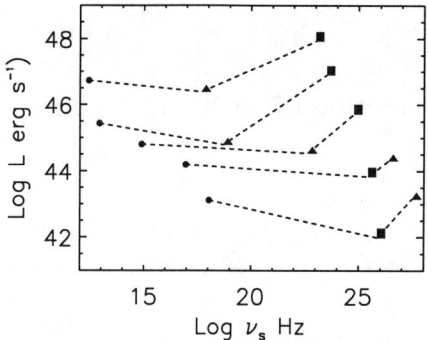

Figure 4. The model luminosity sequence. The synchrotron (circles), SSC (triangles), and EC (squares) luminosity as a function of frequency for 5 model sources of different intrinsic power. The broken lines link points from a single model.

chrotron power increases. The Compton peak frequency in weak sources is in the TeV regime ($\approx 10^{25-26}$ Hz) and it is due to SSC, while in powerful sources it is in the GeV regime ($\approx 10^{23}$ Hz) and it is due to EC. According to the model, although in these bright sources the energy losses are dominated by EC, the hard X-ray emission is due to SSC, in agreement with resent observations (Kubo et al. 1998).

We note here that Fig. 4 corresponds to sources that are oriented at an angle $\theta = 1/\Gamma$. In reality one should expect a range of angles, which will give rise to significant scattering around the presented trend. This scattering is visible in the work of Fossati et al. (1998) and indicates that a proper unification scheme for blazars should include the effects of orientation.

References

Dermer, C. D. & Schlickeiser, R. 1994, ApJS, 90, 945
Fossati, G., et al. 1998, MNRAS, 299, 433
Georganopoulos, M. & Kirk, J. G. 2000, in preparation
Kaspi, S., et al. 2000, ApJ, 533, 631
Kubo, H., et al. 1998, ApJ, 504, 693
Pian, E., et al. 1999, ApJ, 521, 112
Rees, M. J. 1978, MNRAS, 184, 61
Sambruna, R. M., Maraschi, L., & Urry, C. M. 1996, ApJ, 463, 444
Scarpa, R. & Falomo, R. 1997, A&A, 325, 109

Chandra Detection of an X-ray Jet in 3C 371

R. M. Sambruna

George Mason University, Dept. of Physics and Astronomy and School of Computational Sciences, 4400 University Drive, M/S 3F3, Fairfax, VA 22030

C. Megan Urry, R. Scarpa

Space Telescope Science Institute, 3700 San Martin Drive, Baltimore, MD 21218

F. Tavecchio, L. Maraschi

Osservatorio Astronomico di Brera, Via Brera 28, 20121 Milano, Italy

J. E. Pesce

Eureka Corporation

Abstract. We report the detection at X-rays of the optical jet of 3C 371, from a short (10 ks) *Chandra* exposure in AO1. Despite the limited signal-to-noise ratio of the data, an interesting morphology is suggested. Most of the X-ray photons are collected in correspondence of a relative minimum in the radio and optical. The X-ray flux is below the extrapolation from the radio-to-optical continuum, suggesting synchrotron as a plausible emission mechanism. However, this interpretation is in contrast to the observed multiwavelength morphology of the jet. Deeper *Chandra* observations are needed to confirm the jet detection at X-rays and measure more accurately its properties.

1. Introduction

How relativistic jets form and evolve in Active Galactic Nuclei (AGN) is of fundamental astrophysical importance to ultimately understand the physical processes near the central black hole, whereby energy is extracted and channeled into highly collimated structures. Traditionally, the first studies of jets on kpc-scales were at radio wavelengths, where hundreds of jets were mapped and studied (e.g. Bridle & Perley 1984). A handful of jets were detected at optical/UV wavelengths with ground-based telescopes and *HST* (presently less than 20; Sparks et al. 1994; Scarpa et al. 1999). The radio through UV emission is generally consistent with synchrotron radiation.

In the X-rays, only very few jets were detected at the limited sensitivities of *Einstein* and *ROSAT* (M87, 3C 273, and Cen A; Feigelson et al. 1981; Harris et al. 1987; Biretta et al. 1991). Plausible emission mechanisms for the

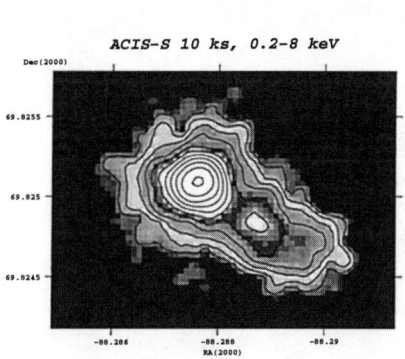 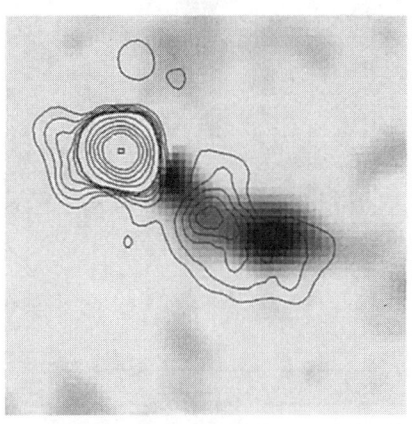

Figure 1. *Chandra* observations of the optical jet of 3C 371. [*Left, (a)*]: X-ray image in 0.2–8 keV; contours from the same image are superimposed. A faint elongation is apparent in the S-W direction, coincident with the optical and radio jet. A total of 330 counts were collected in the X-ray jet in 0.2–8 keV, in a short exposure of 10 ks. [*Right, (b)*]: Superposition of the X-ray contours on the *HST* image of the jet (Scarpa et al. 1999). The central source was subtracted in the *HST* data. Most of the X-ray counts were collected in the first knot, coincident with a minimum of the optical and radio intensity.

X-rays included synchrotron and/or inverse Compton from relativistic electrons, and thermal bremsstrahlung from shocked ambient gas. However, at the poor resolutions of the previous X-ray experiments, no detailed study of the X-ray morphology and spectrum was possible.

This situation is now changing thanks to the advent of the *Chandra* X-ray Observatory. With its high angular resolution (0.5″/pixel) and improved sensitivity, *Chandra* is ideally suited to the study of relatively faint, complex structures as extragalactic X-ray jets. Indeed, after only one year of operations, surprising results were already obtained for a handful of radio and optical jets (Kraft et al. 2000; Chartas et al. 2000; Schwartz et al. 2000; Wilson et al. 2000). As part of our ongoing program of multiwavelength study of jets (e.g. Sambruna et al. 2000a; Tavecchio et al. 2000), we report here on *Chandra* observations of 3C 371, host of a known optical jet. The X-ray counterpart of the optical jet is detected for the first time, with an interesting morphology. We discuss possible emission mechanisms and highlight key future observations. More details on the data and interpretation will be presented in a forthcoming paper (Sambruna et al. 2000b).

2. *Chandra* Observations of 3C 371: a New X-ray Jet

The nearby (z=0.051) radio source 3C 371 exhibits properties intermediate between a BL Lacertae object and a radio galaxy. Its core emission is rapidly variable at all observed wavelengths, and has a spectral distribution typical of a

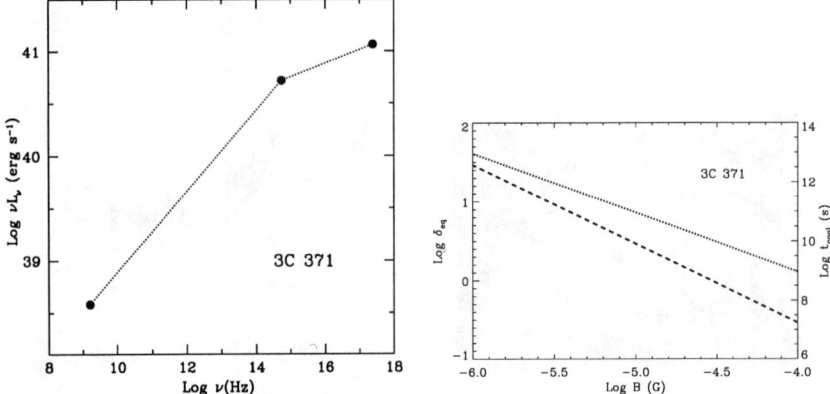

Figure 2. [*[Left, (a)]*]: Spectral Energy Distribution for the brighter X-ray knot in the jet. The X-ray flux lies below the extrapolation from the radio-to-optical continuum, suggesting that the X-rays originate via synchrotron off the same population of relativistic electrons as at the longer wavelengths. [*Right, (b)*]: Example of estimated physical quantities using the multiwavelength emission (radio, optical, and X-ray) from knot B in 3C 371. Left-hand scale and dashed line: permitted values of the Doppler factor δ and magnetic field B assuming equipartition. Right-hand side and dotted line: cooling time (logarithm of the time in seconds) for synchrotron emission of electrons producing 1 keV photons for different values of B and assuming equipartition.

Low-energy Peaked BL Lac (Sambruna, Maraschi, & Urry 1996). On the other hands, emission lines are observed in its optical and UV spectrum (e.g. Worrall et al. 1984), and its radio morphology is characterized by two giant lobes and a 25″-long, one-sided jet (Wrobel & Lind 1990), typical of FRII galaxies.

The radio jet exhibits bends and curvature as if interacting with the external medium (Gomez & Marscher 2000; Wrobel & Lind 1990). No superluminal motion has been observed, with an upper limit of $\sim 1.4h^{-1}c$ to the plasma motion (Gomez & Marscher 2000). The jet-to-counterjet ratio from the radio is high, 1700:1 (Gomez & Marscher 2000), consistent with a viewing angle $\theta \lesssim 18°$ and Lorentz factor $\Gamma \gtrsim 3.2$. The radio properties suggest that the jet knots are stationary features such as shocks, likely triggered by interaction with the external medium. The presence of energetic electrons and high magnetic fields is confirmed by the recent detection of the jet at optical, in ground-based and our *HST* observations (Nilsson et al. 1997; Scarpa et al. 1999). The optical observations probe only the 5″-long, brightest part of the jet, with at least three knots at 1.7″ (knot B), 3.1″ (knot A, the brightest), and 4.5″ (knot D, the faintest) from the nucleus. Overall, the optical morphology tracks the radio one closely in the region of overlap (Scarpa et al. 1999).

Chandra observed 3C 371 on 21 March 2000, as part of our AO1 program aimed at imaging the extended X-ray environment of BL Lacs (Pesce et al. 2000), with the source at the nominal aimpoint of the ACIS-S3 chip. The exposure time was 10 ks. Figure 1a shows the 0.2–8 keV image of 3C 371. A faint elongation is apparent in the S-W direction, at the same position angle on the sky as the optical jet. The X-ray jet has a bright structure at $\sim 1.7''$ from the core, coincident with the position of optical knot B, after which the X-ray emission extends to $\sim 5''$. A total of 330 counts are collected from the X-ray jet, of which ~ 240 are in the first knot, where the optical and radio data have a relative minimum. This is illustrated in Figure 1b, which shows the superposition of the *Chandra* X-ray contours on the *HST* image. The width of the knot is $\sim 1''$ (Scarpa et al. 1999).

Despite the limited numbers of counts, we extracted the X-ray spectrum from the brighter knot B, and analyzed it using currently available calibration files. The data are consistent in 0.5–8 keV with a power law spectrum, absorbed by Galactic N_H only (4.9×10^{20} cm^{-2}), and photon index $\Gamma = 3.1^{+0.4}_{-0.3}$ (errors at 90% confidence for one parameter of interest). The observed 2–10 keV flux is $F_{2-10\ keV} \sim 8 \times 10^{-15}$ erg cm^{-2} s^{-1}.

The Spectral Energy Distribution (SED) from radio to X-rays of knot B is shown in Figure 2a. The X-ray flux is below the extrapolation from the radio-to-optical continuum. The most natural interpretation of the SED is that the X-rays are due to synchrotron emission from the same population of relativistic electrons responsible for the longer wavelengths, with synchrotron losses responsible for the steep cutoff implied by the X-ray continuum. Applying a synchrotron homogeneous model, and assuming equipartition, we derive a magnetic field of 10^{-5} G and a Doppler factor of a few (for a knot radius of $\sim 10^{21}$ cm, or $1''$). The derived electron lifetimes at 1 keV are long enough for the electrons to cross the knot and accumulate locally.

One problem with the synchrotron interpretation is that the jet morphology should be strictly similar at all wavelengths, contrary to what we observe. It is possible that alternative/additional models are required. For example, part of the X-rays could originate via thermal bremmstrahlung ($kT \lesssim 1$ keV) of shocked ambient gas. Unfortunately, due to the limited number of counts we can not discriminate in the spectrum between a thermal and non-thermal model. Another possibility is that the X-rays originate via inverse Compton scattering. In this case, the X-ray flux would lie above the optical continuum.

We can not exclude the possibility that the X-ray counts in knot B are contaminated by pileup effects. The *Chandra* observations of 3C 371 were primarily designed to study of the larger-scale (kpc) X-ray properties of the source, and no precaution was taken to minimize pileup. A deeper *Chandra* exposure, optimally designed, is needed to confirm the X-ray structure of the jet and measure more accurately its X-ray properties. An additional benefit would be multi-band *HST* images, which would provide an estimate of the slope of the optical continuum and help discriminate among models (e.g. synchrotron vs. inverse Compton).

3. Conclusions

A new X-ray jet has been detected in 3C 371 in a short *Chandra* exposure. Despite the limited signal-to-noise ratio of the X-ray data, an intriguing morphology is apparent, with most of the X-ray counts coming from a region of relative minimum radio and optical emission. The broad-band SED is consistent with a synchrotron origin for the X-rays; however, this interpretation poses problems for the multiwavelength morphology of the jet. Deeper *Chandra* observations, and additional *HST* images at various wavelengths, are needed to confirm the X-ray morphology of the jet and measure accurately its X-ray spectrum.

Acknowledgments. RMS acknowledges partial financial support from NASA grant NAG5–7925, and from NASA contract NAS–38252 while at Pennsylvania State University. JEP acknowledges support from NASA grant NAS8–39073 from a *Chandra* GO1 program.

References

Biretta, J. A., et al. 1991, AJ, 101, 1632
Bridley, A. H. & Perley, R. A. 1984, ARA&A, 22, 319
Chartas, G., et al. 2000, ApJ, in press (astro-ph/0005227)
Feigelson, E. D., et al. 1981, ApJ, 251, 31
Gomez, J.-L. & Marscher, A.P. 2000, ApJ, 530, 245
Kraft, R., et al. 2000, ApJ, 531, L9
Harris, D. E. & Stern, C. P. 1987, ApJ, 313, 136
Nilsson, K., et al. 1997, ApJ, 484, L107
Pesce, J. E., et al. 2000, in prep.
Sambruna, R. M., et al. 2000a, ApJ, submitted
Sambruna, R. M., et al. 2000b, in prep.
Sambruna, R. M., Maraschi, L., & Urry, C. M. 1996, ApJ, 463, 444
Scarpa, R., Urry, C. M., Falomo, R., & Treves, A. 1999, ApJ, 526, 643
Sparks, W. B., Biretta, J. A., & Macchetto, F. 1994, ApJS, 90, 909
Schwartz, D. A., et al. 2000, ApJ, 540, L69
Tavecchio, F., Maraschi, L., Sambruna, R. M., & Urry, C. M. 2000, ApJ, submitted (astro-ph/0007441)
Wilson, A. S., Young, A. J., & Shopbell, P. L. 2000, ApJ, in press (astro-ph/0008467)
Worrall, D. M., et al. 1984, ApJ, 278, 521
Wrobel, J. M. & Lind, K. R. 1990, ApJ, 348, 135

Continuous Long-look Observations of TeV Blazars using ASCA

Chiharu Tanihata, Tadayuki Takahashi

Institute of Space and Astronautical Science, 3-1-1, Yoshinodai, Sagamihara, Kanagawa, 229-8150, Japan

Jun Kataoka

Department of Physics, Kyoto University, Sakyo-ku, Kyoto, 606-8502, Japan

C. Megan Urry

Space Telescope Science Institute, 3700 San Martin Drive, Baltimore, MD, 21218

Abstract. Two uninterrupted, 10-day *ASCA* observations of the TeV blazars Mrk 501 and PKS 2155−304 both show continuous strong X-ray flaring. Despite the relatively faint intensity states of each source, there was no identifiable quiescent period. Structure function analysis shows both blazars have a characteristic time scale of ~ 1 day, and there is little variability below that time scale. Together with symmetry of flares, we suggest the dominant time scale is the light crossing time rather than acceleration or cooling, in which case the emission region size is $\sim 10^{16-17}$ cm.

1. Introduction

The recent *ASCA* long-look observation of Mrk 421, a 7-day uninterrupted continuous observation in April 1998, showed rapid daily flares not previously resolved with sparser sampling (Takahashi et al. 2000). This rapid variability (or flaring) is one of the important characteristic of blazars, providing various information on conditions in the jet. Variability in the synchrotron spectral component is known to be most rapid above the νF_ν peak of the emission (Ulrich, Maraschi & Urry 1997). For "TeV blazars," the X-ray range coincides with the high energy end of the synchrotron component, where the highest energy electrons contribute.

To optimize *ASCA* science in its eighth year, especially as a new generation of X-ray telescopes was becoming available, the AO8 program was explicitly devoted to long observations of interesting targets. We therefore proposed two 10-day campaigns for the next two X-ray brightest blazars following Mrk 421; Mrk 501 and PKS 2155−304. Here we describe the *ASCA* observations and data analysis, and discuss the temporal characteristics of the X-ray light curves.

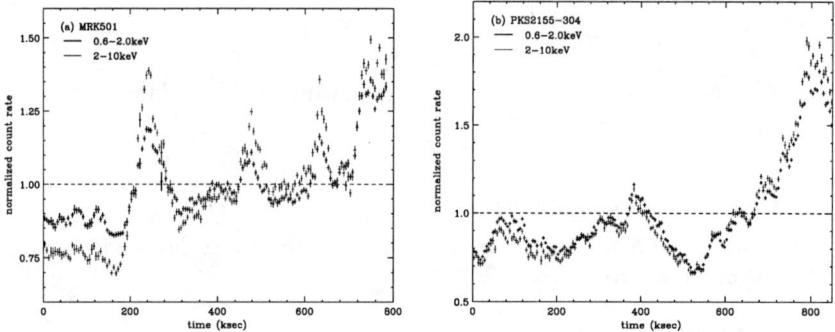

Figure 1. The 10-day light curves of the TeV blazars Mrk 501 and PKS 2155−304, taken by *ASCA* in spring 2000. The dark and light points represent the soft and hard X-rays (Tanihata et al. 2000)

2. Observations

We observed Mrk 501 and PKS 2155−304 with *ASCA* during 2000 Mar 1.50–11.00 UT and 2000 May 1.55–11.50, respectively. Using the normal screening criteria, source photons were extracted from circular regions centered at the target with 3′ and 6′ radii for SIS and GIS. Confirmed from the source-free regions that the background count rates and their fluctuations were negligible, we did not perform any background subtraction. For the X-ray light curves, SIS and GIS data were combined to maximize the signal-to-noise ratio in each temporal bin.

3. X-Ray Light Curves

3.1. Mrk 501

The 10-day observation of Mrk 501 started with a moderately faint state compared to historical observations with 2–10 keV flux at $\sim 6.0 \times 10^{-11}$ erg cm^{-2} s^{-1}, After 2 days of small variations, there was a large flare, followed by 2 more clear flares, and a further rise at the end. Between the large flares there is additional, significant variability, so that Mrk 501 is never in a truly quiescent state. The 10-day light curve is shown in Figure 1(a), where the dark points represent the soft X-ray band (0.6–2 keV), and the light points represent the hard X-ray band (2–10 keV), in 5 ks bins. Each light curve was normalized to its mean.

The amplitude of variability in the hard X-ray is clearly larger, indicating a hardening of the spectrum during high states. The flares in the two bands are well correlated, with apparently no lags larger than a few temporal bins (i.e. lags $\lesssim 10^4$ s). The spectra integrated over short time periods were well fitted by power laws with photon index varying within 1.8–2.2, suggesting that the synchrotron peak is around the *ASCA* range. Importantly, all flares were nearly symmetric, as those seen in the long-look observation of Mrk 421.

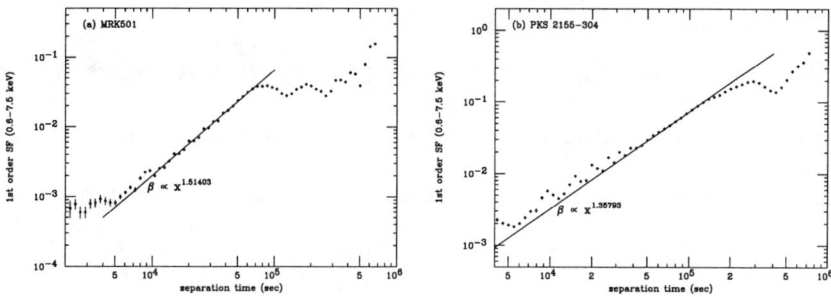

Figure 2. Structure functions for the X-ray light curves of Mrk 501 and PKS 2155−304. Both BL Lac objects show a clear break at a characteristic time scale.

3.2. PKS 2155−304

The observation of PKS 2155−304 also started in a relatively faint state, with 2–10 keV flux at $\sim 3.0 \times 10^{-11}$ erg cm^{-2} s^{-1}. The *ASCA* light curve of PKS 2155−304 is shown in Figure 1(b). In this case, the intensity increased strongly towards the end of the observation by nearly a factor of 3, and again, there is continuous symmetric flaring throughout the whole observation, each having amplitude $\sim$ 30–50%.

Here the variability amplitudes in the hard and soft X-ray bands are very similar, indicating little spectral change during even the large flare. The spectra was well fitted by a power law with photon index $\Gamma = 2.6$–2.8. Here the *ASCA* range must be above the synchrotron peak, and observed X-rays represent the high-energy tail of the electron energy distribution.

3.3. Variability Time Scale

The continuous coverage and long duration of the *ASCA* observations provide the best opportunity to date for studying variability in blazars, enabling us to do a more sophisticated analysis than simply estimating doubling times. We calculated the structure functions (Simonetti et al. 1985) for Mrk 501 and PKS 2155−304, shown in Figure 2. The duration and sampling of the observation limit the useful part of the structure function to time scales in the range $5 \times 10^3 \lesssim t \lesssim 3 \times 10^5$ s.

Both structure functions show a clear break from steep slope at short time scales to flatter slope at long time scales. This break represents a characteristic time scale of variability. At shorter time scales, the slope is much steeper even than a "random walk" case, meaning there is a rapid decrease in variability amplitude. On longer time scales, the slope flattens and there is little increase in variability amplitude. The observed breaks are at $\sim$ 1 day for Mrk 501 and $\sim$ 2.5 days for PKS 2155−304 (the latter would be better determined with a longer observation). The steep slopes are similar, $\sim$ 1.5 for Mrk 501 and $\sim$ 1.3 for PKS 2155−304.

4. Discussion

The long-look observations of 3 bright TeV blazars, including Mrk 421 in 1998 (Takahashi et al. 2000), have revealed that daily flares are common for all sources, and a particular high state is not required for flaring to occur.

The time scales of variability can be interpreted in terms of the time scales for electron injection, escape, acceleration, and cooling, taking into account propagation of light across the emitting volume and from the source to us (e.g. Chiaberge & Ghisellini 1999; Kusunose, Takahara, & Li 2000; Fossati et al. 2000). For example, soft lags have been interpreted as signatures of synchrotron cooling (Takahashi et al. 1996; Urry et al. 1997; Kataoka et al. 2000). Intensive X-ray monitoring of blazars has indicated not only soft lags but in some cases hard lags (Sembay et al. 1993; Takahashi et al. 2000), which can be interpreted in terms of acceleration time scales. The results in Takahashi et al. (2000) suggest that lags can differ flare by flare.

A notice is all flares being nearly symmetric. This implies that the flare time scales are dominated by the light crossing time, and any faster time scale is smeared out. This is consistent with the electron cooling times of the high energy electrons, emitting the X-ray synchrotron photons, usually calculated to be shorter than the observed flare time scales, when using the magnetic field and beaming factor estimated from the spectral energy distribution. If this is the case, the fact that flat tops are hardly seen in light curves would indicate that the injection time must be comparable to the light crossing time (see Chiaberge & Ghisellini 1998). The structure function analysis showed that variability time scales faster than ~ 1 day is strongly suppressed. This also supports the crossing time filtering all variations. Assuming that the typical time scale is the light crossing time, we can estimate the emission region size to be $R \sim c t_{var} \delta \sim 10^{16-17}$ cm. Since the opening angle of the jet is assumed to be $\sim 1/\delta$, we can also estimate the position of the emission region to be $D \sim 10^{17-18}$ cm from the center of the AGN. Further discussion of characteristic time scales are presented in Kataoka et al. (2000).

References

Chiaberge, M. & Ghisellini, G. 1999, MNRAS, 306, 551
Fossati, G., et al. 2000, ApJ, 541, 166
Kataoka, J., et al. 2000, ApJ, 528, 243
Kataoka, J., et al. 2000, submitted
Kusunose, M., Takahara, F., & Li, H. 2000, ApJ, 536, 299
Sembay, S., et al. 1993, ApJ, 404, 112
Simonetti, J. H., Cordes, J. M., & Heeschen, D. S., 1985, ApJ, 296, 46
Takahashi, T., et al. 1996, ApJ, 470, L89
Takahashi, T., et al. 2000, ApJL, in press (astro-ph/0008505)
Tanihata, C., et al. 2000, in preparation
Ulrich, M. H., Maraschi, L., & Urry, C. M. 1997, ARA&A, 35, 445
Urry, C. M., et al. 1997, ApJ, 486, 799

Spectral Energy Distribution of Low Power FR I Radio Galaxies

E. Trussoni, A. Capetti

Osservatorio Astronomico di Torino, Pino Torinese, Italy

A. Celotti, M. Chiaberge

SISSA, Trieste, Italy

Abstract. We discuss the broad band properties of two low luminosity radio galaxies: 3C 66B and 3C 346. Their broad band Spectral Energy Distributions (SED) require the existence of a peak at energies below the infrared band and have an X-ray emission with a flat spectrum. The SED of both sources are thus similar to those of Low Energy peaked BL Lacs (LBL), as expected from the unified scheme for AGN. More quantitatively, the values of their spectral indices are also similar to those of LBL, once the relativistic effects are properly taken into account. While BL Lac objects show a clear trend between luminosity and position of the SED peak, we find FR I with luminosities spanning over three orders of magnitude, with very similar shape of their SED.

1. Introduction

Low luminosity radio galaxies (type Fanaroff-Riley I, FR I) are usually identified as the parent population of BL Lac objects with misoriented jets (Urry & Padovani 1995). This unified model found support in these last few years by the similarities between the properties of the extended radio emission and host galaxies of the two classes. From the point of view of their nuclear properties, optical and X-ray observations of FR I showed the presence of correlations between the radio core luminosity with their X-ray and optical luminosities (Canosa et al. 1999; Chiaberge, Capetti, & Celotti 1999) which support a dominant non thermal origin for their nuclear emission, similarly to what is observed in BL Lac objects.

More insights on the unified scheme can be clearly obtained by comparing the overall Spectral Energy Distribution (SED) of the FR I objects with the SED of BL Lacs. This kind of analysis has been recently performed on a few objects and the results are basically consistent with the unified scheme (Capetti et al. 2000, hereafter C2000). In particular, the broad band spectral indices of radio galaxies are compatible, once the effects of the relativistic beaming are properly taken into account, with those of BL Lacs. Nonetheless, the simplest one-zone jet model is not sufficient to explain quantitatively for the difference in luminosity between FR I and BL Lac objects, as the FR I nuclei appear to be overluminous by a factor of 10–10^4. Following the suggestion by Chiaberge et al. (2000), these results can be reconciled with the unification scheme if a strong velocity structure is present across the jets in BL Lac.

Here we extend this analysis by presenting the data on the SED of two radio galaxies (3C 66B and 3C 346) and discussing their implications on the validity of the unified model.

2. The Sources and the Data

In Table 1 we summarize the main data on these two radio-galaxies, both associated with isolated elliptical galaxies. The core radio luminosities ($L_{c,r}$) have been deduced by the VLA observations of Giovannini et al. (1988). At optical and ultraviolet wavelengths, we used the results of Chiaberge et al. (1999) and the data of the public archive of HST to obtain the values of $L_{c,o}$ and $L_{c,UV}$.

For the SED construction it is crucial to determine the value of the slope of the spectrum at X-ray energies. We have then analyzed the pointed observations of the PSPC instrument on the Rosat satellite, isolating the non-thermal (power-law) component from the thermal (coronal) emission. In both the radio galaxies the nuclear emission has quite flat, non thermal spectrum with photon index ≈ 1.3 (3C 66B) and ≈ 1.7 (3C 346). The X-ray luminosity ($L_{c,X}$) is consistent with the correlation $L_{c,r}$–$L_{c,X}$ found by Canosa et al. (1999) for low luminosity radio galaxies.

Table 1. The Main Data on the Sources ($H_o = 75$ km s^{-1} Mpc^{-1})

source	z	m_v	$\nu L_{c,r}$[a]	$\nu L_{c,o}$[b]	$\nu L_{c,UV1}$[c]	$\nu L_{c,UV2}$[d]	$\nu L_{c,X}$[e]
3C 66B	0.021	15.0	0.7e38	3.7e41	1.4e41	–	3.8e41
3C 346	0.162	17.5	5.6e41	1.1e43	–	3.3e42	2.6e43

Luminosity are in erg s^{-1} at: [a] 5 GHz, [b] 7020 Å, [c] 3400 Å, [d] 1400 Å and [e] 1 keV rest frame.

3. Discussion

The SED of both 3C 66B and 3C 346 show a clear decline in the optical/infrared band, as can be seen in Fig. 1 (where we plot also the SED of 3C 264 and 3C 270 discussed in C2000). This indicates that an emission peak is present and it is located at energies below the infrared band. The overall shape of these SED is thus remarkably similar to what is observed in Low Frequency Peaked BL Lacs (LBL). In these latter class of sources the X-ray emission is interpreted as a (self) Compton component, produced by the same population of relativistic electrons which radiates the synchrotron emission observed at lower energies. This is consistent to what is seen for the two FR I, which indeed show rather flat X-ray spectra. These results are in overall agreement with what is expected from the unified model.

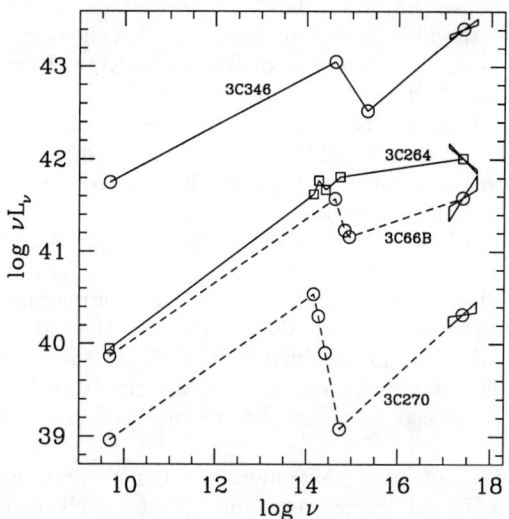

Figure 1. Plot of the SED (νL_ν vs. ν) of 3C 66B and 3C 346 (the SED of 3C 264 and 3C 270 are also plotted, Capetti et al. 2000).

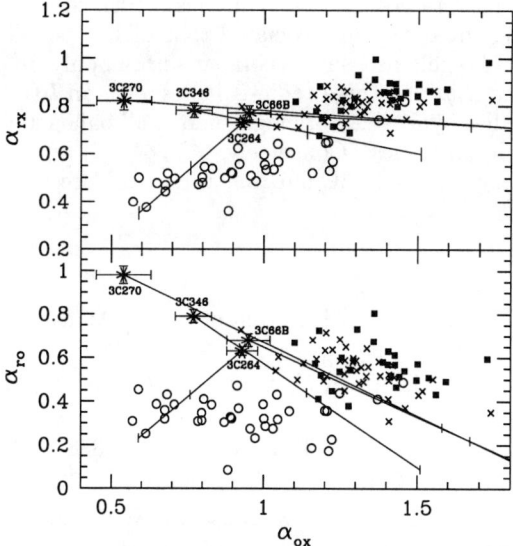

Figure 2. Broad band spectral indices of 3C 66B and 3C 346 (the indices of 3C 264 and 3C 270 are also plotted, Capetti et al. 2000). The circles are HBL, while the crosses and squares are LBL and FSRQ, respectively. The solid lines are the beaming tracks, and the vertical ticks correspond to an increase of δ by a factor of 10 and 100, respectively.

On the other hand, the broad band spectral indices for these sources (also including those presented by C2000) are not located in the same region of the α_{rx} vs. α_{ox} and α_{ro} vs. α_{ox} planes proper of BL Lac objects. However a consistent comparison of the SED of radio galaxies and BL Lacs must include also the relativistic effects. Accordingly we have built the 'beaming track' followed by each FR I sources for an increasing beaming factor δ. These tracks represent the variations of the spectral indices induced by the upward energy shift associated to the relativistic beaming when the jet is seen closer to the line of sight. The representative point of both radio galaxies reaches the region in which LBL are located as soon as the value of δ is increased by a factor as small as ~ 3 (corresponding to a luminosity increase of ~ 2 orders of magnitude). This solves the apparent contrast between the SED shape and the values of the spectral indices, confirming the results obtained by C2000. In this case, however, a jet velocity structure is not strictly required as the resulting beamed luminosity does not necessarily exceed the observed luminosities in BL Lac with similar extended properties.

In blazars, the broad band phenomenology can be accounted for by a continuous distribution in the energy position of the synchrotron and Compton emission peaks and their relative luminosity (Padovani & Giommi 1995; Fossati et al. 1998). This trend seems to be connected with the source (bolometric and radio) power, increasing from HBL to LBL and FSRQ (Ghisellini et al. 1998). HBL are also known to have extended radio luminosities significantly smaller than LBL. Together with the data presented by C2000, there are now four well described SED of FR I radio-galaxies. One has a SED shape similar to HBL, while the remaining three can be associated to a LBL type. Given the limited data available for radio galaxies, such a comparison can only be preliminary, but we can already note that the three radio-galaxies with an LBL type SED have luminosities that differ by 3 orders of magnitude in all bands and by a factor 100 in extended radio luminosities. Conversely, 3C 264 has nuclear and extended radio luminosity very similar to 3C 66B but their SED are radically different.

References

Canosa, C. M., Worrall, D. M., Hardcastle, M. J., & Birkinshaw, M. 1999, MNRAS, 310, 30
Capetti, A., Trussoni, E., Celotti, A., Feretti, L., & Chiaberge, M. 2000, MNRAS, in press (C2000)
Chiaberge, M., Capetti, A., & Celotti, A. 1999, A&A, 358, 104
Chiaberge, M., Celotti, A., Capetti, A., & Ghisellini, G. 2000, A&A, 358, 104
Fossati, G., Maraschi, L., Celotti, A., & Ghisellini, G. 1998, MNRAS, 299, 433
Ghisellini, G., Celotti, A., Fossati, G., Maraschi, L., & Comastri, A. 1998, MNRAS, 301, 451
Giovannini, G., Feretti, L., Gregorini, L., & Parma, P. 1988, A&A, 199, 73
Padovani, P. & Giommi, P. 1995, ApJ, 444, 567
Urry, C. M. & Padovani, P. 1995, PASP, 107, 803

BL Lacs at the Blue End of the Blazar Sequence

L. Costamante

Univ. of Milan, Milan, Italy; Osservatorio di Brera, Milan, Italy

G. Ghisellini, A. Wolter, G. Tagliaferri

Osservatorio di Brera, Milan, Italy

G. Fossati

CASS/UCSD, La Jolla, CA, USA

P. Padovani

ESA/Space Telescope Science Institute, Baltimore, MD, USA

P. Giommi

BeppoSAX Science Data Center, ASI, Roma, Italy

Abstract. We present the main results of seven *Beppo*SAX observations performed with the aim to find and study more objects with "extreme" synchrotron peak frequencies ($\gtrsim 1$ keV). Five sources have been confirmed as "extreme," with one, 1ES 1426+428, showing a peak energy at or above 100 keV. Our results seem also to confirm the higher spectral variability of "high peak" objects compared to the "low peak" ones.

1. Introduction

Among blazars, BL Lacertae objects are the sources which show the highest variety of synchrotron peak frequencies, ranging from the IR to UV—soft X energies (i.e. LBLs and HBLs, respectively, see Padovani & Giommi 1995), and even up to 100 keV, as demonstrated by the cases of Mkn 501 and 1ES 2344+514 (Pian et al. 1997; Giommi et al. 2000). Although recent surveys (like DXRBS, Perlman et al. 1998; and RGB, Laurent-Muehleisen et al. 1998) are now sampling quite well most of this sequence, very little is known about the high energy branch: few objects have shown peak energies $\gtrsim 1$ keV, and the most extreme values have been observed, up to now, only during exceptional flares. The high synchrotron frequencies displayed by these sources, flagging the presence of high relativistic electrons, makes them the most interesting objects for studying the particle acceleration mechanism at its limits, and good candidates for TeV emission through the inverse Compton process. Here we report the main results of an observational campaign with the X-ray satellite *Beppo*SAX performed with the aim to find and study other sources with these properties.

The candidates have been selected from the Einstein Slew Survey and the bright source catalog of the Rosat All Sky Survey (RASSBSC) on the basis of properties suggesting high peak frequencies: a) very high F_x/F_{radio} ratio ($> 3\times 10^{-10}$ erg cm^{-2} s^{-1}/Jy, at [0.1–2.4] keV and 5 GHz respectively); b) strong ($> 10^{-11}$ erg cm^{-2} s^{-1}) and flat ($\alpha_x \lesssim 1$, when available) 0.1–2.4 keV X-ray spectra, connecting smoothly with the flux at lower frequencies; c) with appropriate values of α_{ro}, α_{ox} and α_{rx}, similar to other extreme BL Lacs.

Tab. 1 lists the seven sources observed by *Beppo*SAX in this program. The data have been reduced and analyzed according to the SDC Cookbook instructions (details in Costamante et al. 2000, in preparation)

The spectra have been fitted both with single and broken absorbed power-law models: the best fits parameters are reported in Table 1, and Fig. 3 shows the Spectral Energy Distribution (SED) for all sources.

Table 1. Best fits parameters

Source	N_H 10^{20} cm^{-2}	α_1	E_{break} keV	α_2	F_{1keV} μJy	F_{2-10} ergs/cm^{2}s	χ^2_r/d.o.f.
1ES 0120+340	5.2 $gal.$	$0.8^{-1.1}_{+0.3}$	$1.4^{-0.7}_{+1.0}$	$1.32^{-0.08}_{+0.08}$	$4.5^{-0.6}_{+2.1}$	1.3×10^{-11}	0.92/93
1ES 0033+595	61^{-12}_{+12}	$0.82^{-0.33}_{+0.12}$	$2.9^{-0.6}_{+2.4}$	$1.08^{-0.05}_{+0.11}$	$13.2^{-2.5}_{+2.5}$	6.0×10^{-11}	1.05/151
PKS 0548–322	$4.2^{-0.9}_{+1.1}$	$0.91^{-0.16}_{+0.10}$	$4.5^{-2.3}_{+1.8}$	$1.4^{-0.3}_{+0.6}$	$5.7^{-0.5}_{+0.5}$	2.3×10^{-11}	0.95/82
GB 1114+203	1.36 $gal.$	$1.23^{-0.11}_{+0.10}$	$1.2^{-0.2}_{+0.4}$	$1.90^{-0.09}_{+0.10}$	$5.0^{-0.5}_{+0.5}$	6.2×10^{-12}	0.95/81
1ES 1218+304	1.73 $gal.$	$1.02^{-0.10}_{+0.09}$	$1.4^{-0.2}_{+0.4}$	$1.56^{-0.05}_{+0.05}$	$7.3^{-0.7}_{+0.4}$	1.5×10^{-11}	0.89/93
1ES 1426+428	$1.5^{-0.3}_{+0.4}$	$0.92^{-0.04}_{+0.04}$	—	—	$4.6^{-0.2}_{+0.2}$	2.0×10^{-11}	1.00/89
H 2356–309	1.3 $gal.$	$0.78^{-0.09}_{+0.06}$	$1.8^{-0.6}_{+0.6}$	$1.10^{-0.05}_{+0.05}$	$6.2^{-0.5}_{+0.5}$	2.5×10^{-11}	0.94/35

Errors at $\Delta\chi^2 = 4.61$

2. Results

All sources, except GB 1114+203, have been detected also in the PDS instrument. For five sources the peak of the synchrotron emission is in the X-ray band, near or above 1 keV, thus confirming their "extreme" nature. GB 1114+203 and 1ES 1218+304, instead, present a curved spectrum with both the spectral indices steep (i.e. > 1), which locates the peak energy of the synchrotron emission below the observed X-ray band, thus qualifying these objects as typical HBLs. These results confirm the good efficiency of the adopted selection criteria at finding high ν_{peak} sources, and we are now beginning to populate the high energy branch of the synchrotron peak sequence.

With more objects, it is interesting to compare the properties of this type of BL Lacs with those of HBLs and LBLs. In Fig. 1 α_x and the broad band spectral index α_{rx} are plotted vs. the frequency of the synchrotron peak. As shown in the left panel, the "extreme" BL Lacs data seem to suggest a flattening of the correlation between α_{rx} and ν_{peak} at large values of ν_{peak}. This is in agreement with the scenario of a synchrotron peak moving smoothly from lower to higher

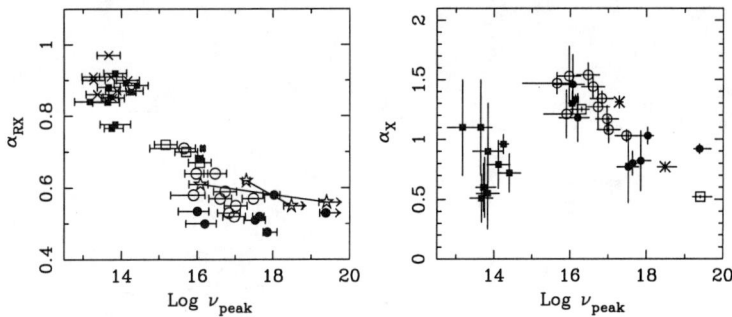

Figure 1. α_{rx} and α_x vs. ν_{peak} for our 7 sources (filled circles) together with the HBL and LBL in Wolter et al. 1998 (open marks and crosses) and the 1 Jy BL Lacs data (Padovani et al. 2000, in preparation, filled squares). The "quiescent" and "flaring" states of Mkn 501 and 1ES 2344+514 are represented, on the left as stars with connecting lines, and on the right as open squares and asterisks, respectively.

energies: as long as radio and X-ray fluxes are produced by different branches of the synchrotron emission (before and after the peak), they change differently as the peak shifts, thus changing α_{rx}. When the peak moves into the X-ray band and beyond, both fluxes come from the same branch of the synchrotron emission, and so begin to change similarly as the peak moves at still higher frequencies, stabilizing α_{rx} at a common (flat) value.

The "moving peak scenario" also nicely accounts for the shape of the relation α_x–ν_{peak}. In this case, the spectral index just traces the upcoming of the synchrotron emission in the X-ray band: it steepens from LBLs to HBLs, when the main contribution in the X-ray band passes from the flat inverse Compton emission to the steep tail of the synchrotron emission, and then starts flattening again as the synchrotron peak moves into the X-ray band and beyond, eventually reaching a "stable" flat value corresponding to the typical slope of the synchrotron emission well before the peak.

Among the five extreme sources observed, the most interesting one turned out to be 1ES 1426+428. This object has shown a flat spectral index ($\alpha_x = 0.92$) up to the PDS band, after taking into account other contaminating objects in the PDS f.o.v. (details in Costamante et al. 2000, in preparation). **This constrains the synchrotron peak to lie near or above 100 keV**, and makes this object the third source ever found with such extreme peak energies. Quite interestingly, however, as opposed to the other two "over 100 keV" sources, the X-ray flux during the observation was not particularly high (see Fig. 2, left): it is then likely that 1ES 1426+428 is the first source found in a "quiescent extreme" state. Note that all three sources, in previous observations, were characterized by steep spectral indices, i.e. with ν_{peak} below the X-ray band, like typical HBLs. These objects have undergone a shift in the synchrotron peak frequencies of two orders of magnitude or more, with or without luminosity changes (see Fig. 2, right). It is still uncertain if this extreme spectral variability is common to all blazars or only to HBLs ($\nu_{peak} \gtrsim 10^{15}$ Hz). However, it is interesting to note that

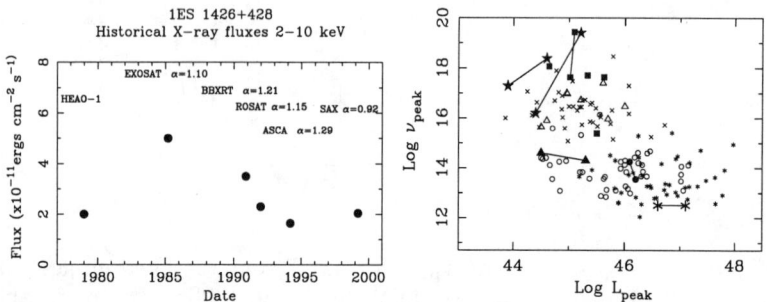

Figure 2. Left: Historical 2–10 keV fluxes for 1ES 1426+428. Above each point are also reported the instrument and the measured 2–10 keV spectral index. Right: peak frequencies vs. luminosity at the peak. Crosses, circles and asterisk are BL Lacs and FSRQs data from Fossati et al. 1998. Open triangles are the HBLs from Wolter et al. 1998, filled squares are our data. The lines connect different states for Mkn 501 and 1ES 2344+514 (stars), 1ES 1426+428 (filled squares), BL Lac (filled triangles), OJ 287 (filled circles) and 3C279 (large asterisks).

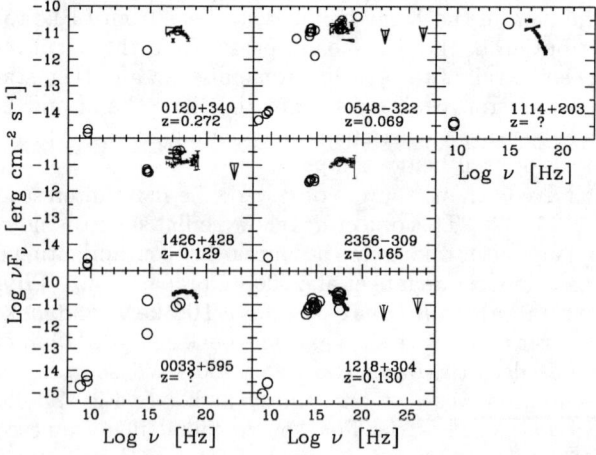

Figure 3. The SEDs of the seven sources, made with *Beppo*SAX and literature data.

for some well studied "low peak" sources (for example, BL Lac itself, OJ 287 and 3C279, see Fig. 2) the value of $\nu_{\rm peak}$ seems to remain much more constant.

Acknowledgments. L.C. thanks the ST ScI visitor program and the CARIPLO Foundation for support.

References

Fossati, G., et al. 1998, MNRAS, 299, 433
Giommi, P., Padovani, P., & Perlman, E. 2000, MNRAS, 317, 743
Laurent-Muehleisen, S. A., et al. 1998, ApJS, 118, 127
Padovani, P. & Giommi, P. 1995, ApJ, 444, 567
Perlman, E., et al. 1998, AJ, 115, 1253
Pian, E., et al. 1997, ApJ, 492, 17L
Wolter, A., et al. 1998, A&A, 335, 899

A Polarization Flare in 3C 273: A Clue to Jet Physics

Lara L. Cross, Beverley J. Wills

Astronomy Department, University of Texas at Austin, Austin, TX, 78712

J. H. Hough

Department of Physical Sciences, University of Hertfordshire, Hatfield, Hertfordshire AL10 9AB

J. A. Bailey

Anglo-Australian Observatory, P.O. Box 296, Epping, NSW 2121, Australia

Abstract. We present *UBVRIJHK* polarization and flux density observations of the quasar 3C 273 obtained during a time of outburst over two weeks in 1988 February. We have modelled these data with two power law components, each with wavelength-independent position angle. These components are roughly perpendicular. The steeper-spectrum component has higher infrared polarized flux density, with the electric vector approximately transverse to the projected direction of the VLBI jet. The K-band polarized flux density and position angle, and possibly the spectral index of the two components, are correlated with a time lag of less than a day. We explain our results in terms of a shocked jet model with two nearly co-spatial components: a shock component with magnetic field approximately perpendicular to the jet and the other with magnetic field approximately parallel to the jet.

1. The Data

Our data consist of polarization position angle (E-field), percentage polarization, and total flux at *UBVRIJHK* for 3C 273 for each night of UT 10, 12–18 February 1988. The data for a representative night (12 Feb.) are shown in Figure 1, where the solid lines represent the best-fitting model for that night (see §2). The data were taken at UKIRT with the Mark II Hatfield Polarimeter (Hough, Peacock, & Bailey 1991), which makes simultaneous optical and near-infrared observations, important for this variable blazar. The data have been corrected for interstellar polarization and Galactic extinction. A brief discussion of these data was presented by Wills (1991).

At the time of our observations 3C 273 was in an outburst state (Courvoisier et al. 1988). This outburst has since been associated with the ejection of VLBI component C9 ($t_0 = 1988.1 \pm 0.1$, Bååth et al. 1991). At this time the polarization was higher than usual.

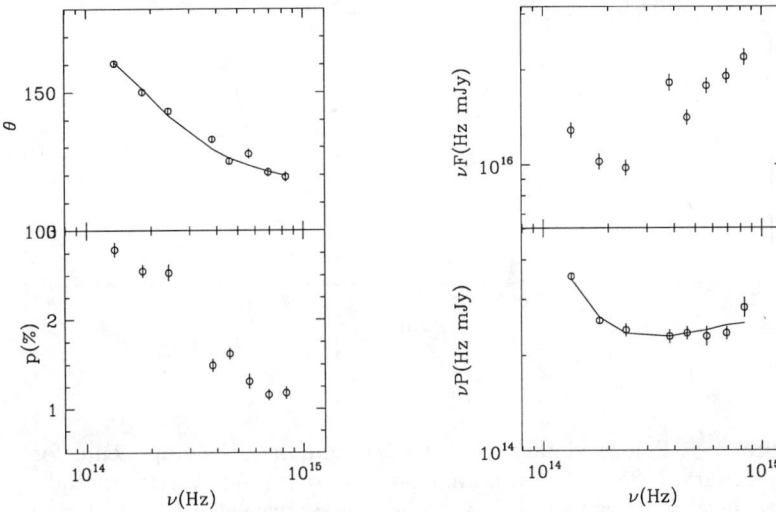

Figure 1. Frequency-dependence of polarization position angle and percentage, total and polarized flux (in mJy), corrected for Galactic extinction and interstellar polarization, for 12 February, 1988. Solid line represents the best-fitting model. The high value of the total flux in I-band is due to the presence of the Hα line in this band.

In our data both the percentage polarization and the position angle are very frequency-dependent, and this dependence changes from night to night. The frequency dependence of θ, particularly a 90° flip on 14 Feb., is suggestive of the presence of two perpendicular polarized components of similar amplitude but different spectral shape.

2. Modelling

We have modelled our data with two power laws in polarized flux density that vary from night to night in amplitude, spectral index, and position angle. The addition of two such components for 12 Feb. is illustrated in Figure 2. In all cases we find that the two components are approximately perpendicular, with the E-vector of the steeper component transverse to the projected jet direction. The model reproduces the data well. Slight changes in the model parameters (for example, $< 3°$ in position angle, < 0.05 in spectral index, and < 0.05 mJy in amplitude) produce noticeably worse fits to the data. The two-component models are thus well-constrained by the data. We find that the components are correlated in amplitude and position angle, and probably spectral index (significance levels of $> 99.75\%$, $> 99\%$, and $> 93\%$ respectively) with a time lag of much less than one day. The changes of the amplitudes and position angles of the components with time are illustrated in Figure 3.

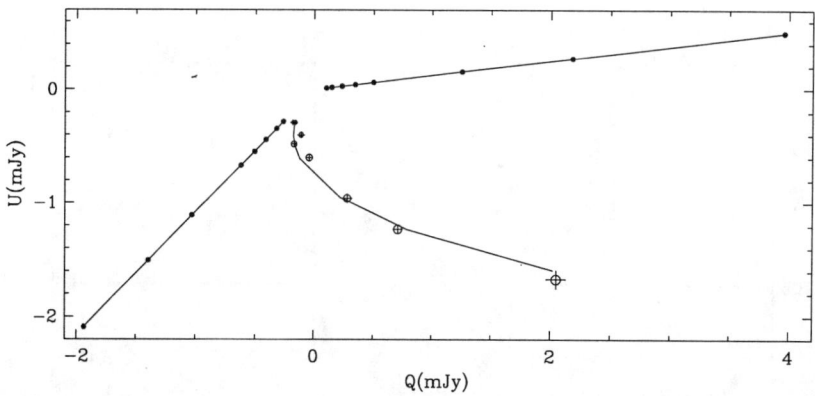

Figure 2. Polarized flux, best model, and model components for 12 February, 1988 in Stokes parameter representation. Circle size indicates wavelength, with larger circles for longer wavelength. Filled dots represent Q and U of the model components at the wavelengths of the passbands. The components add vectorially to give the total model.

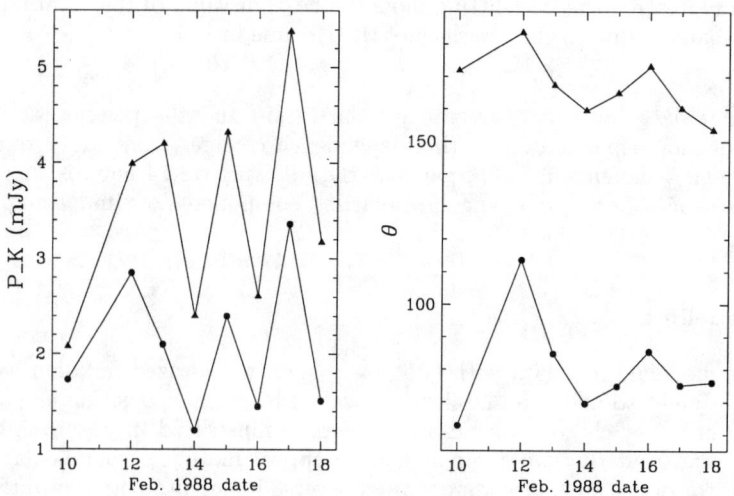

Figure 3. Changes of model K-band amplitude and position angle for the two model components as a function of time. Circles represent component 1, which is approximately parallel to the VLBI jet, and triangles represent component 2, perpendicular to the jet. The parameters for the two components generally rise and fall together. Results are similar for the model spectral indices. For clarity data for each component are connected by straight lines.

3. Discussion

We explain the polarization of 3C 273 in outburst with two synchrotron components. Since these components are nearly perpendicular they tend to cancel each other out, resulting in low polarized flux. This blazar-like polarization is further diluted by a strong, unpolarized non-blazar continuum, resulting in the low observed percentage polarization for this object.

The outburst emission of blazars is often explained in terms of a shocked jet model like that of Marscher & Gear (1985). Such models normally involve a quiescent component and a shock component. We instead require two components associated with the shock. These components must be nearly co-spatial to account for the very short time lag between their polarizations. The simplest model would involve an underlying magnetic field that is parallel to the jet, but which is compressed so as to be perpendicular in the region of the shock. One of our components, which has polarization parallel to the jet and a flatter spectrum, would originate from the region of the perpendicular B-field, as expected from the Marscher & Gear model. The other component, with perpendicular polarization, would be due to high energy electrons that have leaked from the shocked region and are accelerated where the B-field is parallel to the jet. We see these two components superimposed.

References

Bååth, L. B., et al. 1991, A&A, 241, L1
Courvoisier, T. J.-L., et al. 1988, Nature, 335, 330
Hough, J. H., Peacock, T., & Bailey, J. A. 1991, MNRAS, 248, 74
Marscher, A. P. & Gear, W. K. 1985, ApJ, 298, 114
Wills, B. J. 1991, in Variability of Active Galactic Nuclei, eds. H. R. Miller & P. J. Wiita (New York: Cambridge University Press), 87

Simultaneous Optical and X-Ray Observations of BL Lacertae

R. Nesci, E. Massaro, F. Montagni, S. Sclavi

Univ. "La Sapienza," Roma, Italy

T. Balonek, M. Caler, C. Tremonti

Foggy Bottom Observatory, Colgate University, USA

F. D'Alessio

Osservatorio Astronomico di Roma, MontePorzio, Italy

S. Catalano, A. Frasca, E. Marilli

Catania Astrophysical Observatory, Italy

G. Tagliaferri, G. Ghisellini, M. Ravasio

Brera Astronomical Observatory, Italy

P. Giommi

BeppoSAX Scientific Data Center, ASI, Roma, Italy

L. Chiappetti

Istituto di Fisica Cosmica CNR, Milano, Italy

T. Kato, M. Uemura

Kyoto University, Japan

O. M. Kurtanidze, M. G. Nikolashvili

Abastumani Observatory, Abastumani, Republic of Georgia

M. T. Carini, J. C. Noble

Western Kentucky University, USA

G. Tosti, G. Nucciarelli

Perugia University Observatory, Italy

J. Mattox

Frances Marion University, Florence, SC, USA

Abstract. We present the results of simultaneous X-ray and optical observations of BL Lacertae performed in the periods 1999, June 5–7 and December 5–6. The *Beppo*SAX satellite (0.2–100 keV) and the WEBT (Whole Earth Blazar Telescope) collaboration (B,V,R,I bands) were involved in this campaign. During the first observation the optical behavior of the source showed an oscillating pattern with a total observed amplitude of ∼ 0.5 mag. A marked variability was detected on hour time scale in the soft X-rays. Two power-law components were required to fit the X-ray spectrum, which we interpret as the Synchrotron (soft) and the Inverse Compton (hard) components. During the second run only the Inverse Compton component was detected in the X-rays, without appreciable variability neither in the X-rays nor in the optical.

1. Introduction

Simultaneous multiband observations of blazars are a key way to better understand the physical processes responsible for the emission of these puzzling objects. According to our current knowledge there are two main processes at work:

a) Synchrotron emission by relativistic electrons in a somewhat ordered magnetic field, mainly parallel to the relativistic jet, which is responsible for the emission from radio (8 GHz) to soft X-rays (2 keV);

b) Inverse Compton emission from upscattered low-energy photons, either coming from the above mentioned Synchrotron process (Self Compton) or from external origin.

A very effective way to recognize if the X-ray emission is just a tail of the Synchrotron process responsible for the optical one, or is produced by a different mechanism, is to check for its variability simultaneously with the optical emission.

With this purpose in mind, at the IAU General Assembly at Kyoto in 1997 the Whole Earth Blazar Telescope (WEBT) collaboration was started, including several observers scattered all over the world, aiming at performing a 24 hours coverage of some BL Lac objects simultaneously with X-ray or γ-ray observations.

The observations presented in this paper are the result of the interaction between an X-ray team (P.I. G. Tagliaferri) using the *Beppo*SAX satellite and the WEBT collaboration, coordinated in this occasion by R. Nesci.

The first satellite pointing was made on July 5–7 1999, and a second one was performed five months later on December 5–6.

2. Optical Observations

In the first campaign, useful observations were collected by the following groups: Roma University, Abastumani Observatory, Foggy Bottom Observatory (Colgate University), Lowell Observatory (Western Kentucky University), Kyoto University and Perugia Observatory. In the second campaign data were obtained by the groups of Roma, Abastumani, Kyoto, Perugia, and also by the Catania Astrophysical Observatory, while bad weather conditions prevented any contribution from the USA. Photometric observations were performed in the standard bandpasses U, B, V (Johnson) and R, I (Cousins). A list of the involved telescopes and filters is given in Table 1. Magnitudes were evaluated using IRAF (Roma, Abastumani, Foggy Bottom, Lowell) or locally developed codes (Catania, Kyoto, Perugia).

Table 1. WEBT Optical Telescopes

Group	Bands	telescope	detector
Abastumani	B,V,R,I	70 cm	Texas TC241
W. Kentucky (Lowell)	V, I	180 cm	SITe 501A
Catania	U,B,V	91 cm	EMI 9893QA/350 photomultiplier
Foggy Bottom	R,I,V	40 cm	CCD
Kyoto	R	25 cm	Kodak KAF 400
Perugia	R, I	40 cm	Texas TC211
Roma (Vallinfreda)	V	50 cm	Texas TC241
Roma (Greve)	I	35 cm	SITe 501A

Each optical group performed differential photometry of BL Lac with respect to 4 nearby reference stars (Bertaud et al. 1969; Fiorucci and Tosti 1996). The data, sent to the campaign coordinator, were then checked for systematic differences in the instrumental magnitudes of the reference stars; a general light curve of BL Lac was obtained and the color indices were computed when possible.

Since not all the involved observers could use the same filters (see Table 1), to build an optical light curve as complete as possible in time coverage, conversions to the same I band were made using appropriate color indices for each Observatory (typical V−I = 1.3). The resulting light curve is plotted in Fig. 1 (lower panel) for comparison with the X-ray one (upper panel).

It is apparent from Fig. 1 that BL Lac varied continuously during the *Beppo*SAX pointing with an overall amplitude of 0.55 mag. After the maximum at 1334.8, followed by a decay of about 0.4 mag in less than 10 hours, the source behavior was characterized by an increasing trend for about one day, superposed onto small oscillations; the brightness then decreased down to I = 13. These data show how important is the availability of several observers distributed in longitude to achieve a good coverage of the light curve. Indeed, given the shortness of the night near the summer solstice, each telescope could monitor BL Lac only for a few hours.

A color change with the source luminosity (0.06 in V−I for a change of 0.2 R mag) was clearly detected in the Abastumani data set, which provided nearly simultaneous BVRI photometry for June 6, but it is not critical for the transformation of the observed magnitudes from the other bands into the I band,

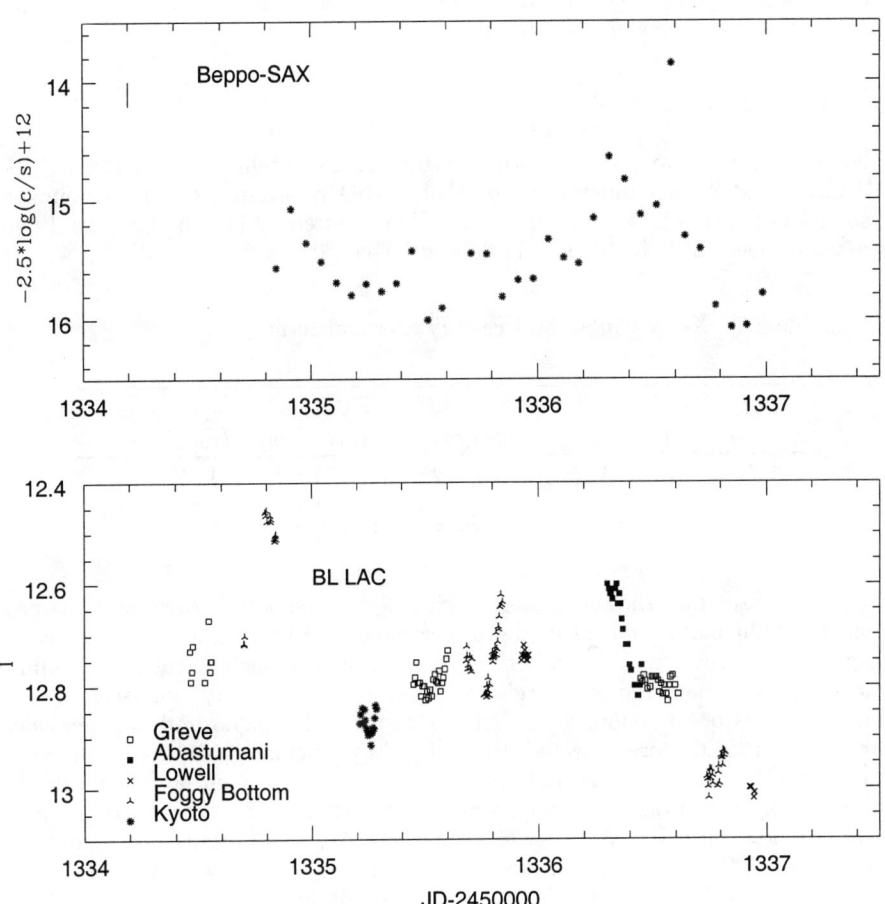

Figure 1. Composite optical I-band light curve (lower panel) and LECS x-ray light curve averaged over 30 m intervals (upper panel), for the June pointing. Notice that the ordinates are in magnitudes for both plots. The typical x-ray error bar is shown at the upper left corner. Perugia data, not plotted for sake of clarity, overlap some Greve points.

so that the long-term behavior of the source during the run shown in Fig. 1 is reliable.

The second run (December) was less lucky from the meteorological point of view, so coverage from the ground was less complete. The source showed an overall variation of only ~ 0.2 mag. The (V−I) color index was about 1.6, substantially steeper than in June, while the average flux level in the I band (12.6) was about the same at both epochs.

3. X-ray Observations

The observations were made with the *Beppo*SAX satellite, using the LECS, MECS and PDS instruments, as part of a ToO program. The first pointing was on June 5, 1999, starting at 08:05 UT and lasted 52 hours; the second one started on December 5, 1999 at 11:06 and lasted 30.5 hours.

Table 2. X-ray Fluxes and Energy Spectral Indices

Date	α_1 (energy)	α_2 (energy)	$F_{2-10\text{keV}}$ (erg/cm^2/s)	I (mag)	α_{opt}
5–7 June	1.6 ± 0.25	0.15 ± 0.22	0.6×10^{-11}	12.7	1.4
5–6 Dec.	—	0.60 ± 0.05	1.2×10^{-11}	12.6	1.9

The source flux showed a large variability in the soft X-rays, as apparent from the light curve plotted in the upper panel of Fig. 1; a logarithmic scale, magnitude like, was used for a better comparison with the optical one. No flux variations were detected at energies above 4 keV on the day time-scale. Two main episodes of variation, a smaller one around JD 1336.3 and a larger one, around JD 1336.6, were detected; the latter, in particular was quite fast having a total duration of about 20 minutes only. Despite the much larger amplitude of the X-ray variations, it is apparent a fair match of the overall optical and X-ray trends until JD 1336.4, while the large X-ray outburst has no optical counterpart. This outburst, however, was real because it was detected both by the LECS and the MECS instruments and the image analysis showed that the photons' coordinates were fully compatible with the position of BL Lac. A spectral fit to the LECS, MECS and PDS data required a two-power-law model ($F_\nu = A_1 \nu^{-\alpha_1} + A_2 \nu^{-\alpha_2}$), with a steep low energy component and a much harder one above about 3 kev, to give a satisfactory χ^2. The best-fit value of the (energy) indices are given in Table 2.

In the second run (December) only the hard component was detected at a level (in the 2–10 keV range) brighter than in June (see Table 2); no significant variability was observed.

4. Conclusions

The main results of this campaign may be summarized as follows:

a) The optical data are well fitted by a single power law, while two power laws are necessary for the X-ray data, at least for the June observations, with a spectral slope of the hard X-ray component flatter than the optical one, indicating that different radiation processes are involved.

b) Variability on intraday time scale was detected both in the optical and in the soft X-rays in June and the variations were fairly correlated, while the hard X-ray flux was stable on this time scale. This result supports the conclusion, derived from the spectral slopes, of the existence of two physical processes responsible for the emission.

c) On a several months time scale the X-ray flux of the hard component was different by a factor ~ 2, while the optical flux in the I band was practically unchanged.

d) The optical spectrum in the December observation was steeper than in June, consistently with the non detection of the soft X-ray component.

The most straightforward interpretation of all these findings is that the soft X-ray flux observed in June is essentially the high energy tail of the Synchrotron emission, having the peak of its spectral energy distribution in the optical. The hard X-ray flux is likely due to the Inverse Compton process, but the lack of variability on intraday time scale suggests that the seed photons of this component are not the optical ones. The upscattered photons could either have a much lower frequency (from the far IR or millimetric range), where the emission is expected to be less variable than around the peak, or be originated in a region external to the jet.

We remark that BL Lac is the third case, after ON 231 (Tagliaferri et al. 2000), and S5 0716+714 (Giommi et al. 1999), of a *Beppo*SAX detection of the emission from both Synchrotron and Inverse Compton processes in a BL Lac object.

Some observational points however still requires a deeper understanding:

a) Despite the source was at the same optical level in the two runs, both the optical and the X-ray spectral slopes were different. Therefore there is not a simple relation between the flux level and the spectral shape when different episodes of variation are considered, while a fair correlation is present within a single episode.

b) The nature of the large and fast burst detected in the soft X-rays during the June observation without an optical counterpart is unclear. It is the only episode of this type observed so far. Further simultaneous observational campaigns in the optical and X-rays are therefore necessary to search for other similar events.

References

Bertaud, C., et al. 1969, A&A, 3, 436
Fiorucci, M. & Tosti, G. 1996, A&AS, 116, 403
Giommi, P., et al. 1999, A&A, 351 59
Tagliaferri, G., et al. 2000 A&A, 354 431

The Physics of Blazar Optical Emission Regions. I. Alignment of Optical Polarization and the VLBI Jet

Michael J. Yuan

University of Texas at Austin

Hien Tran

The Johns Hopkins University

Beverley Wills, Derek Wills

University of Texas at Austin

Abstract. We collected optical and near IR linear polarization data obtained over 20–30 years for a sample of 51 blazars. For each object, we calculated the probability that the distribution of position angles was isotropic. The distribution of these probabilities was sharply peaked, with 27 blazars showing a probability $< 15\%$ of an isotropic distribution of position angles. For these 27 objects we defined a preferred position angle. For those 17 out of 27 blazars showing a well-defined radio structure angle (jet position angle) on VLBI scales (1–3 mas), we looked at the distribution of angle differences—the optical polarization relative to the radio position angles. This distribution is sharply peaked, especially for the BL Lac objects, with alignment better than $15°$ for half the sample. Those blazars with preferred optical position angles were much less likely to have bent jets on 1–20 mas scales. These results support a shock-in-jet hypothesis for the jet optical emission regions.

1. Introduction

Polarization observations have long been a very important probe of the internal structure of blazar jets. Bright spatially resolved knots often show radio polarization (**E** vector) aligned with the projected jet direction, indicating a perpendicular magnetic field. This suggests that shocks are responsible for compressing the jet magnetic field and accelerating the synchrotron-emitting electrons (e.g. Aller, Aller, & Hughes 1985).

The relation between optical polarization and VLBI structure provides a unique tool for investigating the regions of jet formation on $\lesssim$ parsec scales. While previous statistical investigations have shown a tendency for optical polarization to be aligned with the jet (Impey et al. 1991; Rusk & Seaquist 1985), the interpretation of optical polarization is less clear because the emitting regions in blazars are not resolved, blazars often show violent short-term optical polarization variability, and the old radio observations did not have sufficient angular resolution to probe the region near the optically emitting core. A few quasi-simultaneous optical-VLBI observations indicate that the optical polarization is aligned with the direction of newly-ejected blobs at the highest VLBI resolutions (Gabuzda & Sitko 1996; Lister & Smith 2000). The optical polarization may originate in shocks at the base of the jet.

We re-address the question of optical alignment, taking advantage of a more extensive optical polarization database, and more and improved VLBI maps.

2. Data and Derived Parameters

Our sample consists of 31 BL Lac type objects (BLLs) and 20 high polarization QSOs (simply called QSOs) with extensive optical linear polarization data and high quality VLBI maps taken from the literature. Optical polarization data from observations spanning 20–30 years were collected from the literature and McDonald Observatory archives.

We determined the following parameters for each blazar:

1. The probability that the measured optical polarization vectors are drawn from an isotropic distribution.
2. The preferred optical polarization position angle. This is the angle of the vector average of the unit vectors corresponding to each polarization measurement. We calculate this for objects with an isotropic distribution probability < 15%. In these cases, our data are consistent with a single preferred angle (for an exception with two preferred angles, see Cross & Wills, these proceedings).
3. The position angle for the VLBI inner structure. While some blazar jets are straight, many are curved even on very small scales (Gomez et al. 1999; Kellermann et al. 1998). Therefore, we measure position angles on both 1–3 mas and 5–20 mas scales, and determine a jet bending angle (the difference between them).

3. Results

1. Most BLLs show long-term, preferred optical-polarization angles despite their violent short-term variability (Figure 1–left). The probability of this distribution arising by chance is $\ll 10^{-4}$ for BLLs alone, and for BLLs and QSOs combined. The QSOs' distribution is significantly different from the BLLs' (0.5% chance for the two to arise from the same underlying distribution) and consistent with an isotropic angle distribution.
2. When we look only at the objects with preferred optical polarization position angles most BLLs have preferred optical polarization aligned with the VLBI 1–3 mas jet. For BLLs, or BLLs and QSOs combined, the probability that Figure 1—right represents an isotropic distribution of angles is < 0.1%.
3. The objects with preferred optical polarization angles show a strong tendency to have straight VLBI jets (bending angle < 15°) compared with objects with no preferred optical polarization angles (Figure 2). The probability for the objects with preferred optical angles to have the same VLBI bending angle distribution as the ones with no preferred angles, is less than 1%. Objects with preferred optical polarization angle and small VLBI bending are mostly BLLs.

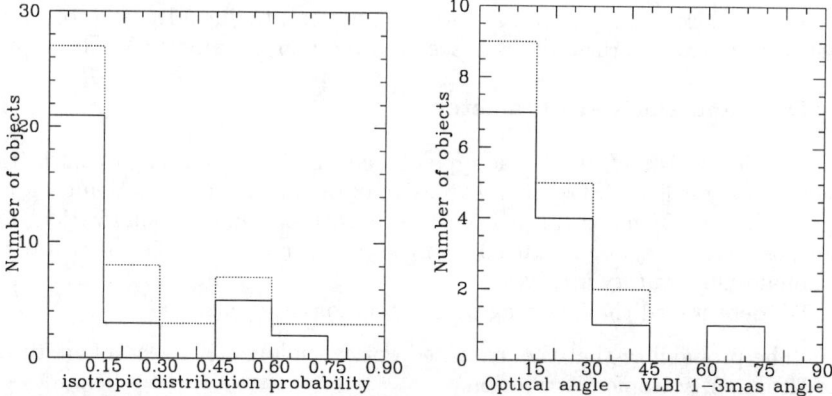

Figure 1. Histograms of (left) isotropic distribution probability of optical polarization position angles; (right) the difference between preferred optical polarization and VLBI 1–3 mas jet angle. (Solid bars are for BL Lac objects and dashed bars are for BLLs and QSOs combined).

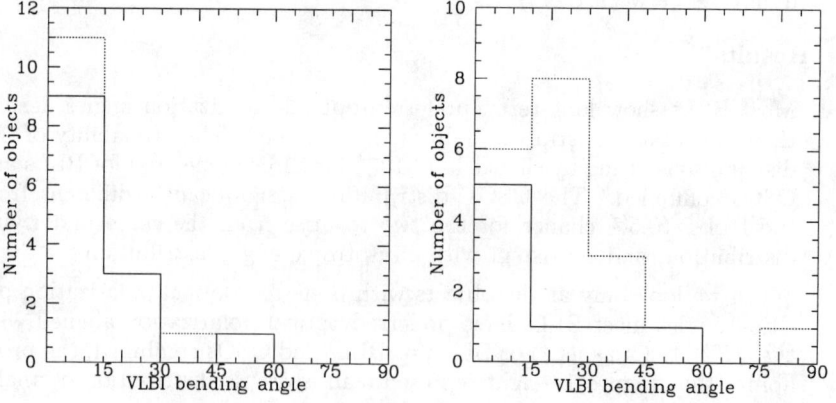

Figure 2. Histograms of VLBI bending angle for (left) objects with preferred optical polarization angles and for (right) objects without preferred optical polarization angles. (Solid bars are for BL Lac objects and dashed bars are for BLLs and QSOs combined).

4. Discussion

A natural explanation for the result that optical polarization tends to align with the jet, is that the optical synchrotron emission arises from a shock front in which the jet magnetic field has been compressed, on average, perpendicular to the jet. The large scatter in the optical polarization angles for a given blazar suggests that, in the inner jet region, the compressed magnetic field changes direction with time. Possible explanations are that the inner jet is internally unstable, or shocks may form via interaction with gas surrounding the central engine. The jets of QSOs may be affected by gas in the NLR and BLR, present in QSOs but absent in BLLs. This may explain why QSOs show preferred optical polarization angles less frequently. The variations may be enhanced by the effects of foreshortening and relativistic beaming.

The tendency that objects with well-defined preferred optical polarization directions also have very small VLBI scale jet bending indicates that a well-behaved straight jet on parsec to Kpc scales corresponds to a well-behaved jet on sub-parsec (optical) scales. Large curvature is likely to be the effect of projection of small jet curvature at very small viewing angles (Gower et al. 1982). Possible causes of jet curvature are [1] an interaction with the environment, or [2] an apparent curvature. In the first case, how does the base of the jet know about the environment on much larger scales? The angular resolution of optical observations is at best a factor of 100 worse than VLBI, often > 100 mas. So we do not have direct evidence to test our assumption that the more energetic optical photons arise near the base of the jet. The optical emission could arise in the same shocks giving rise to cm-wavelength emission. The observation of rapid polarization variation at cm wavelengths, outside the core, gives credence to this idea (Gabuzda et al. 2000). In the second case, the direction of particle ejection may vary with time, for example, via a precession jet (e.g. Hummel et al. 1997). Present data are inadequate to address changes in optical polarization position angles on precession time scales.

References

Aller, H. D., Aller, M. F., & Hughes, P. A. 1985, ApJ, 298, 296
Gabuzda, D. C., Kochenov, P., et al. 2000, MNRAS, 313, 627
Gabuzda, D. C., Sitko, M. L., & Smith, P. S. 1996, AJ, 112, 1877
Gomez, J., Marscher, A. P., et al. 1999, ApJ, 519, 642
Gower, A. C., Gregory, P. C., Unruh, W. G., & Hutchings, J. B. 1982, ApJ, 262, 478
Hummel, C. A., Krichbaum, T. P., et al. 1997, A&A, 324 857
Impey, C. D., Lawrence, C. R., & Tapia, S. 1991, ApJ, 375, 46
Lister, M. L. & Smith, P. S. 2000, ApJ, 541, 66
Kellermann, K. I., Vermeulen, R. C., et al. 1998, AJ, 115, 1295
Rusk, R. & Seaquist, E. R. 1985, AJ, 90, 30

The Physics of Blazar Optical Emission Regions. II. Magnetic Field Orientation, Viewing Angle and Beaming

Michael J. Yuan

University of Texas at Austin

Hien Tran

The Johns Hopkins University

Beverley Wills, Derek Wills

University of Texas at Austin

Abstract. For a sample of 51 blazars with extensive optical polarization data, we used circular statistics to calculate the scatter among the polarization position angles for each object. We found that this scatter is correlated with the radio core dominance. We compared this relationship with the predictions of a simple transverse shock model. The result suggests that blazar jets are likely to cover a wide range of speeds consistent with those derived from observations of superluminal motion.

1. Introduction

Jets are very important for AGN unification. All radio loud objects have powerful jets. In the orientation Unification Scheme, blazars, with their extreme flux and polarization variability, have the jets beamed towards us. However viewing angle unification is not the complete story. Do all the jets have similar physical properties on average, so that orientation is able to explain most of the variations of observed properties among blazars?

Optical polarization of blazars provides magnetic field information very close to the jet formation region and is therefore a very important tool for probing jet physics. We have investigated a sample of 51 blazars with many optical polarization measurements made over the past 20–30 years. In paper I, we have investigated the relationship between the optical polarization position angle and the VLBI radio structure. The results are consistent with a shock model. In this paper, we focus on the variability of optical polarization position angles. Assuming a simple shock model, we link the angle variability with jet orientation and speed to explore the jet physical parameters among blazars.

2. Data and Analysis

2.1. Optical Polarization

Since a blazar's optical polarization swings over a wide range of angles, there exists a 180° ambiguity. Linear statistics based on normal distributions is not strictly correct. Therefore we use circular statistics—a technique little used in astronomy but commonly used for example, to analyze directions of returning homing pigeons, or wind directions. Thus, from the polarization position angles θ_i, we calculate the circular standard deviation $c(\theta)$ about the mean angle, to represent their scatter (Fisher 1993). Polarization data are axial, i.e., only cover the range 0° to 180°. However, the $2\theta_i$ are truly circular data. Therefore, we first calculate $c(2\theta)$, then divide by 2 to get $c(\theta)$:

$$c(\theta) = \left[-\ln \left(\frac{\sqrt{(\sum_{i=1}^n \cos 2\theta_i)^2 + (\sum_{i=1}^n \sin 2\theta_i)^2}}{n} \right) \right]^{\frac{1}{2}}$$

where n is the total number of measurements. If the sample has a large intrinsic scatter, the determination of the real scatter depends on the accurate shape of the wrapped distribution between $-\pi/2$ and $\pi/2$. If the sampling of data is not enough to reveal the detailed shape of the distribution, the $c(\theta)$ calculation algorithm gives a maximum $c(\theta) \sim 66°$.

2.2. Radio Core Dominance

In the unification scheme of powerful radio sources, the core-dominated sources are the Doppler-boosted jet, viewed end-on. At larger angles the core becomes much weaker and collimated jets link the well resolved double lobes. Doppler boosting in core-dominant sources is supported by their very high brightness temperatures and by superluminal motions. Statistical analyses of radio source samples support the unified scheme (Chiaberge, these proceedings). The ratio of core-to-lobe luminosity, or the core dominance, R, is therefore a good indicator of viewing angle. However, in blazar samples, viewing angle is not the only parameter affecting Doppler boosting. From work of other authors (Orr & Browne 1982), we know that R depends on three physical parameters and can be expressed as

$$\log R = \log(R_T \times (1 - \beta \times \cos(\phi))^{-3})$$

where ϕ is the viewing angle of the jet, β is the bulk speed of the emitting electrons in the jet and R_T is the tangential value of R measured for an average edge-on jet in radio galaxies. We used $R_T = 0.0024$ at 5 GHz rest frequency (e.g. Figure 1 of Hoekstra, Barthel & Hes 1997). While R is of course sensitive to ϕ, for smaller ϕ (e.g. < 20°), it is increasingly sensitive to β.

For each object we calculated R at 5 GHz rest frequency, using core and lobe flux densities from literature. In Section 4, we will calculate $\log R$ from models using the formula above and compare the models with observed values.

3. Correlation

Figure 1a shows that there is a correlation between $c(\theta)$ and $\log R$. The correlation is significant at the 99.9% level. However, we note that the correlation is

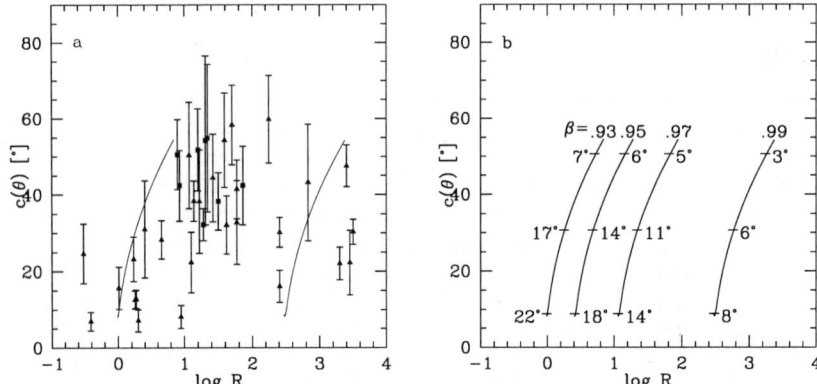

Figure 1. (a) Correlation between log R and the scatter of optical polarization position angle. Triangles are BL Lacs and squares are QSOs. The two thin curves represent $\beta = 0.93, 0.99$ models. (b) Model simulations. Each represents a model with fixed β and variable ϕ. The β for each model is noted on top of each curve. The markers along each curve denote the corresponding ϕ in the observer's frame.

not linear. For the log R<2 region, which includes most blazars, $c(\theta)$ increases with log R, but for log R>2, there is a lot of scatter. No object has $c(\theta)$ larger than 66°, as expected from the algorithms for large scatter and small sample size.

4. Model

The model we use makes the following assumptions:

1. Optical emission comes from a transverse shock propagating along the jet.

2. The shock compresses the magnetic field so much that the magnetic component along the jet direction is negligible.

3. In the plane of the shock, there is a dominant magnetic field direction that changes randomly with time. This magnetic field direction corresponds to the emission region whose radiation happens to be boosted towards us. Different ejection directions observed for emerging blobs on VLBI scales and complex VLBI structure suggests that these dominant regions change rapidly.

4. The shock is optically thin.

When the shock is viewed face on, the magnetic field direction projected on the sky will vary randomly with time. As the viewing angle is increased, the projected magnetic field direction will lie increasingly perpendicular to the projected jet direction.

A simplification is afforded by the Lorentz invariance of the ratios of Stokes parameters, so our calculations of projected magnetic field direction in the co-moving frame (viewing angle ϕ') can be referred to the observer's frame using the relation (Bjornsson 1982):

$$\sin(\phi') = \frac{\sin(\phi)}{1 - \beta\cos(\phi)} \cdot \sqrt{1 - \beta^2}$$

For a given set of β and ϕ, we generate 100 randomly oriented magnetic field directions to simulate the random changes over time inside the shock plane and project them onto the sky in the co-moving frame. The circular standard deviation of those angles can then be compared with real observations, using the relation between R and ϕ given earlier. R_T is fixed by observations, so the only free parameter in this model is β.

The results are shown in Figure 1b. Each curve represents a model with fixed β, with ϕ decreasing from bottom left to top right. The values of ϕ are marked along the curve. From left to right, $\beta = 0.93, 0.95, 0.97, 0.99$. The upper limit on $c(\theta)$ is the result of the computational limit. The lower limits are caused by the fact that we only have 100 simulated vectors to do statistics and any one vector that happens to be parallel to the jet direction contributes a lot to the scatter. In principle, if we had an increasing number of vectors, the lower limit for $c(\theta)$ would approach 0. On our model, the $c(\theta)$ vs. log R correlation arises because $\beta \sim 0.95$ for most blazars, and so the viewing angle dependence dominates for log R< 2. For log R> 2, the scatter is the result of a tail to higher jet speeds.

Our simple model with only one free parameter, β, accounts for the correlation we find between the scatter in optical polarization angles, $c(\theta)$, and the core-dominance, log R. Figure 1 shows that the average jet speed differs significantly amongst the powerful blazars, with a range of values comparable with those derived from superluminal motion and from the statistics of core-dominance in radio-source surveys (Urry & Padovani 1995). In Unified Schemes it is important to take into account that log R is not simply a measure of orientation. Jet speed introduces considerable scatter, especially at log R > 2.

References

Bjornsson, C.-I. 1982, ApJ, 260, 855
Fisher, N. I. 1993, Statistical Analysis of Circular Data, (Cambridge: Cambridge University Press)
Hoekstra, H., Barthel, P. D., & Hes, R. 1997, A&A, 319, 757
Orr, M. J. L. & Browne, I. W. A. 1982, MNRAS, 200, 1067
Urry, C. M. & Padovani, P. 1995, PASP, 107, 803

On the AGN Storage Ring Central Engine Model for Blazars

Howard D. Greyber

10123 Falls Road, Potomac, MD 20854, USA

1. Blazar Central Engine Model

The physics of the central engine of blazars is still an object for research but simultaneous observations from the Chandra X-ray Observatory, the Hubble Space Telescope and the VLBI (VSOP) are giving clues. The Chandra study of the X-ray jet found in quasar PKS 0637−752 concludes that all the standard scenarios are eliminated, leaving only models with inhomogeneous magnetic field structures and/or extreme departures from equipartition.

The X-ray and radio jets coincide remarkably well. For the inner portion of the radio jet, the X-ray morphology closely matches that of the latest radio images. The outer portion of the radio jet, and a radio component to the east, show no X-ray emission to a limit of about 100 times lower flux. However the magnetic field is longitudinal where the X-ray emission is strong and perpendicular to the jet where no X-ray emission is detected.

This author's "Strong" Magnetic Field model (SMF) for the nuclei of AGN postulates that the structure of the jet is not homogeneous, but has a high magnetic field parallel to the jet direction in the outside surface of the cylindrical jet with a low magnetic field inside the jet (Greyber 1988; 1989a,b; 1990). The plasma, being diamagnetic, tends to exclude the field from the jet. However a transition region of annular ring of plasma around the jet, adjacent to the high longitudinal magnetic field, produces the X-rays observed. Thus the relativistic particles making up the jet that are inside the annular ring do not radiate X-rays greatly, and can travel very long distances forming the megaparsec radio jets observed for many quasars.

Figure 1 shows the SMF model for the central engine. The unusual feature of the model is clearly the intense, highly relativistic current loop or Storage Ring around the black hole. The thin toroid of plasma, surrounding and anchored to the current loop by the Maxwell "frozen-in" field condition, is gravitationally bound to the black hole. The gravitational force on the thin toroid by the black hole balances the bursting force of the current loop in a stable equilibrium.

As matter accretes into the originally dipole magnetic field, the magnetic field shape is distorted producing two topological mirrors. Eventually the matter pressure near one throat gets larger than the plasma pressure, forcing the throat to open, and a huge blob of hot plasma is explosively ejected. The blob is also accelerated for a short time by the squeezing effect as the dipole field increases; the process is repeated as long as accretion persists, and the succession of ejected blobs forms the clumpy relativistic jets observed.

On the AGN Storage Ring Central Engine Model for Blazars 159

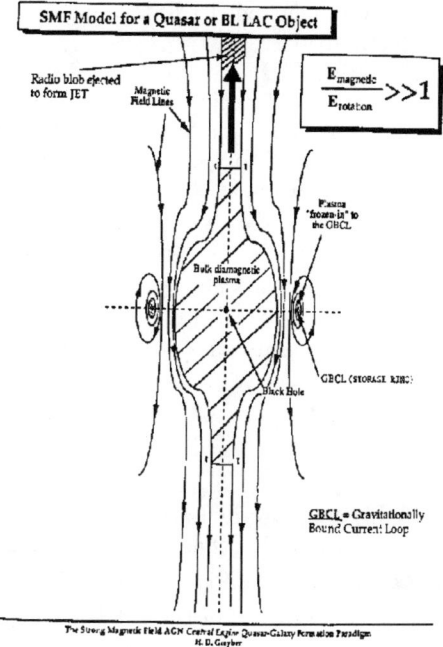

Figure 1. The "Strong" Magnetic Field model for blazars.

2. Origin of the Magnetic Field at Combination Time

Philipp Kronberg (Univ. of Toronto) and collaborators have recently discovered significant magnetic fields that appear to be pervasive in space. Their results show that the "Coma" cluster of galaxies, and even all "normal" rich clusters in the nearby Universe, have several microgauss magnetic fields permeating them. Weaker fields on megaparsec scales have even been detected recently in the extragalactic medium outside of clusters, i.e. very distant from any galaxy.

These observations fit perfectly with the processes invoked in the SMF model that create a significant primordial magnetic field, at and after Combination Time (Greyber 2000). This field is formed precisely in the thin spatially curved sheets of clusters of galaxies around huge voids, which have been reported by de Lapparent(1996) and by Einasto et al. (1997). It is this interplay of gravitation and magnetic fields that explains the wide variety in the dynamics and topology of objects of the dimension of galaxies. After all, the electromagnetic force is the only other long range force in physics besides gravitation, and is more than 10^{36} times stronger than Gravity.

It is clear from the SMF model that the magnetic field will be aligned, almost uniform, over the diameter of a pregalactic plasma cloud, which, as it collapses under gravity, forms the galaxy/quasar. Around the central object, presumably a black hole, a intense, highly relativistic current loop (Storage Ring) forms which evolves into the magnetic field topology shown in Fig. 1. In 1961 for astronomical reasons, Greyber (1961; 1962) postulated such a current ring (or gravitationally bound current loop) forming during galaxy formation in the presence of a magnetic field.

However, in a pioneering MHD calculation, Mestel & Strittmatter (1967) showed how the gravitational contraction of a plasma sphere in a uniform magnetic field, with a small but finite plasma resistivity, drastically changes the magnetic field topology from that for perfect conductivity, forming an equatorial current loop. As Nobelist Hannes Alfven emphasized several decades ago, the lifetime of an electric current, and its associated magnetic field in space, will be much larger than the Hubble time, i.e. far larger than the age of the Universe.

Acknowledgments. Thanks are due to the late Donald H. Menzel for advice and warm encouragement, to Martin D. Kruskal for a valuable conversation, and to Gart Westerhout for permission to use the U. S. Naval Observatory library.

References

de Lapparent, V. 1996, in Mapping the Large-Scale Structure in Cosmology, Les Houches Session LX, eds. R. Schaeffer et al., (Elsevier Press)

Einasto, J., Starobinsky, A., et al. 1997, Nature, 385, 139

Greyber, H. D. 1961, in Transactions of the I.A.U., 11B, 332

Greyber, H. D. 1988, in Supermassive Black Holes, ed. M. Kafatos, (Cambridge University Press), 360

Greyber, H. D. 1989, in The Center of the Galaxy, ed. Mark Morris, (Kluwer Acad. Press), 335

Greyber, H. D. 1989, Comments On Astrophysics, 13, 201

Greyber, H. D. 1990, in Fourteenth Texas Symposium, Annals New York Acad. Sciences, 571, 239

Greyber, H. D. 2000, in The Greatest Explosions Since the Big Bang: Supernovae and Gamma Ray Bursts, eds. M. Livio et al., Poster Book

Mestel, L. & Strittmatter, P. 1967, MNRAS, 137, 95

Schwartz, D. A., et al. 2000, ApJ, 540, L69

U. S. Air Force Office of Scientific Research, Report no. 2958 (1962)

Part 3

New Surveys, Number Counts, Luminosity Functions

Deep Blazar Surveys

Paolo Padovani

Space Telescope Science Institute, 3700 San Martin Drive, Baltimore, MD, 21218, USA

Affiliated to the Astrophysics Division, Space Science Department, European Space Agency

On leave from Dipartimento di Fisica, II Università di Roma "Tor Vergata," Via della Ricerca Scientifica 1, I-00133 Roma, Italy

Abstract. I address the need for deep blazar surveys by showing that our current understanding of blazars is based on a relatively small number of intrinsically luminous sources. I then review the on-going deeper surveys, addressing in particular their limits and limitations. Finally, I present some preliminary results on the evolutionary properties of faint blazars as derived from the Deep X-ray Radio Blazar Survey (DXRBS).

1. Introduction

Blazars are the most extreme variety of Active Galactic Nuclei (AGN) known. Their signal properties, discussed in detail in this volume, include irregular, rapid variability; high optical polarization; core-dominant radio morphology; apparent superluminal motion; flat ($\alpha_r \lesssim 0.5$) radio spectra; and a broad continuum extending from the radio through the gamma-rays (e.g. Urry & Padovani 1995). Blazar properties are consistent with relativistic beaming, that is bulk relativistic motion of the emitting plasma at small angles to the line of sight (as originally proposed by Blandford & Rees in 1978), which gives rise to strong amplification and collimation in the observer's frame. It then follows that an object's appearance depends strongly on orientation. Hence the need for "Unified Schemes," which look at intrinsic, isotropic properties, to unify fundamentally identical (but apparently different) classes of AGN.

The blazar class includes flat-spectrum radio quasars (FSRQ) and BL Lacertae objects. These are thought to be the "beamed" counterparts of high- and low-luminosity radio galaxies, respectively. The main difference between the two blazar classes lies in their emission lines, which are strong and quasar-like for FSRQ and weak or in some cases outright absent in BL Lacs. The current view is that there is actually a continuity of at least some properties between the two classes, so the distinction between a BL Lac and an FSRQ can be somewhat blurred (Landt et al., these proceedings).

Due to their peculiar orientation with respect to our line of sight, blazars represent a very rare class of objects, making up considerably less than 5% of all AGN (Padovani 1997). As a consequence, all existing blazar samples were, until very recently, relatively small and, due also to the difficulty in identifying them, at high fluxes. It then follows that *our understanding of the blazar phenomenon is mostly based on a relatively small number of intrinsically luminous sources, which means we have only sampled the tip of the iceberg of the blazar population.* For example, the radio luminosity function (LF) of FSRQ derived by Urry & Padovani (1995), although based on 52 sources (the best that could be done at the time), included only one source at $L_r < 10^{26.5}$ W Hz^{-1}, the power which coincides roughly with the predicted flattening of the LF based on unified schemes (see §7). Moreover, only in the limited range $10^{26.9} < L_r < 10^{27.7}$ W Hz^{-1} were the statistics good enough to have more than one source per bin! The need for deeper, larger blazar samples is obvious.

2. The "Classical" Blazar Samples

Before I discuss the on-going, deeper blazar surveys, I summarize here the basic facts about the "classical" blazar samples, the ones we all know and love and on which our knowledge of blazars is based.

BL Lacs

- 1 Jy, radio flux-limited, $f_{5\text{GHz}} > 1$ Jy, with radio spectral index cut $\alpha_r \leq 0.5$, $V < 20$; complete sample includes 34 objects (Stickel et al. 1991);

- EMSS, X-ray flux-limited, $f_{0.3-3.5\text{keV}} \gtrsim 2 \times 10^{-13}$ erg/cm^2/s; complete sample includes 41 objects (Stocke et al. 1991; Rector et al. 2000);

- IPC Slew, X-ray flux-limited, $f_{0.3-3.5\text{keV}} \gtrsim 2 \times 10^{-12}$ erg/cm^2/s; complete sample includes 51 objects (Perlman et al. 1996).

Flat-spectrum Radio Quasars

- 2 Jy, radio flux-limited, $f_{2.7\text{GHz}} > 2$ Jy; complete sample includes 52 objects (Wall & Peacock 1985; di Serego Alighieri et al. 1994).

3. New Blazar Samples

Many groups are tackling the problem of assembling deeper, sizable blazar samples, for the reasons discussed above: number statistics and limiting fluxes. Most of these samples take advantage of the fact that blazars are relatively strong radio and X-ray sources and use a double radio/X-ray selection method (unlike the "classical" samples). Another difference lies in the identification process. When dealing with catalogs of up to $\sim 1,000$ sources, one can obtain an optical spectrum of all of them and identify the blazars. With the deeper, larger catalogs available today, with numbers $\gtrsim 100,000$ and reaching the millions, this becomes impossible without access to dedicated facilities or unlimited resources. Hence the need to increase the efficiency (e.g., via cross-correlation methods) to restrict the number of blazar candidates down to a manageable number.

Since not all groups could present their results at this conference, I summarize here the main on-going surveys, in chronological order. I make no claim of completeness, but I have tried my best to include the largest, deepest samples.

BL Lacs

- DXRBS (Deep X-ray Radio Blazar Survey); uses radio (GB6, PMN)/X-ray (WGA [ROSAT PSPC]) selection; survey limits are $f_{5GHz} \gtrsim 50$ mJy, $f_{0.1-2.4keV} \gtrsim 2 \times 10^{-14}$ erg/cm^2/s, with a cut in radio spectral index $\alpha_r \leq 0.7$; complete sample includes 37 objects (43 in whole sample) and is $\sim 90\%$ identified as of October 2000 (Perlman et al. 1998; Landt et al. 2001; and papers in preparation);

- RGB (ROSAT All Sky Survey [RASS]-Green Bank) sample; uses radio (GB6)/X-ray (RASS) selection, with an optical limit; survey limits are $f_{5GHz} \gtrsim 20$ mJy, $f_{0.1-2.4keV} \gtrsim 3 \times 10^{-13}$ erg/cm^2/s, $B < 18$; complete sample includes 33 objects (127 in whole sample) and is $\sim 94\%$ identified (Laurent-Muehleisen et al. 1998; 1999);

- REX (Radio-Emitting X-ray) sample; uses radio (NVSS)/X-ray (ROSAT PSPC) selection; survey limits are $f_{1.4GHz} > 5$ mJy, $f_{0.1-2.4keV} \gtrsim 3 \times 10^{-14}$ erg/cm^2/s; sample includes 72 objects, $\sim 30\%$ identified; subsample of ~ 40 objects with $f_{0.1-2.4keV} \gtrsim 4 \times 10^{-13}$ erg/cm^2/s is $\sim 90\%$ identified (Caccianiga et al. 1999; 2000; Caccianiga et al., these proceedings);

- "Sedentary" Survey; uses radio (NVSS)/X-ray (RASS)/optical (APM, COSMOS) selection; survey limits are $f_{1.4GHz} > 3.5$ mJy, $f_{0.1-2.4keV} \gtrsim 10^{-12}$ erg/cm^2/s; two-point spectral index selection as well, $\alpha_{rx} \lesssim 0.56$, $\alpha_{ro} > 0.2$, to select a region populated by high-energy peaked BL Lacs (HBL) at $\sim 85\%$ level. Sample includes 155 candidates, $\sim 70\%$ identified, but high efficiency expected (Giommi, Menna & Padovani 1999; Giommi et al., these proceedings);

- FIRST Flat Spectrum sample; uses radio (FIRST, GB6) selection, with an optical limit; survey limits are $f_{1.4GHz} > 35$ mJy, $f_{5GHz} > 20$ mJy, $B < 19$, with a cut in radio spectral index $\alpha_r < 0.5$. Sample includes 87 sources and is $\sim 84\%$ identified (Laurent-Muehleisen et al., in preparation).

Flat-spectrum Radio Quasars

- Parkes 0.25 Jy sample; uses radio selection (PKS); survey limit is $f_{2.7GHz} > 250$ mJy, with a cut in radio spectral index $\alpha_r \leq 0.4$; 444 sources, sample is 100% identified, in the process of being published (Shaver et al. 1996; Hook et al. 1999; Jackson & Wall, these proceedings);

- DXRBS (Deep X-ray Radio Blazar Survey); uses radio (GB6, PMN)/X-ray (WGA [ROSAT]) selection; survey limits are $f_{5GHz} \gtrsim 50$ mJy, $f_{0.1-2.4keV} \gtrsim 2 \times 10^{-14}$ erg/cm^2/s, with a cut in radio spectral index $\alpha_r \leq 0.7$; complete sample includes 187 objects (193 in whole sample) and is $\sim 90\%$ identified as of October 2000 (Perlman et al. 1998; Landt et al. 2001; and papers in preparation);

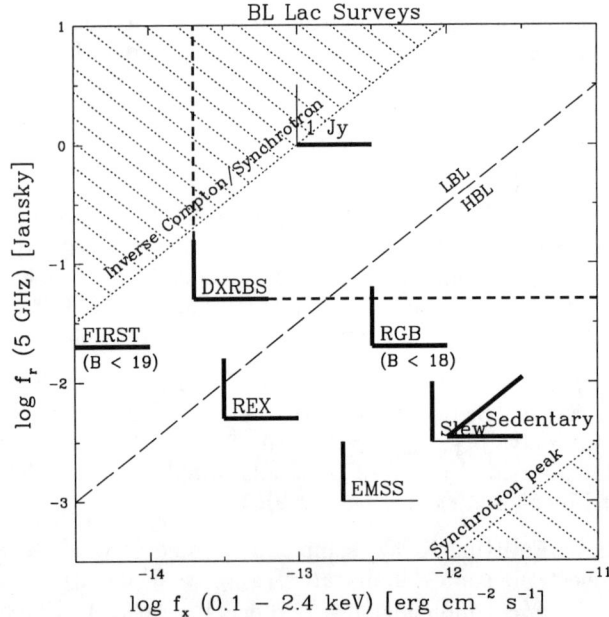

Figure 1. The sampling of the radio flux–X-ray flux plane by different BL Lac surveys. Thick lines represent "hard" survey limits, while thin lines are the fluxes reached in a band other than the one of selection. Sources belonging to a given survey occupy a region of the plane whose bottom-left corner is indicated by the thick/thin lines, as exemplified for DXRBS (short-dashed lines). The long-dashed line divides HBL from LBL, while the hatched regions represent the "forbidden" zones, where no known BL Lacs have been found so far. See text for more details.

- FIRST Flat Spectrum sample; uses radio (FIRST, GB6) selection, with an optical limit; survey limits are $f_{1.4GHz} > 35$ mJy, $f_{5GHz} > 20$ mJy, $B < 19$, with a cut in radio spectral index $\alpha_r < 0.5$; 332 sources, sample is $\sim 84\%$ identified (Laurent-Muehleisen et al., in preparation).

4. Parameter Space Coverage

It is important to assess what regions of parameter space these various surveys are sensitive to, in order to understand what constraints they can or cannot put on blazar demographics. Given the double (radio/X-ray) selection criteria of most of the new surveys and the fact that the "classical" blazar samples were either radio or X-ray selected, I analyze how these samples cover the radio–X-ray flux plane.

This is shown in Figure 1 for BL Lacs. Every survey is characterized by one (or two) flux limits (thick lines), while the smallest flux reached by a sample in a band other than the one of selection is given by a thin line. For example, the

EMSS BL Lacs reach $f_x \sim 2 \times 10^{-13}$ erg/cm^2/s (thick line), by default the X-ray selection limit (actually, the faintest of various limits, due to the nature of the survey). A limit in one band translates into a limit in the other and in this case the radio faintest EMSS BL Lac has a flux $f_r \sim 1$ mJy (thin line). The sources of a given survey occupy a region of the flux-flux plane whose bottom-left corner is shown in the figure.

The long-dashed line in the figure (X-ray-to-radio flux ratio $f_x/f_r = 10^{-11.5}$ erg/cm^2/s/Jy or $\alpha_{rx} \sim 0.78$) divides HBL from low-energy peaked BL Lacs (LBL). Although this distinction might sound arbitrary, there is convincing evidence that HBL are synchrotron-dominated in the X-ray band, unlike LBL where two components (or only one, inverse Compton emission) might coexist (Padovani & Giommi 1996). The parallel dotted lines (lines of constant f_x/f_r) represent the known range in f_x/f_r for BL Lacs, which I derived from available X-ray and radio data. This is $10^{-13} \lesssim f_x/f_r \lesssim 10^{-8.5}$ erg/cm^2/s/Jy (or $0.4 \lesssim \alpha_{rx} \lesssim 1$). No known BL Lacs occupy the hatched regions. I believe that this is not mainly a selection effect, but that there are physical reasons for this. The limit at the low end of the f_x/f_r range (marked "Inverse Compton/Synchrotron" in the figure) is likely due to the fact that in extreme LBL sources the X-ray band is dominated by inverse (synchrotron self-) Compton emission, the radio emission is synchrotron, and the ratio of the two is proportional to the ratio of photon density, W, to B^2, where B is the magnetic field strength. There are probably physical reasons why W/B^2 cannot reach indefinitely low values in blazars (although I cannot exclude that sources with smaller f_x/f_r exist). At the other end, the higher f_x/f_r, the larger the peak frequency of the synchrotron emission, $\nu_{\rm peak}$, in extreme HBL sources. And even in this case there are plausible physical reasons that limit $\nu_{\rm peak}$, which depends on the maximum electron energy (Ghisellini 1999).

A few interesting points can be made about the position of the various surveys on the $f_r - f_x$ plane. First, it is clear why we came to think of radio-selected (RBL) and X-ray selected (XBL) BL Lacs as different types of sources: the 1 Jy and EMSS surveys sample vastly different regions of parameter space. Based on these two "classical" surveys it was hard to see that there was a distribution of synchrotron peak frequencies of which the 1 Jy and EMSS samples represented the two extremes. With the Slew sample we started to bridge the gap, as a few Slew BL Lacs are LBL and "intermediate," but it was not until the more recent surveys (DXRBS, REX, RGB), whose limits straddle the HBL/LBL division, that we realized that intermediate BL Lacs indeed existed in sizable numbers. Second, it is important to realize the limitations of surveys with double (radio/X-ray) flux limits. A survey whose limits fall quite far from the two dotted lines will not provide a complete picture of the BL Lac population. For example, the REX survey cannot provide BL Lac radio number counts to be compared with the predictions of a beaming model based on the 1 Jy sample, simply because it does not include all the BL Lacs above its radio limit (as it misses all those above the radio limit but below the X-ray limit). For the complementary reason, neither can REX provide X-ray number counts to be compared with the predictions from a beaming model based on the EMSS sample. REX will provide radio number counts for HBL, given its proximity to the HBL/LBL dividing line, and X-ray number counts for LBL (as it detects all LBL above its X-ray flux limit). The same arguments apply to the RGB survey, which has the

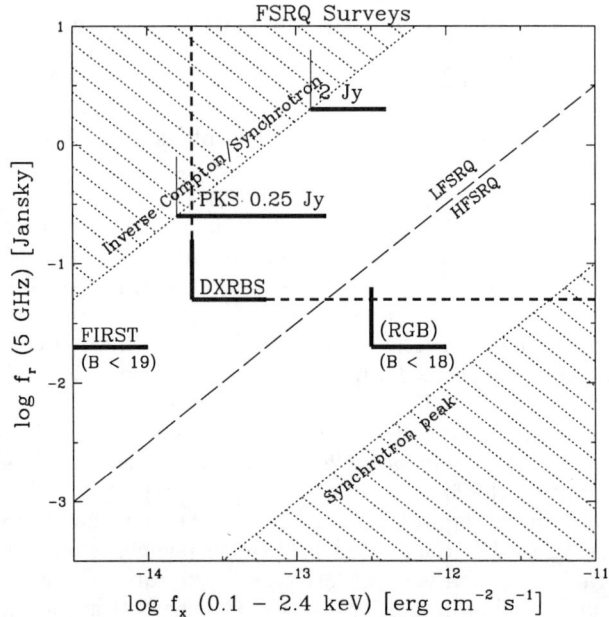

Figure 2. The sampling of the radio flux–X-ray flux plane by different FSRQ surveys. Thick lines represent "hard" survey limits, while thin lines are the fluxes reached in a band other than the one of selection. Sources belonging to a given survey occupy a region of the plane whose bottom-left corner is indicated by the thick/thin lines, as exemplified for DXRBS (short-dashed lines). The long-dashed line divides HFSRQ (high-energy peaked FSRQ) from LFSRQ (low-energy peaked FSRQ), while the hatched regions represent the "forbidden" zones, where no known FSRQ have been found so far. See text for more details.

further problem of an optical limit ($B < 18$). This implies that only BL Lacs with radio-optical spectral index $\alpha_{\rm ro} < \alpha_{\rm ro}({\rm lim})$, where $\alpha_{\rm ro}({\rm lim})$ depends on radio flux (and is ~ 0.4 at the survey limit, for example) will be included. In the case of DXRBS, on the other hand, being relatively close to the leftmost boundary of the BL Lac region, the X-ray flux limit is not as important and can therefore be considered "almost" radio flux-limited only. The ideal sample, of course, has only one, faint, flux limit. FIRST does not have any X-ray cut but unfortunately the optical limit ($B < 19$) implies, as for RGB, that only BL Lacs with radio-optical spectral index $\alpha_{\rm ro}$ flatter than a given value (which depends on radio flux) will be included.

The coverage of the radio–X-ray flux plane for FSRQ surveys is shown in Fig. 2. As in Fig. 1, the parallel dotted lines represent the known range in $f_{\rm x}/f_{\rm r}$. For FSRQ I find $10^{-13.2} \lesssim f_{\rm x}/f_{\rm r} \lesssim 10^{-10}$ erg/cm^2/s/Jy (or $0.6 \lesssim \alpha_{\rm rx} \lesssim 1.05$). The long-dashed line divides high-energy peaked and low-energy peaked FSRQ (HFSRQ and LFSRQ respectively) at $f_{\rm x}/f_{\rm r} = 10^{-11.5}$ erg/cm^2/s/Jy (or $\alpha_{\rm rx} \sim 0.78$). The existence of HFSRQ, flat-spectrum quasars with synchrotron peak in the UV/X-ray band, was not suspected until the first results of DXRBS (Perlman et al. 1998; Landt et al. 2001; Perlman et al., these proceedings; Padovani et al., in preparation). As shown in Fig. 2, in fact, only by reaching relatively faint radio fluxes and by having X-ray information one can sample the HFSRQ region of the plane. The 2 Jy sample had too high of a radio flux limit to include a sizable number of FSRQ above the LFSRQ/HFSRQ line (only two sources in the sample, in fact, have $\alpha_{\rm rx} < 0.78$, both of them with $f_{\rm x} \sim 10^{-11}$ erg/cm^2/s). Note, however, that the region of the plane occupied by HFSRQ is *smaller* than that occupied by HBL (compare the position of the rightmost dotted line labeled "Synchrotron peak" in Fig. 1 and 2). For reasons we still do not understand, there seem to be no FSRQ with synchrotron peak at energies as high as those reached by HBL (Padovani et al., in preparation).

Turning to the surveys themselves, I first note that RGB does not include information on the radio spectral index, which is why it is in parentheses in the figure. Padovani et al. (in preparation) have cross-correlated the RGB sample with the NVSS to obtain radio spectral indices and extract the FSRQ. The limitations of the RGB sample described above (due to its position in the plane) still apply but given its radio/X-ray flux limits this is the survey which is most suited to find HFSRQ. As discussed above DXRBS, by being close to the leftmost dotted line, can be considered "almost" radio flux-limited only and therefore provides a sample of FSRQ which can be used to test the predictions of unified schemes at low radio fluxes. FIRST reaches even deeper fluxes but the optical limit ($B < 19$) implies that it will not give a complete picture of the FSRQ population.

In summary, it is vital to understand what the various surveys can and cannot provide and have their limits and limitations clear. In particular, *surveys with more than one flux limit can provide a complete picture of the blazar population only if the additional limits are relatively close to one edge of the region of parameter space occupied by blazars.* I now turn to analyze the preliminary results of DXRBS, the survey which I am directly involved with and for which I have direct access to the data, in terms of blazar demographics.

5. The Evolutionary Properties of DXRBS Blazars

The basic idea behind the Deep X-ray Radio Blazar Survey (DXRBS) is quite simple: blazars are relatively strong X-ray and radio emitters so selecting X-ray and radio sources with flat radio spectrum (one of their defining properties) should be a very efficient way to find these rare sources. By adopting a spectral index cut $\alpha_{\rm r} \leq 0.7$ DXRBS: 1. selects all FSRQ (defined by $\alpha_{\rm r} \leq 0.5$); 2. selects basically 100% of BL Lacs; 3. excludes the large majority of radio galaxies.

The survey limits are given in §3, while details on the selection technique and identification procedures can be found in Perlman et al. (1998) and Landt et al. (2001). Here I will just note that DXRBS is currently the faintest and largest flat-spectrum radio sample with nearly complete ($\sim 90\%$ as of October 2000) identification. Redshift information is available for $\sim 95\%$ of the identified sources.

The simplest way to study the evolutionary properties of a sample is through the $V/V_{\rm max}$ test or, since the X-ray flux limit is a function of the area, the $V_{\rm e}/V_{\rm a}$

test (Avni & Bahcall 1980). Values of V_e/V_a significantly different from 0.5 indicate evolution, which will be positive (i.e. sources were more luminous and/or more numerous in the past) for values > 0.5, or negative (i.e. sources were less luminous and/or less numerous in the past) for values < 0.5. Moreover, one can fit an evolutionary model to the sample by finding the evolutionary parameter which makes $V_e/V_a = 0.5$.

The DXRBS sky coverage (the area of sky surveyed as a function of X-ray flux) has been derived by Paolo Giommi and Matteo Perri and will be used to derive the evolutionary properties of the sample. I present here some preliminary results based on the sample as of July 2000 ($\sim$ 30 more sources have been identified in August 2000 but are not included in this analysis). The sky coverage is difficult to determine in the regions of the ROSAT PSPC field of view affected by the rib structure ($13' <$ offset $< 24'$) That area, and the sources within, have therefore been excluded from this analysis. Moreover, only sources with $f_r > 51$ mJy have been included since we still have not computed the sky coverage of the PMN survey below this flux (this excludes however only a handful of objects). Table 1 gives the sub-sample, the mean V_e/V_a value, $\langle V_e/V_a \rangle$, the number of objects, and the best fit parameter τ assuming a pure luminosity evolution of the type $P(z) = P(0)exp[T(z)/\tau]$ (where $T(z)$ is the look-back time). The values $H_0 = 50$ km/s/Mpc and $q_0 = 0$ have been adopted.

Table 1. DXRBS Evolutionary Properties

Sample	$\langle V_e/V_a \rangle$	N	τ
All FSRQ	0.58 ± 0.03	119	$0.35^{+0.14}_{-0.08}$
HFSRQ	0.71 ± 0.05	32	$0.17^{+0.03}_{-0.02}$
BL Lacs	0.57 ± 0.05	30	
HBL	0.65 ± 0.09	11	
LBL	0.52 ± 0.07	19	
Unclassified ($z = 1.5$)	0.80 ± 0.05	39	

The main results are the following:

1. DXRBS FSRQ evolve; however, their $\langle V_e/V_a \rangle$ value and evolutionary parameter τ reflect the fact that the sample is not completely identified (incompleteness decreases $\langle V_e/V_a \rangle$). By restricting the analysis to the HFSRQ (defined here by $\alpha_{rx} \leq 0.78$), a basically complete sub-sample as most unidentified sources have $\alpha_{rx} > 0.78$ (pending the effect of the k-correction on their α_{rx} values) $\langle V_e/V_a \rangle$ increases and the value of τ becomes consistent (within $\sim 2\sigma$) with that of 2 Jy FSRQ (Urry & Padovani 1995).

2. DXRBS BL Lacs do not evolve, i.e. their $\langle V_e/V_a \rangle$ value is not significantly different from 0.5 (and consequently $\tau \gtrsim 1$). (The results for BL Lacs are however more uncertain because of the smaller number statistics and the fact that $\sim 30\%$ of them have no redshift; $z = 0.4$ was assumed in this case).

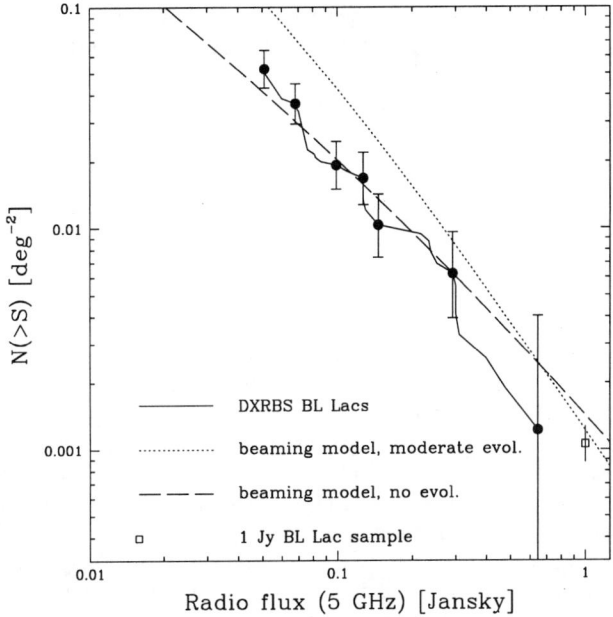

Figure 3. The (preliminary) radio integral number counts at 5 GHz of DXRBS BL Lacs (solid line and filled points) compared to the predictions of a beaming model with the moderate evolution of the 1 Jy BL Lacs (dotted line) and no evolution (dashed line). The empty square represents the surface density of the 1 Jy BL Lacs. Error bars correspond to 1σ Poisson errors and are shown only for a few selected points for clarity. The drop in the observed counts at high fluxes is due to the serendipitous nature of DXRBS. All the ROSAT targets, in fact, have been excluded and these were mostly well-known, high-flux sources.

3. The $\langle V_e/V_a \rangle$ values for HBL and LBL are not significantly different. This is a new result, which contradicts the commonly accepted fact that HBL and LBL have different evolutionary properties. Notice that *for the first time* we can study the evolution of HBL and LBL *within the same sample*. Previous comparisons had been made between the 1 Jy (radio-selected) and the EMSS samples (X-ray-selected). Admittedly, the errors on the $\langle V_e/V_a \rangle$ values are rather large but since, as noticed above, the still unidentified sources are mostly of the LBL type, completion of the identification process will likely *decrease* the difference between the HBL and LBL values.

4. $\langle V_e/V_a \rangle$ for the still unclassified sources which, based on our results so far, will be for the most part FSRQ, is quite high (assuming $z = 1.5$, the mean value for the FSRQ); this implies that when these sources will be identified and included in the whole FSRQ sample, the FSRQ $\langle V_e/V_a \rangle$ will likely reach that of the HFSRQ.

The V_e/V_a test is a simple way to study the evolutionary properties of a sample. To move to the demographics one needs number counts or, if complete redshift information is available, the luminosity function. I will address these in turn for the DXRBS BL Lacs and FSRQ.

6. DXRBS BL Lac Number Counts

For the past 10 years or so the only sizable, complete, radio-selected sample of BL Lacs has been the 1 Jy sample (Stickel et al. 1991). The predictions of relativistic beaming have been tested and tuned to this sample and constraints on beaming parameters (Lorentz factor distribution, angles) have been derived. We can now test unified schemes on a sample which reaches $\sim$ 20 times fainter radio fluxes. Given the fact that redshifts are still missing for $\sim$ 30% of the DXRBS BL Lacs we start by deriving the radio number counts.

Figure 3 shows the (preliminary) integral number counts at 5 GHz for the DXRBS BL Lacs down to $\sim$ 50 mJy, compared to the predictions of unified schemes based on a fit to the 1 Jy LF (Urry & Padovani 1995). The DXRBS counts have been corrected for incompleteness by scaling them up by 15%. The dotted line assumes the best-fit 1 Jy evolution ($\tau = 0.32^{+0.27}_{-0.08}$). Note that the V/V_m value for the 1 Jy BL Lacs is 0.60 ± 0.05, i.e. a departure from the non evolutionary case significant only at the 2σ level. For this reason, and because the V_e/V_a results for DXRBS are at present consistent with no evolution, I also show the surface density of BL Lacs predicted assuming no evolution. Fig. 3 shows that the preliminary number counts agree with the no evolution case, in agreement with the V_e/V_a results. There are a couple of caveats, however, which should be kept in mind. First, the identification process of the DXRBS sample is not complete yet. This has been taken into account by scaling the counts up appropriately but most of the unidentified sources are the faint end so the shape of the counts could change. Second, the definition of a BL Lac for the 1 Jy and DXRBS samples is different, the latter being less restrictive following Marchã et al. (1996). This implies that the comparison between predictions and observations should be restricted to the DXRBS BL Lacs which fulfill the 1 Jy definition. As these make up $\sim$ 70% of the sample, however, this should not make much of a difference.

7. DXRBS FSRQ Luminosity Function

The situation for FSRQ is better, both because of the better statistics and the fact that redshifts are available for all objects. In this case we can then derive directly the luminosity function and compare it with what expected from unified schemes. I take into account the fact that the identification is not complete yet by applying the best-fit evolution derived from the complete subsample of HFSRQ to the whole sample. Keeping this in mind, Figure 4 presents the (preliminary) local radio luminosity function (de-evolved to zero redshift using the best-fit evolution) for the DXRBS FSRQ. The predictions of unified schemes based on a fit to the 2 Jy LF (Urry & Padovani 1995) are also shown (solid line). A few interesting points can be made: 1. the 2 Jy and DXRBS LFs are in good agreement in the region of overlap; 2. DXRBS has much better statistics: the

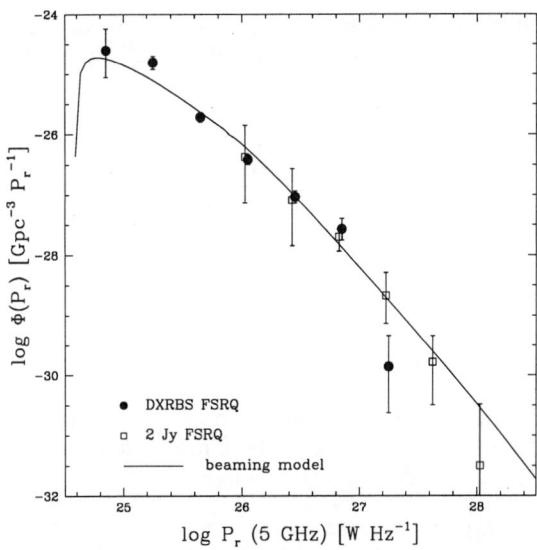

Figure 4. The (preliminary) radio luminosity function of DXRBS FSRQ (filled points) compared to the predictions of a beaming model based on the 2 Jy luminosity function and evolution (solid line). The open squares represent the 2 Jy luminosity function. Error bars correspond to 1σ Poisson errors.

two lowest bins of the 2 Jy LF contain only one object each, while the number of DXRBS sources in the same bins is ~ 20–30; 3. the DXRBS LF reaches powers more than one order of magnitude smaller than those reached by the 2 Jy LF, as expected given the much fainter (~ 30) flux limit; 4. the DXRBS LF is in (amazingly!) good agreement with the predictions of unified schemes; 5. we are getting close to the limits of the FSRQ "Universe"; as FSRQ are thought to be the beamed counterparts of high-power radio galaxies, their luminosity function should end at relatively high powers. Assuming that the value inferred from the fit to the 2 Jy LF is correct (solid line in the figure, based on the 2 Jy LF of Fanaroff-Riley type II radio galaxies; see Urry & Padovani 1995), then DXRBS is approaching that value.

8. Even Deeper Surveys?

What is in store for the future? Will we be able to go even deeper in our quest for blazars, to probe the even less powerful sources? It will not be easy. Consider in fact a radio survey reaching ~ 1 mJy. A typical radio-loud source (with a two-point radio-optical spectral index $\alpha_{\rm ro} \sim 0.6$) will have $V \sim 24$, beyond the reach for spectroscopy of 4m class telescopes even in the presence of strong, broad lines, let alone if one is dealing with a BL Lac! Similarly, at the Chandra/XMM fluxes $f_{\rm x} \sim 10^{-15}$ erg/cm^2/s a typical radio-loud source (with $\alpha_{\rm ox} \sim 1.2$) will reach $V \sim 26$. These magnitudes are starting to become problematic for spectral identification even for 8–10m class telescopes, especially in the absence of strong features. I stress that these problems will plague all radio-loud AGN and not only blazars!

This means that we will need to be very efficient in our pre-selection of candidates, as optical identification will require large resources. Statistical identification of sources based on their location in multi-parameter space, which will imply a smaller need for optical spectra (similar to the method employed for the "Sedentary" survey; §3), will also have to become more common.

9. Summary

The main conclusions are as follows:

1. "Classical" blazar samples are small and at relatively high fluxes; it then follows that our understanding of the blazar phenomenon is based mostly on the intrinsically most powerful sources.

2. A number of on-going, deeper surveys are probing the more common, less luminous blazars, and will reveal the bulk of the blazar population. Before drawing conclusions about blazar demographics, however, care has to be taken to assess the limitations of these surveys and what regions of blazar parameter space they are sampling.

3. Preliminary results of the Deep X-ray Radio Blazar Survey (DXRBS) in terms of evolution, number counts, and luminosity functions agree with the predictions of unified schemes (based on samples having flux limits ~ 20 times larger).

4. Even deeper blazar surveys will face daunting identification problems, due to the faintness of the optical counterparts. The good news is, however, that due to the relatively high radio powers of flat-spectrum quasar, we might be approaching the limits of their Universe.

Acknowledgments. The work on DXRBS reported here has been done in collaboration with, amongst others, Paolo Giommi, Hermine Landt, and Eric Perlman.

References

Avni, Y. & Bahcall, J. N. 1980, ApJ, 235, 694
Blandford, R. D. & Rees, M. J. 1978, in Pittsburgh Conference on BL Lac Objects, ed. A. N. Wolfe (Pittsburgh: University of Pittsburgh Press), 328
Caccianiga, A., et al. 1999, ApJ, 513, 51
Caccianiga, A., et al. 2000, A&AS, 144, 247
di Serego Alighieri, S., Danziger, J., Morganti, R., & Tadhunter, C. 1994, MNRAS, 269, 998
Giommi, P., Menna, M. T., & Padovani, P. 1999, MNRAS, 310, 465

Ghisellini, G. 1999, Astroparticle Physics, 11, 11

Hook, I. M., McMahon, R. G., & Shaver, P. A. 1999, in Looking Deep in the Southern Sky, eds. R. Morganti & W. J. Couch (Heidelberg: Springer-Verlag), 211

Landt, H., Padovani, P., Perlman, E. S., Giommi, P., Bignall, H., & Tzioumis, A. 2001, MNRAS, in press (astro-ph/0012356)

Laurent-Muehleisen, S. A., Kollgaard, R. I., Ciardullo, R., Feigelson, E. D., Brinkmann, W., & Siebert, J. 1998, ApJS, 118, 127

Laurent-Muehleisen, S. A., Kollgaard, R. I., Feigelson, E. D., Brinkmann, W., & Siebert, J. 1999, ApJ, 525, 127

Marchã, M. J. M., Browne, I. W. A., Impey, C., & Smith, P. S. 1996, MNRAS, 281, 425

Padovani, P. 1997, in Very High Energy Phenomena in the Universe, eds. Y. Giraud-Héraud & J. Trân Thanh Vân (Paris: Ed. Frontières), 7

Padovani, P. & Giommi, P. 1996, MNRAS, 279, 526

Perlman, E. S., et al. 1996, ApJS, 104, 251

Perlman, E. S., Padovani, P., Giommi, P., Sambruna, R., Laurence, R. J., Tzioumis, A., & Reynolds, J. 1998, AJ, 115, 1253

Rector, T. A., Stocke, J. T., Perlman, E. S., Morris, S. L., & Gioia, I. M. 2000, ApJ, in press (astro-ph/0006215)

Shaver, P. A., Wall, J. V., Kellermann, K. I., Jackson, C. A., & Hawkins, M. R. S. 1996, Nature, 384, 439

Stickel, M., Padovani, P., Urry, C. M., Fried, J. W., & Kühr, H. 1991, ApJ, 374, 431

Stocke, J., et al. 1991, ApJS, 76, 813

Urry, C. M. & Padovani, P. 1995, PASP, 107, 803

Wall, J. V. & Peacock, J. A. 1985, MNRAS, 216, 173

The BLEIS Project

Ilaria Cagnoni, Annalisa Celotti, Davide Poccecai

SISSA-ISAS, Via Beirut 2-4, 34014 Trieste, Italy

Abstract. The BLEIS project's aim is to select samples of blazars, radio galaxies and quasars from the ESO Imaging Survey (EIS) Wide, a deep optical survey designed to provide a database of VLT targets. The BLEIS samples will be useful not only in understanding the physics of the selected objects, but also in cosmological and statistical studies. The optical selection will give us new means of testing unified and evolution models of blazars. We present the pilot sample of 15 candidates (m_V = 20.5–24.05) obtained cross-correlating the NVSS radio survey and the EIS Patch B.

1. Introduction

We started the BLEIS (Blazars + EIS) project (Cagnoni et al. 2000), a search for blazars from the faint optical images of the ESO Imaging Survey Wide (EIS Wide, e.g. Nonino et al. 1999), with the aim of testing beaming and evolution models by selecting the FAINTEST sample ever in the OPTICAL band. This will be the *first* blazar sample selected from optical images and colors at such faint fluxes (B = 24.6, V = 24.4 and I = 23.7, 80% completeness).

1.1. Optical

The "classical" blazars samples were selected either from the radio (the RBL and quasars from the 1 and 2 Jy samples, Stickel, Meisenheimer & Kuhr, 1994; Wall & Peacock 1985) or from the X-ray band (the XBL and radio loud (RL) quasars from the EMSS, e.g. Wolter et al. 1994; Wolter & Celotti 2000), and the selected objects show different properties.

One of the main issues regarding BL Lacs is their evolution: while the X-ray selected BL Lacertae objects show negative evolution, the radio selected ones are consistent with no evolution. It is unclear whether this difference is due to selection effects related to the blazars SED or to the existence of two different populations.

Another open issue regards the density of X-ray selected RL quasars: the ones selected from the EMSS seem to have a higher density than predicted from radio samples. The BLEIS project, looking at an energy band in between the radio and X-ray energies, could shed light on the "evolutionary puzzle" and on the quasar density problem.

1.2. Faintness

The BLEIS sample, being the faintest sample available, can, even with its small area of 16 deg^2, place strong constraints on number-counts predictions, highly uncertain at the BLEIS flux level. As an example, at 1.4 GHz down to 5 mJy, the model of Jackson & Wall (1999) predicts $\sim$ 4–5 BL Lacs deg^{-2}, while Urry, Pàdovani & Stickel (1991) predict $\sim$ 0.5 deg^{-2}; this means 74 and 8 BL Lacs respectively in the EIS Wide fields.

Other results from the BLEIS project will be e.g. the selection of:

- a sample of high redshift quasars ($z > 3.5$, 3 good candidates already found from patch B)

- a sample of radio galaxies.

These samples will be important on their own to obtain an independent estimate on the very high redshift population of radio-loud quasars, as a test for evolutionary and unification models, and for other statistical studies.

2. The Pilot Sample

We use for our project the EIS Wide, a survey covering 4 regions of the southern sky for a total of $\sim$ 16 deg^2 in up to two colors. We started our analysis from Patch B, being the region covered with all the 3 filters (B, V and I). Since blazars are all radio sources, we cross-correlated each EIS Patch B filter catalog with the NRAO VLA Sky Survey (NVSS, Condon et al. 1998) sources. We then excluded all the optical sources fainter than the catalogs 80% completeness limits and cut at a radio flux of 5 mJy (twice the NVSS limit). We considered only the radio sources with a counterpart in all the 3 filters and we obtained a final sample of 15 sources. The NVSS radio error circle of few arc seconds is such that more than one optical counterpart is present for $\sim$half of the 15 selected radio sources. This corresponds to 24 V-band counterparts.

3. Properties of the BLEIS Sources

3.1. Optical Properties

To understand the nature of the BLEIS sources, we applied the classification based on optical colors presented in Zaggia et al. (1999) and shown in Figure 1 (left) to the 24 V-band counterparts of the pilot sample. We added to the Zaggia et al. classification (regions 1, 3, 4 and 5 in Fig. 1 (left)) an area (region 2) describing the colors ranges of the BL Lacertae objects measured by Moles et al. (1985).

Figure 1 (right) shows the colors of the BLEIS V-band counterparts compared to the Zaggia et al. classification: 5 of them fall in the region occupied by high redshift quasars ($z > 3.5$), 3 in the low redshift quasar region ($z < 3$), 4 in the BL Lac region and 4 fall out of the known regions. 7 counterparts are not visible in the I and/or B filters and, obviously, are not included in the plot.

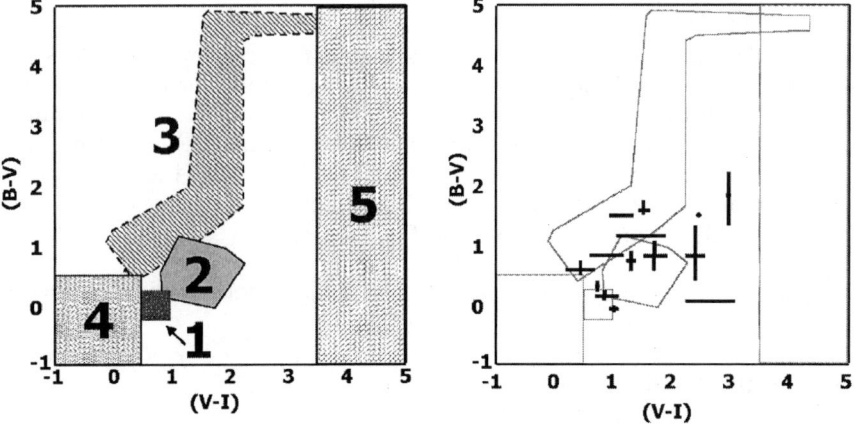

Figure 1. (Left) Classification based on optical colors: in region 1 the low redshift ($z < 3$) and in region 3 the high redshift ($z > 3.5$) quasars are expected; regions 4 and 5 are occupied by very low mass stars and white dwarfs respectively (Zaggia et al. 1999). Region 2 represents the colors of BL Lacertae objects measured by Moles et al. (1985). (Right) Optical classification of the V band counterparts of the BLEIS pilot sample.

3.2. α_{RO} Distribution

By extrapolating the NVSS flux (assuming a flat slope to 5 GHz) and using the EIS V magnitudes, we estimated the α_{RO} distribution for our sources and compared it to various classes of sources, namely: the X-ray selected BL Lacertae objects of the *Einstein* Slew Survey (Elvis et al. 1992); the radio selected BL Lacs of the 1 Jy sample (Stickel et al. 1994); the radio galaxies of the 3CR sample (whose flux was extrapolated to 5 GHz assuming an average slope for the radio spectrum of $\alpha = 0.7$); the Flat Spectrum Radio Quasar sample selected by Padovani & Urry (1992) from the '2 Jy sample' of Wall & Peacock (1985). The α_{RO} distribution of the BLEIS sources is compatible with those of radio selected BL Lacs and FSRQ.

3.3. Future work

We recently applied for VLA mapping of the EIS patches B, C and D[1] to obtain good positions ($\sim 1''$) and faint flux limits (~ 1 mJy). We expect that these new fainter radio data will double the number of the BLEIS sources in Patch B and provide a unique match between radio and optical sources, not possible with the NVSS positional accuracy of a few arcsec. VLA mapping at two frequencies will also provide information on the radio spectral slopes and will allow the selection of the radio galaxies and of the Steep Spectrum Radio Quasars.

---

[1] Patch A is not accessible to VLA and was already observed with the Australia Telescope Compact Array, Prandoni et al. 2000

We have a total of 24 V-band counterparts for the 15 NVSS sources, 19 of which with $m_V \leq 23.6$ and thus observable at VLT. We applied for observing time to get a spectral classification and a redshift estimate for these sources. With these information we will be able to:

1. determine the counts of blazars at low fluxes and compare them to the predictions of evolution and beaming models (e.g. there is a factor of 10 difference in the BL Lacs density predictions of Jackson & Wall (1999) and of Urry et al. (1991) at the BLEIS flux limits and a turnover in quasar counts at low radio flux level is expected from radio samples (Padovani & Urry 1992) and not found in X-ray ones);

2. estimate a luminosity function, a redshift distribution and compare the results with the predictions of the unified models and with the extrapolation of brighter samples; and

3. assess the goodness of the classification presented in Zaggia et al. (1999) based on optical colors.

We will also extend the work to the other EIS patches (A, C and D) using the existing radio data (NVSS and SUMMS, e.g. Bock, Large, & Sadler 1999) and hopefully the VLA data we proposed for.

References

Bock, D. C. J., Large, M. I., & Sadler, E. M. 1999, AJ, 177, 1578
Cagnoni, I., Celotti, A., & Poccecai, D. 2000, Mem. SAIt, in press (astro-ph/0006258)
Condon, J. J., et al. 1998, AJ, 115, 1693
Elvis, M., et al. 1992, ApJS, 80, 257
Jackson, C. A. & Wall, J. V. 1999, MNRAS, 304, 160
Nonino, M., et al. 1999, A&AS, 137, 51
Prandoni, I., et al. 2000 A&A, in press (astro-ph/0007395)
Padovani, P., & Urry, C. M. 1992, ApJ, 387, 449
Stickel, M., Meisenheimer, K., & Kuhr, H. 1994, A&AS, 105, 211
Urry, C. M., Padovani, P., & Stickel, M. 1991, ApJ, 382, 501
Wall, J. V. & Peacock, J. A. 1985, MNRAS, 216, 173
Wolter, A., et al. 1994, ApJ, 433, 29
Wolter, A. & Celotti, A. 2000, A&A, submitted
Zaggia, S., et al. 1999, A&AS, 137, 75

The Reddest BL Lacs?

Sónia Antón, Ian Browne

Jodrell Bank Observatory, Cheshire SK11 9DL, U.K.

Abstract. We have been investigating the Spectral Energy Distribution (SED) of a sample of nearby radio-loud flat-spectrum objects. The synchrotron peak frequencies, ν_{peak}, are distributed approximately uniformly from 10^{12} to 10^{15} Hz. There is no evidence of bimodality separating the other objects from BL Lacs. Our results suggest that all these objects are similar non-thermal emitters, and they only get classified as BL Lacs if their ν_{peak} fall in the near IR/optical range.

1. Introduction

Apart from the radio-loudness and flatness of the broadband spectrum, an object is traditionally classified as BL Lac if has weak emission lines, EW < 5 Å (Stickel et al. 1991) and strong non-thermal optical component, the later is best estimated by the 4000 Å break contrast, C, which is required to be < 0.25 (Stocke et al. 1991). Both the quantities EW and C are naturally dependent on the frequency of cut-off in the synchrotron electron energy distribution— as indicated by the synchrotron peak frequency ν_{peak}. BL Lac objects show synchrotron peaks from the infrared/optical wavelengths, the "red BL Lacs," up to the EUV region, the "blue BL Lacs," red and blue BL Lacs representing the extremes of the same population (e.g. Padovani 1999 and references therein). That the BL Lac population has a wide ν_{peak} distribution is well established, but, its limits are not firmly constrained. Based on new observations of a nearby sample, we try and identify the reddest BL Lacs.

2. The Objects

We have been investigating the 200 mJy sample through a multiwavelength study. This is a nearby ($z < 0.2$) radio-selected sample, that comprises objects with core-dominated morphology at 8.4 GHz, $S_{5GHz} \geq 200$ mJy, flat radio spectra between 1.4 and 5 GHz and R $\leq$ 17 mag (see Marchã et al. 1996). Their extended radio emission is below the FRII range (Dennet-Thorpe & Marchã, in preparation). The objects all have similar radio properties. But they show a remarkable distribution of optical activity, ranging through Seyfert-like objects, BL Lac objects to "normal" elliptical galaxies (Marchã et al. 1996). New data have been gathered in different bands: submillimeter data with SCUBA, infrared data with ISO and optical imaging with NOT. Here we present the results from a sub-sample of 34 objects (out of 66). These are the objects with flat broadband

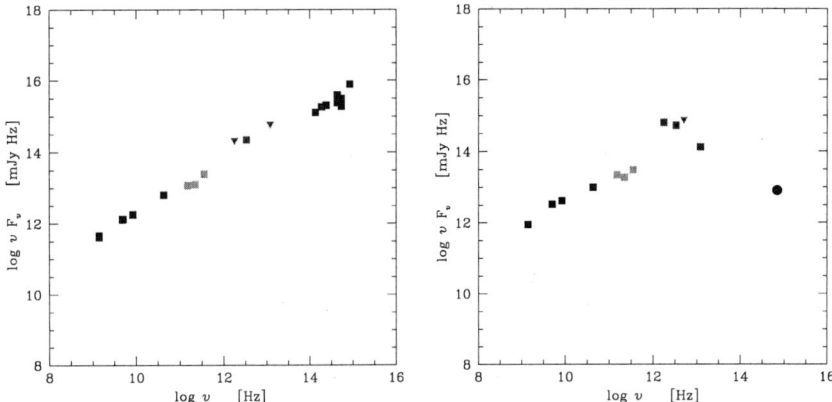

Figure 1. Left—SED of the BL Lac 1133+704. Right—SED of the PEG 1144+352.

spectra between radio and millimeter/sub-millimeter bands, or even beyond. According to their optical properties the objects are divided in the following categories: 15 objects have EW ≤ 5 Å & C ≤ 0.25 and they are classified as BL Lacs, 7 objects have EW ≤ 50 Å & C < 0.35 and they are classified as BL Lac Candidates (BLC), 9 objects have EW ≤ 50 Å & C ≥ 0.35 and they are classified as Passive Elliptical Galaxies (PEGs), 3 objects have broad emission lines with EW ≥ 50 Å and C < 0.35 and are classified as Seyfert 1-like objects.

3. SEDs

The objects show a broadband spectrum that is flat (F$\sim \nu^\alpha$, $\alpha > -0.5$) up to the millimeter/sub-millimeter wavelengths, and even up to higher frequencies (see two examples in Figure 1). When observed with VLBI techniques all the objects show core-jet morphology. Thus, the available information suggests that the emission is synchrotron emission from the nuclear core-jet. Their synchrotron peak frequencies (estimated from the SEDs) are presented in Figure 2. Interestingly enough, the values of ν_{peak} are widely and fairly *continuously* distributed from $10^{10.7}$ to $10^{14.6}$ Hz. That is, there is no clear separation between well known BL Lacs and the other objects. This suggests that the 34 objects are drawn from the same population, one that contains a wide distribution of non-thermal cut-offs, ranging over 3 orders of magnitude.

4. Optical Classification and ν_{peak} distribution

The fact that there is such a wide distribution of ν_{peak} must be relevant for the classification of an object as a BL Lac. In Figure 3–left the values of ν_{peak} are plotted against the 4000 Å break contrasts from Marchã et al. (1996). That figure shows that the C and ν_{cutoff} are correlated: the objects with larger C have lower frequency cut-offs whereas objects that have smaller C have higher

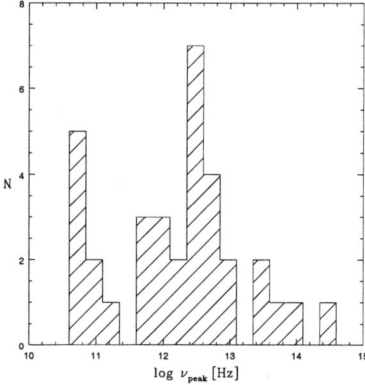

Figure 2. The distribution of the synchrotron peak frequencies.

frequency cut-offs. This shows the obvious, i.e. that $\nu_{\rm peak}$ is important in determining if an object is recognized as a BL Lac or something else. Figure 3–right presents the EW-log ($\nu_{\rm peak}$) plot—the EW are from Marchã et al. (1996). That figure reveals that objects with EW > 7.5 Å all have $\nu_{\rm peak} < 10^{12.5}$ Hz. Also, there is no clear correlation between the EW and ν_{peak}, something that would be expected if what was photoionizing the emission line region was closely related with the non-thermal component. The lack of a clear correlation between EW and non-thermal component and the fact that all objects with EW < 5 Å have $\nu_{\rm peak} > 10^{12.5}$ Hz can be interpreted if the observed differences in the EWs are not due to the differences of emission line strengths, but instead, the observed differences are the fingerprint of different amounts of emission line dilution. Note that if the EWs are mainly tracing differences in the amount of line dilution, then the EW does not seem a good way to distinguish BL Lacs from non-BL Lacs.

5. Discussion

In the framework of unification, both spectral and structural information indicate that these objects have similar properties to those found in FRI/BL Lac population: the 34 objects discussed here are low-luminosity radio-loud coredominated objects, their synchrotron emission extends to high frequencies, their extended radio emission is below the FR II range. Optical imaging shows that the objects are hosted by similar elliptical galaxies which live in similar environments (Antón 2000; Marchã et al., in prep.). The results presented in Figure 2 suggest that this group of objects are from a single population, but one that has a very broad distribution in non-thermal cut-off frequencies. The spread of $\nu_{\rm peak}$ accounts for the observed spread of the strengths of the optical nonthermal emission, and therefore for different optical-classification. In light of our results, the very-low-frequency peaked objects are interpreted as "redder" versions of classical BL Lacs. "Redder" BL Lacs might be "tilted" BL Lacs, the location of their cut-off being reduced in frequency compared with a fully "beamed" BL Lac. But some of the "redder" BL Lacs may be intrinsically red,

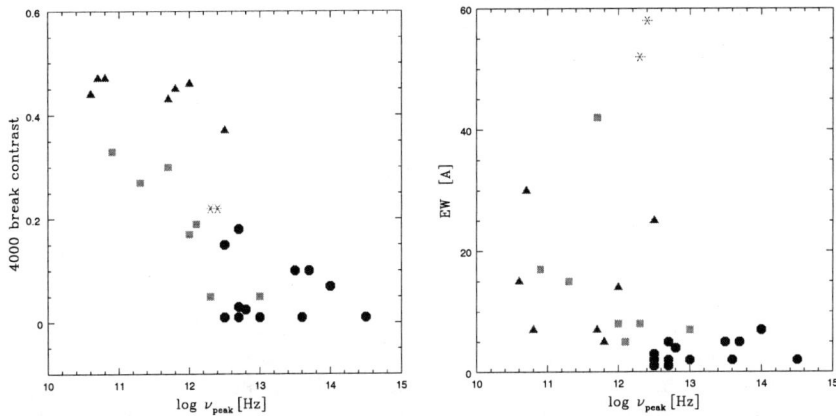

Figure 3. **Left**: 4000 Å break contrast vs. log ν_{peak}. PEGs are represented by triangles, BL Lac Candidates are represented by squares, BL Lacs are represented by circles, and Seyfert 1-like objects are represented by stars. **Right**: EW vs. log ν_{peak}. The symbols are as before.

i.e., their synchrotron cut-offs *intrinsically* happen at lower frequency than those of the "red" BL Lacs.

6. Conclusions

The recognition and classification of a low-luminosity core-jet flat-spectrum object as a BL Lac is heavily dependent on the location of the synchrotron peak frequency. Our results suggest that the classical definition of BL Lacs (EW $\leq$ 5 Å & C < 0.25) is biased towards objects with synchrotron peak frequencies of $\nu_{peak} \geq 10^{12.5}$ Hz. If so, then this definition excludes *a priori* BL Lac objects with lower synchrotron peak frequencies, that is "redder" BL Lacs.

Acknowledgments. This work was supported by the European Commission, TMR Programme, Research Network Contract ERBFMRXCT96-0034 "CERES," and PRAXIS XXI Programme through the grant BD/5532/95.

References

Antón, S. 2000, Ph.D. thesis. University of Manchester

Marchã, M., Browne, I. W. A., Impey, C.D., & Smith, P. S. 1996, MNRAS, 281, 425

Padovani, P. 1999, in ASP Conf. Ser., 159, BL Lac Phenomenon (San Francisco: ASP), 339

Stickel, M., Padovani, P., Urry, C. M., Fried, J. W., & Kuhr, H. 1991, ApJ, 374, 431

Stocke, J. T., Morris, S. L., Gioia, I. M., Maccacaro, T., Schild, R., Wolter, A., Fleming, T. A., & Henry, J. P. 1991, ApJS, 76, 813

Blazar Demographics and Physics
ASP Conference Series, Vol. 227, 2001
Paolo Padovani and C. Megan Urry, eds.

Hidden BL Lacertae Objects Near and Far

John T. Stocke

Center for Astrophysics & Space Astronomy, and Dept. of Astrophysical & Planetary Sciences, University of Colorado, Boulder, CO 80309-0389

Abstract. I describe two difficulties with current research into the cosmic evolution of BL Lacertae Objects: (1). Possible sample incompleteness due to unrecognized (i.e. "hidden") BL Lacs; and (2). The absence of a viable physical model of the evolution, to which current and future observations can be compared.

1. Introduction

The unusual evolutionary trend seen in the number and/or luminosity of BL Lac Objects (a.k.a. "negative" evolution) has been the subject of many recent papers and of several talks at this conference. But two things remain frustrating (for me at least) involving these evolutionary studies:

1. Unlike quasars (or other AGN classes for that matter), the identification of a complete BL Lac sample is fraught with difficulty and uncertainty because BL Lac Objects are defined in "negative" terms (e.g. they lack the strong emission lines of quasars). Other defining qualities may also eliminate true BL Lacs from samples under study, making them incomplete in a systematic way.

2. But like the quasars (and other AGN classes as well), there is no physical model to which the evolutionary trends are compared; i.e. functions are fit and argued over (e.g. pure luminosity evolution vs. luminosity-dependent density evolution) but these are algebraic functions without much physical basis.

In this brief account I will address both of these issues with respect to current samples as well as what has been or will be seen in deep CHANDRA images.

2. Incompleteness in Current Samples

In this Section I will describe three ways in which true BL Lacs can be "hidden" in X-ray and radio-selected samples, potentially causing incompleteness:

1. Despite many X-ray selected BL Lacs (XBLs) having spectral energy distributions (SEDs) which peak in the optical/UV (so-called "high energy peaked BL Lacs": HBLs), their optical spectra often have a significant

contribution from host galaxy starlight. Browne & Marchã (1993) and Marchã & Browne (1995) pointed out that such objects could be confused with FR-1 radio galaxies in clusters so that the X-ray emission would be assigned incorrectly to the cluster. Thus, some XBLs, particularly those with modest optical non-thermal luminosity, could be misidentified as clusters of galaxies.

2. Investigators of radio surveys used to identify radio selected BL Lac (RBL) samples have set a radio spectral index cutoff to eliminate the large number of radio galaxies in these samples (e.g. the 1 Jy sample uses $\alpha \geq -0.5$; Stickel et al. 1991). However, some XBLs have radio spectral indices steeper than this, suggesting that a flat radio spectral index is not required for all BL Lacs. By this spectral index fiat RBLs would be misidentified as radio galaxies.

3. Both RBLs and XBLs could be "hiding" in rich clusters of galaxies where the radio emission would be ascribed to a brightest cluster galaxy (BCGs like NGC 1275 = 3C84 = Perseus A) and the X-ray emission ascribed to dense, cooling gas (i.e. a "cooling flow"). These BL Lacs would be misidentified as FR1 radio galaxies in "cooling flow" clusters.

2.1. Hidden XBLs

A large gallery of XBL spectra can be found in the recent compilation paper of the EMSS XBL sample by Rector et al. (2000). These spectra show that, unlike RBLs which often have weak (low equivalent width and low luminosity) emission lines in their optical spectra, XBLs only very rarely have emission lines but often have significant contributions from host galaxy starlight. In my first examination of XBLs (Stocke et al. 1991), I suggested a BL Lac classification criteria to separate BL Lacs from FR 1 radio galaxies using the strength of CaII H & K break at ≈ 4000 Å rest. A Ca II "break strength" of $\leq 25\%$ was required for BL Lacs, whereas 50% is typical of current epoch ellipticals (e.g. Dressler & Shectman 1987). It is important to understand that, like the radio spectral index limit, the Ca II break limit is arbitrary, and so Maria Marchã in her thesis correctly pointed out that this criteria could prevent us from correctly identifying low luminosity BL Lac Objects in X-ray selected samples (Browne & Marchã 1993; Marchã & Browne 1995). Other, more imaginative, but still arbitrary, criteria have been proposed since then (e.g. Marchã et al. 1996). These investigations led us (Rector, Stocke, & Perlman 1999) to use the ROSAT HRI to image some poor clusters in the EMSS which might actually be BL Lac Objects alá Marchã. She was correct! We found that some of these EMSS sources were pointlike at radio galaxy locations: new EMSS BL Lacs misidentified as clusters and thus "hiding" in these clusters. However, some of these new BL Lacs had optical spectra whose Ca II breaks violated both the old Stocke et al. (1991) and the newer Marchã et al. (1996) criteria. Two are nearly indistinguishable from FR 1 radio galaxies or normal elliptical galaxies in their Ca II breaks, and yet they are still quite luminous X-ray point sources ($L_x \approx 10^{44}$ ergs s^{-1}). Optical spectra of these new BL Lacs can be found in Rector, Stocke, & Perlman (1999). An important result, however, is that even with these new BL Lacs included, the $<V/V_{max}>$ is still significantly less than 0.5, especially for extreme XBLs which have $<V/V_{\text{max}}>= 0.27 \pm 0.08$ (Rector et al. 2000), solidifying "negative" evolution for XBLs.

Are there other "hidden" BL Lacs of this sort? I suspect, yes! Here I mention only two interesting cases:

(1) The Owen et al. (1996) optical spectroscopy of rich cluster radio galaxies which failed to find many BL Lac candidates (3C264 and IC310 in Perseus being two of the four possibilities). Given the above, I suspect that CHANDRA X-ray cluster imaging will find new examples of luminous point X-ray sources associated with cluster radio galaxies without obvious non-thermal continua optically (see Section 2.3).

(2) Numerous investigators have been bewildered to find faint X-ray sources (some with quite hard X-ray spectra) associated with "passive" elliptical galaxies. By "passive" these investigators have meant that they fail to find obvious AGN signatures optically (e.g. non-thermal continuum + strong emission lines). By way of example, I direct the reader to the Nature article by Mushotzky, Cowie, Barger, & Arnaud (2000) on optical identifications in a CHANDRA deep field, in which three relatively bright, "passive" ellipticals have been found. The optical characteristics of these galaxies are quite similar to the low-luminosity BL Lacs in Rector, Stocke, & Perlman (1999). So, I predict that these sources will have SEDs similar to other BL Lac Objects (i.e. radio detections at sub-mJy levels). And one only has to redshift the SEDs displayed in Sambruna, Maraschi, & Urry (1996) to realize that extremely hard soft X-ray spectra are possible for BL Lacs without requiring any internal absorption (as has been proposed for other potential hard-spectrum X-ray background contributors like obscured Seyferts).

2.2. Hidden RBLs

Radio continuum images of XBLs in Perlman & Stocke (1993) and Rector, Stocke, & Perlman (1999) show that many have luminous extended structure, which is likely to be steep in radio spectral index. Single epoch radio spectral indices have been obtained for only 8 XBLs (Stocke et al. 1985), for which only half would pass the -0.5 spectral index cutoff used to define a BL Lac candidate list in the 1 Jy sample (Stickel et al. 1991). Spectral index observations of the entire EMSS sample are now being obtained to see how important the arbitrary spectral index cut imposed upon RBL samples is (Wolter, Rector, & Stocke, in progress). But, given the current data, I suspect that there could be many BL Lacs "hidden" amongst the large number of radio galaxies in the 1 Jy. Some of these could have optical spectra similar to the low-luminosity XBLs in Rector, Stocke, & Perlman (1999), further complicating their identification (e.g. 3C264). Unfortunately, optical polarimetry blueward of the Ca break may be the only viable observation useful in testing this prediction. But, if enough of these new BL Lacs are found in the 1 Jy, their presence could alter the $<V/V_{\text{max}}>$ statistic for that sample.

2.3. Other Hidden BL Lacs

A detailed imaging survey of BL Lacs conducted at the Canada-France-Hawaii 3.6m several years ago (Wurtz et al. 1996, 1997) presented a confusing result: BL Lacs were not found in the brightest cluster galaxies (BCGs) or in the richest clusters (Abell richness classes > 1) in the current epoch. How can this be if the parent population of BL Lacs are FR 1s, which are abundant in nearby rich clusters (e.g. Owen et al. 1996 and references therein)? A possibility, not yet fully explored, but now possible with CHANDRA, is that cluster X-ray emission "hides" some BL Lac Objects. This is especially true in the case of "cooling flow" clusters, whose central X-ray excesses are thought to be due to a dense, cooling ICM. However, if some BCGs in these rich clusters are unrecognized BL Lac Objects, their X-ray emission could contribute to the excess thought to be a "cooling flow." Because the soft X-ray spectra of XBLs are similar to ICM emission (see e.g. the *Beppo*SAX spectra in Wolter et al. 1998), these XBLs could "masquerade" as "cooling flows." But, one might argue that the optical spectra of BCGs in "cooling flow" clusters rules these objects out as BL Lacs, since strong emission lines are present (particularly [OII] and Hα). But these emission lines arise in spatially very extended gas (tens to hundreds of kpc) and so are *not* the nuclear emission lines of an AGN. I predict that CHANDRA ACIS imaging of nearby "cooling flow" clusters will discover luminous point X-ray sources associated with many BCGs (BCGs in "cooling flow" clusters are very often radio galaxies; Burns 1990). The presence of these point sources will indicate previously unknown BL Lac Objects and will reduce substantially the required mass inflow for the "cooling flow." We have searched the EMSS clusters for such circumstances using ROSAT HRI images and have found only one cluster which could contain a BL Lac in its BCG: MS1455.0+2232. While CHANDRA images will test this suggestion, this one "hidden" BL Lac will not alter the statistics of the EMSS sample.

3. Models for AGN Evolution?

So, despite the substantial concerns about completeness described above, the EMSS XBL sample has been very thoroughly scrutinized for "hidden" BL Lacs and continues to exhibit "negative" evolution. Further, recent work on other AGN samples (e.g. Urry & Padovani 1995) find that BL Lac and FR1 evolution can be characterized as $L(z)/L(z=0) = (1+z)^\Gamma$, with $\Gamma = -4.0$, while FR2 samples have $\Gamma = +3.8$. These results suggest that FR2s evolve (i.e. fade to become) FR1s. Other indications that this is the case comes from the investigations into the environment of quasars (Ellingson, Yee & Green 1991), FR2 radio galaxies (Harvanek et al. 2000) and FR1 radio galaxies (e.g. Longair & Seldner 1979; Yee & Lopez-Cruz 1999). Briefly, from these works: quasars are found in clusters only at $z \geq 0.45$, FR2s are found in clusters only at $z \geq 0.2$ and FR1s are found in clusters at all redshifts out to 0.8 (e.g. Stocke et al. 1999).

What can we make of these results? Is a physical model suggested or is any suggestion premature? To me these results suggest two things: (1) the orientation unification advocated by Bartel (1989) is ruled out since some FR2 radio galaxies are found in clusters at lower redshifts than any quasars are found in clusters; and (2) an evolutionary scheme by which quasars fade to become FR2 radio galaxies which fade to become FR1 radio galaxies on an e-folding timescale of 0.9 Gyrs (Harvanek et al. 2000) is quite plausible for cluster AGN. After all, what has become of the quasars which inhabited rich clusters at $z \sim 0.5$? This hypothesis is also consistent with the evolutionary trends for BL Lacs/FR1s and FR2s mentioned above.

The interesting question is: what physical phenomenon drives this rapid fading of cluster AGN (and perhaps AGN in general)? Ellingson, Yee, & Green (1991) list two possibilities: (1) the evolution of the cluster potential well greatly reduces the number of interactions and mergers between gas rich galaxies which could "feed" an AGN (e.g. Roos 1981); and (2) the rapidly increasing ICM density prevents, in some as yet not understood way, the "feeding" of the AGN (Stocke & Perrenod 1981). Both of these hypotheses are supported by the direct X-ray observations of clusters which show that: (1) in the richest clusters of galaxies a dense ICM is already in place at $z \sim 1$ (e.g. Donahue et al. 1999) and (2) in poorer clusters at $z \sim 0.5$ (wherein quasars are found) there is no evidence for a dense ICM (e.g. Rector, Stocke, & Ellingson 1995; Hall et al. 1997; Harvanek et al. 2000), but similar richness clusters at $z \sim 0$ have dense ICMs as well as FR1 radio galaxies. In fact, there is little to distinguish these two hypotheses at low-z, since they both prevent further feeding of the AGN as the ICM thickens and heats due to the deepening of the potential well.

However, at the highest redshifts these two hypotheses predict opposite outcomes. Extrapolating to $z \geq 4$, before the epoch of quasars, the Roos (1981) interaction hypothesis would predict even greater numbers of interactions (because the Universe was smaller and the number of galaxies could only have been larger than today). Thus, by this hypothesis, the highest observed quasars are the first AGN, whose highest observed redshifts are due to supermassive black hole formation timescales. However, by the Stocke & Perrenod (1981) hypothesis, the Universal ICM is denser at high-z, so that an AGN evolution opposite in sense to what we observe at $z \leq 0.5$ is possible; i.e. in this case, the Black Holes are already formed at even higher redshifts as FR1s. Then these BL Lac/FR1s brighten to become FR2s, which brighten to become quasars with cosmic time. This idea was first presented at the Como Conference on BL Lac Objects (Stocke 1987). If this idea is correct, then very high-z BL Lac Objects should be extremely numerous but hard to discover. Using the observed SEDs of typical XBLs redshifted to $z \geq 4$, the Sloan digital sky survey (SDSS) would be the most sensitive search mechanism. And, indeed, one strange AGN at $z = 4.7$, which in many ways resembles a BL Lac Object, has been found in the SDSS commissioning data (Fan et al. 2000)! If this radical hypothesis for quasar evolution at high-z is correct, the SDSS should discover other examples of this new class of high-z BL Lacs, which could contribute significantly to the very faint X-ray source count (and to the X-ray Background). Using the observed SEDs and luminosity functions of low-z BL Lacs, and assuming that all quasars were once BL Lacs at higher redshifts ($z \geq 4$), this new class could contribute up to ~ 1000 X-ray sources deg^{-2} at fluxes of 10^{-15} ergs cm^{-2} s^{-1}. And, if these objects have SEDs similar to low-z BL Lacs, they would be spectrally very hard.

Acknowledgments. JTS would like to thank Drs. R. Wurtz, E. Perlman, & T. Rector for their past work, which contributes so importantly to this presentation, and their current interest in these strange objects. Anything of note and veracity herein is due to them; all misconceptions are the exclusive property of their ex-thesis advisor.

References

Barthel, P. D. 1989, ApJ, 336, 606
Browne, I. W. A. & Marchã, M. J. M. 1993, MNRAS, 261, 795
Burns, J. O. 1990, AJ, 99, 14
Donahue, M., Voit, G. M, Scharf, C. A., Gioia, I. M., Mullis, C. R., Hughes, J. P., & Stocke, J. T. 1999, ApJ,527, 525
Dressler, A. & Shectman, S. A. 1987, AJ, 94, 899
Ellingson, E., Yee, H. K. C., & Green, R. F. 1991, ApJ, 371, 49
Fan, X., et al. 2000, ApJL, in press
Hall, P. B., Ellingson, E., & Green, R. F. 1997, AJ, 113, 1179
Harvanek, M., Ellingson, E., Stocke, J. T., & Rhee, G. H. 2000, AJ, submitted
Longair, M. S. & Seldner, M. 1979, MNRAS, 189, 433
Marchã, M. J. M. & Browne, I. W. A. 1995, MNRAS, 275, 951
Marchã, M. J. M., Browne, I. W. A., Impey, C. D., & Smith, P. S. 1996, MNRAS, 281, 425
Mushotzky, R. F., Cowie, L. L., Barger, A., & Arnaud, K. 2000, Nature, 404, 459
Owen, F. N., Ledlow, M. J., & Keel, W. C. 1996, AJ, 111, 5
Perlman, E. S. & Stocke, J. T. 1993, ApJ, 406, 430
Rector, T. A., Stocke, J. T., & Ellingson, E. 1995, AJ, 110, 1492
Rector, T. A., Stocke, J. T., & Perlman, E. S. 1999, ApJ, 516, 145
Rector, T. A., Stocke, J. T., Perlman, E. S., Morris, S. L., & Gioia, I. M. 2000, AJ, 120, 1626
Roos, N. 1981, A&A, 104, 218
Sambruna, R. M., Maraschi, L., & Urry, C. M. 1996, ApJ, 463, 444
Stickel, M., Padovani, P., Urry, C. M., Fried, J. W., & Kühr, H. 1991, ApJ, 374, 431
Stocke, J. T. & Perrenod, S. C. 1981, ApJ, 245, 375
Stocke, J. T., Liebert, J., Schmidt, G., Gioia, I. M., Maccacaro, T., Schild, R. E., Maccagni, D., & Arp, H. C. 1985, ApJ, 298, 619
Stocke, J. T. 1988, in BL Lac Objects, eds. L. Maraschi, T. Maccacaro & M. H. Ulrich (Heidelberg: Springer-Verlag), 456
Stocke, J. T., Morris, S. L., Gioia, I. M., Maccacaro, T., Schild, R., Wolter, A., Fleming, T. A., & Henry, J. P. 1991, ApJS, 76, 813
Stocke, J. T., Perlman, E. S., Gioia, I. M., & Harvanek, M. 1999, AJ, 117, 1967
Urry, C. M. & Padovani, P. 1995, PASP, 107, 803
Wolter, A., et al. 1998, A&A, 335, 899
Wurtz, R. E., Stocke, J. T., & Yee, H. K. C. 1996, ApJS, 103, 109
Wurtz, R. E., Stocke, J. T., Ellingson, E., & Yee, H. K. C. 1997, ApJ, 480, 547
Yee, H. K. C. & López-Cruz, O. 1999, AJ, 117, 1985

Radio Properties of REX BL Lacs and Galaxies

Anna Wolter

Osservatorio Astronomico di Brera, Milano, Italy

A. Caccianiga

Observatório Astronómico de Lisboa, Lisboa, Portugal

T. Maccacaro, R. Della Ceca

Osservatorio Astronomico di Brera, Milano, Italy

I. M. Gioia

Istituto di Radioastronomia, Bologna, Italy

Institute for Astronomy, Honolulu, HI

F. Cavallotti, M. Minoia

Università degli Studi di Milano, Milano, Italy

Abstract. Detailed VLA observations have been gathered for a number of sources classified as either BL Lacs or galaxies, derived from the REX survey. We focus in particular on the sources identified by us, for which we have in hand homogeneous optical data, to study in more detail than allowed by the NVSS the radio properties of these sources in the framework of AGN unified models.

1. Introduction

The REX survey (Radio Emitting X-ray sources; see Maccacaro et al. 1998 and Caccianiga et al. 1999 for details of the survey) is based on pointed ROSAT/PSPC observations and NVSS (NRAO VLA Sky Survey; Condon et al. 1998) data. About 500 of the $\sim$ 1600 REX sources have an optical identification obtained working down from the highest X-ray fluxes, in order to have larger and larger completely identified subsamples (see Caccianiga et al., these proceedings for the first results on the X-ray–Bright-REX sample of BL Lacs).

AGN and BL Lacs are the most common counterparts, followed by galaxies, either isolated or in clusters. The objects without emission lines are classified according to the amount of the Ca II discontinuity in our optical spectral data. Values larger than 40% define galaxies. BL Lacs are defined as having a Ca II discontinuity smaller than 25%, while objects with Ca II discontinuity between 25% and 40% are defined as BL Lac candidates. Most of the galaxy classification

Table 1. Identification Breakdown as of July 2000

ID	New	Literature	Total
AGN	96	138	234
BL+cand.	40	32	72
Galaxies	55	121	176

from the literature are instead based on the optical extended morphology. In Table 1 we list the updated identification breakdown.

2. Testing Unified Models

Unified models (see Urry & Padovani 1995 for a review) imply that both radio and optical appearance are governed by the proximity of the line of sight to the radio axis, which is supposed to correspond also to the rotation axis of the accretion disk.

We want to test if the angle to the line of sight is really the primary parameter on which the transition between galaxies and BL Lacs rests. To do so we concentrate on the subsample of non-emission line objects (i.e. galaxies, BL Lacs and candidates), in particular those that have been observed by us, since for those objects we have homogeneous optical data. In the radio band, we expect a transition from an extended steep-spectrum source to a dominant flat-spectrum core due to the Doppler boosted foreground jet. In the optical band the direction with respect to the line of sight should be measured by a Ca II absorption less and less prominent as the angle decreases, due to the increasing dominance of the non-thermal (synchrotron) component.

We have thus performed VLA observations at two frequencies (20 and 6 cm) in two different configurations. Forty-one objects have been observed in the most resolved configuration (A) and a subset of 35 in B configuration. The objects are almost equally divided in Galaxies (11), BL Lac candidates (12) and "bona-fide" BL Lacs (18), according, as said before, to the measure of the Ca II depression. The resolution of these observations ranges from $\sim 0.3''$ to $\sim 3''$, depending on the VLA configuration and observing frequency. At the redshift of the observed sample ($z \sim 0.1$–0.2) we are therefore probing structures of the order of a few kpc.

2.1. Core Dominance

One of the best indicators of the orientation of the beamed radiation with respect to the line of sight in the radio band is the so-called "core dominance," R, defined as the ratio between the *core* flux (F_C) and the *extended* flux. In practice, the core flux is defined as the peak of the emission, while the extended flux is the difference between the total integrated flux (F_T) and the peak flux: $R = \frac{F_C}{(F_T - F_C)}$

We compute the core dominance R at both 6 cm and 20 cm for the A configuration data. The first frequency gives the best resolution available in our data, however in many literature papers—from the VLA and comparable size telescopes—the 20 cm data are used.

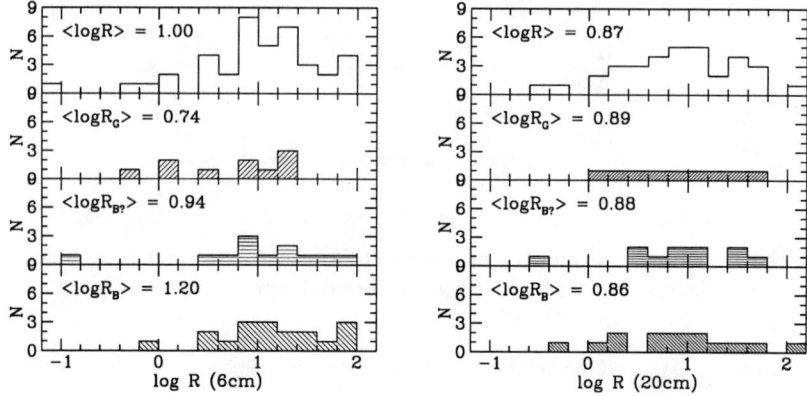

Figure 1. *Left*: The core dominance measured at 6 cm in configuration A—i.e. from the most resolved data. From top to bottom: the total sample, the galaxies, the BL Lac candidates and the "bona-fide" BL Lacs. The mean values for each subsamples are listed in each panel. *Right*: The core dominance measured at 20 cm—as is commonly found in literature.

In Figure 1 we plot the distribution of the core dominance at the two frequencies for the different subsamples and for the total sample. Mean values are also reported in the figure. We can compare them with values found in the literature (Morganti et al. 1997; Laurent-Muehleisen et al. 1993) for analogous samples of objects. We have: $\langle \log R \rangle = -1.6$ (for a sample of classical FRI); $\langle \log R \rangle = 0.04$ (for a sample of FRI galaxies, with an extension to fainter objects from the B2 survey); $\langle \log R \rangle = 0.05$ (Steep Spectrum Quasars); $\langle \log R \rangle = 0.5$ (1 Jy BL Lacs); $\langle \log R \rangle = 1.28$ (Flat Spectrum Quasars); $\langle \log R \rangle = 1.3$ (EMSS and HEAO 1 LASS BL Lacs). The fact that different values of R are found for the same objects by using the data sets at the two frequencies implies that caution should be taken when comparing different resolution data.

We can explore a larger sample by using catalog data, in particular the FIRST and NVSS data, although not simultaneous, to compute the core dominance for a subset of REX galaxies, BL Lacs and candidates that have not been observed by us at the VLA. We derive somewhat lower values of R, since we include in this set also some of the well-known nearby objects. In any case, in this larger subsample there is a trend between isolated galaxies, galaxies in clusters, and BL Lacs ($\langle \log R \rangle = -0.04$; 0.28; 0.86, respectively), possibly due to an environment effect because the denser environment in a cluster could constrain the radio emitting gas. However, the overlap in the distribution is large, and the radio galaxies in the REX are definitely more core-dominated than classical FRI.

If R and the Ca II depression are indicators of orientation, then we expect to find a strong correlation between the two. We plot such function in Figure 2. As evident from the figure, the trend in R with Ca II break is very marginal (regression coefficient = -0.4) and also the slope of the regression fit is very flat:

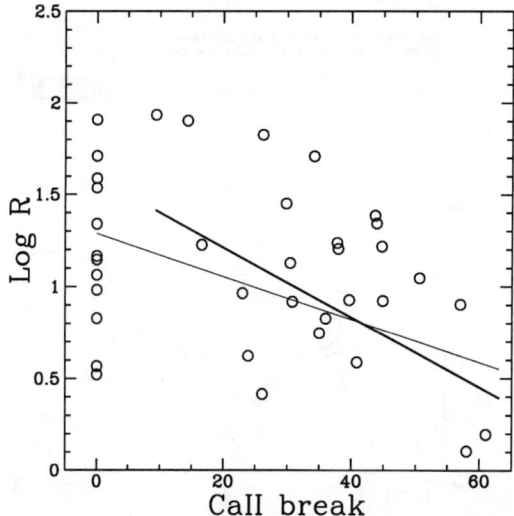

Figure 2. The correlation between the core dominance from VLA data at 6 cm and the Ca II depression as measured from optical data. The two lines are regression fits, see text for details.

we obtain −0.01 (thin line) or −0.02 (thick line) by including or excluding the objects in which the Ca II depression is not visible in the spectrum, that are set at 0.

2.2. Radio Morphology

We expect a more complex radio morphology for objects that preferentially lie in the plane of the sky. It was therefore a bit of a surprise to find that most of the resolved objects are classified as BL Lacs or candidates, and not as galaxies. Actually only ∼ 10% of the objects have been detected with an extended structure in the most resolved configuration, and so the number of objects under consideration is very small. Furthermore, one of the resolved object, 1REX J181335+3144, is consistent with the picture of being the core of a BL Lac jet pointed at us, surrounded by a very faint halo, due to the lobe seen in projection (see Figure 3).

The most striking and unexpected morphology comes from the object shown in Figure 4. This object, 1REX J074722+0905, as can be seen from the featureless optical continuum, is classified as a BL Lac, with no redshift determination. The prominent structure observed is ∼ 20″ long. The radio spectral index (defined by $F_\nu \sim \nu^\alpha$) is −0.6 for the total source, and −0.3 for the core, implying that the extended part is indeed steeper. The extent was already suggested by the NVSS data (with a restoring beam of 45″), where the source, although described by a single Gaussian component, has an integrated flux about 10% larger than the peak flux. Similar morphologies have been found for a few 1 Jy BL Lacs (Cassaro et al. 1999), in particular for 1147+245, for which no emission or absorption lines are seen in the spectrum as in 1REX J074722+0905.

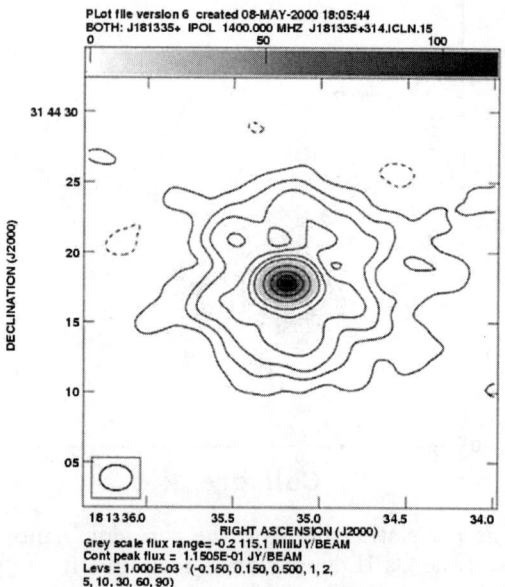

Figure 3. The 20 cm radio map of 1REXJ 181335+3144. The halo around the central pointlike source is only a few percent of the total flux.

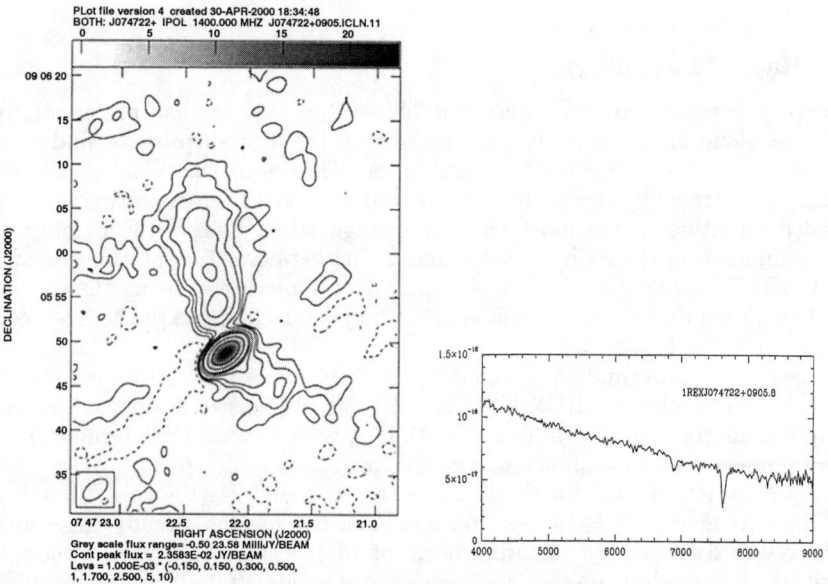

Figure 4. *Left*: The 20 cm radio map of 1REX J074722+0905; *Right*: the featureless optical continuum of the same source.

The other 1 Jy objects with extended radio structure show both absorption and narrow emission lines in the optical spectrum.

3. Conclusions

From the REX survey we have extracted a number of BL Lacs and galaxies, to study their radio properties in the context of Unified Models. Our results seem to imply that objects classified spectroscopically as galaxies, i.e., that do not have emission lines and have a large predominance of stellar light, as measured form the Ca II absorption, are not the same objects that are defined as classical radio galaxies, i.e., with a large radio structure of jets and lobes in the plane of the sky. The resolved morphology contrasts with the BL Lac classification from the optical spectrum in at least one striking case.

The two measures of orientation: core dominance and Ca II depression, from the radio and optical band respectively, seem to be only weakly correlated, suggesting that a revision of the unifying models, in which not only the orientation contributes to change one class into the other, might be necessary.

Acknowledgments. It is a pleasure to thank Greg Taylor for his precious help in reducing the VLA data. This work has received partial financial support from the Italian Space Agency (ASI) and the Italian MURST under grant COFIN98-02-32.

References

Caccianiga, A., et al. 1999, ApJ, 513, 51

Cassaro, P., Stanghellini, C., Bondi, M., Dallacasa, D., Della Ceca, R., & Zappalà, R. A., 1999, A&AS, 139, 601

Condon, J. J., et al. 1998, AJ, 115, 1693

Laurent-Muehleisen, S. A., Kollgaard, R. I., Moellenbrock, G. A., & Feigelson, E. D. 1993, AJ, 106, 875.

Maccacaro, T., et al. 1998, Astron. Nachr., 319, 15

Morganti, R., Oosterloo, T. A., Reynolds, J. E., Tadhunter, C. N., & Migenes, V. 1997, MNRAS, 284, 541

Urry, C. M. & Padovani, P. 1995, PASP, 107, 803

The EMSS Radio Loud Quasar Sample

Anna Wolter

Osservatorio Astronomico di Brera, Via Brera 28, 20121 Milano, Italy

Annalisa Celotti

Sissa, Via Beirut 2-4, 34014 Trieste, Italy

Abstract. We construct the first X-ray selected sample of radio-loud quasars from the EMSS survey. The X-ray selection allows us to explore the properties of radio-loud quasars 10–100 weaker than classical samples in the radio band. There are no radio-loud quasars whose synchrotron peak reaches the UV-soft X-ray band at these (radio) flux levels (as occurs for BL Lac objects selected in the X-ray band), but they appear to have comparatively stronger optical-UV emission. We suggest that this can be ascribed to a significant contribution from a quasi-thermal optical-UV component, emerging due to the comparatively weak non-thermal emission. The detection of sources at low radio fluxes also shows the presence of a large population of steep spectrum quasars, and the lack of the predicted turnover in the quasar counts.

1. Introduction

To understand the properties of a population it is crucial to disentangle the intrinsic features from the selection induced ones. To this aim it is necessary to compare the characteristics of sources selected in different spectral bands. We have therefore considered radio-loud quasars—which so far have been extracted from radio surveys—and studied an X-ray selected sample. We are mainly interested in determining whether the relation between the Spectral Energy Distribution (SED) and the source power proposed by Ghisellini et al. (1998) and Fossati et al. (1998)—a blazar sequence from High-peaked BL Lacs (HBL), to Low-peaked BL Lacs (LBL) to Flat Spectrum Radio quasars (FSRQ)—holds also for fainter X-ray selected quasars. The trend in the SED can be physically accounted for by an increase in the external radiation field along the sequence.

2. The Sample

The sample under study comprises the radio-loud quasars detected in the *Einstein* Medium Sensitivity Survey (EMSS; Gioia et al. 1990; Stocke et al. 1991). It is the first statistically complete X-ray selected sample of (39) radio-loud quasars. Details can be found in Wolter & Celotti (2000, A&A submitted). The sample is cut at $\delta \geq -40°$, and excludes narrow line objects. We dub the sam-

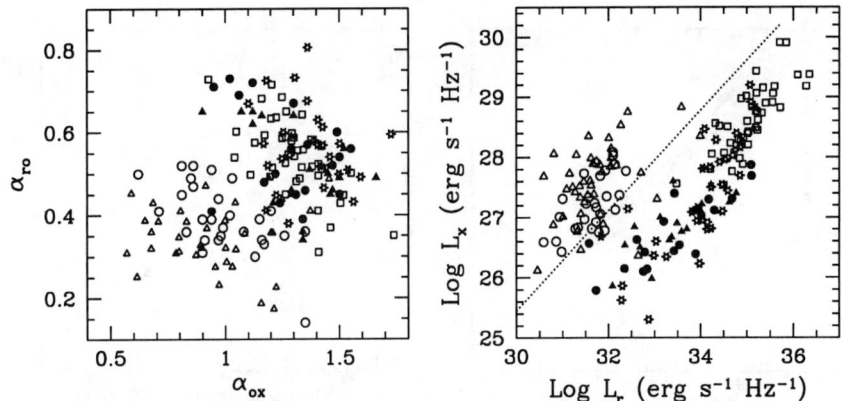

Figure 1. *Left*: Broad band spectral index $\alpha_{\rm ro}$ vs. $\alpha_{\rm ox}$ for the ERL (filled triangles and circles for FS and SS, respectively) compared to: EMSS BL Lacs (empty circles), Slew BL Lacs (empty triangles), 1 Jy BL Lacs (empty stars), 2 Jy FSRQ (empty squares). Just for graphical purposes, blazars selected by Perlman et al. (1998) and Laurent-Muehleisen et al. (1999) are not included. *Right*: X-ray vs. radio luminosity for the same samples. The dotted line indicates the "nominal separation" between HBL and LBL in the luminosity plane (e.g. Fossati et al. 1998). The ERL source in the HBL region is MS0815.7+5233, a weak lined AGN, defined as "BL Lac-like" in Stocke et al. (1991).

ple the EMSS Radio Loud quasar sample (ERL). We divide the sample in flat (FS) and steep (SS) spectrum radio-loud objects ($F_\nu \sim \nu^{-\alpha}$). The distribution of α_r is not bimodal and therefore the choice of a dividing value is somewhat arbitrary. We formally consider it at $\alpha_r = 0.7$.

3. Broad Band Properties

We study the luminosities in the radio, optical and X-ray bands (L_r, L_o, L_x) and the corresponding broad band spectral indices ($\alpha_{\rm ro}$, $\alpha_{\rm ox}$, $\alpha_{\rm rx}$). Through X-ray selection we detect objects with a rather limited range in optical luminosity, but a large span in the radio one, extending the range of sampled L_r towards fainter sources compared to radio-selected quasars.

The radio and X-ray luminosities seem to be correlated at the 99% level even taking into account the common redshift dependence. No differences are found between the distributions of FS and SS, neither in luminosities (KS probability $p \geq 20\%$) nor in broad band spectral indices ($p \geq 47\%$) nor in the trends, suggesting a behavior of the compact beamed component independent of the large scale radio flux dominance.

The shape of the SED appears to be related to the radio luminosity: in fact, the only statistically significant trends found are between $\alpha_{\rm ro}$, $\alpha_{\rm rx}$ and L_r, consistent with being caused by a change in the synchrotron peak energy: the flatter is $\alpha_{\rm ro}$, the higher is the peak energy.

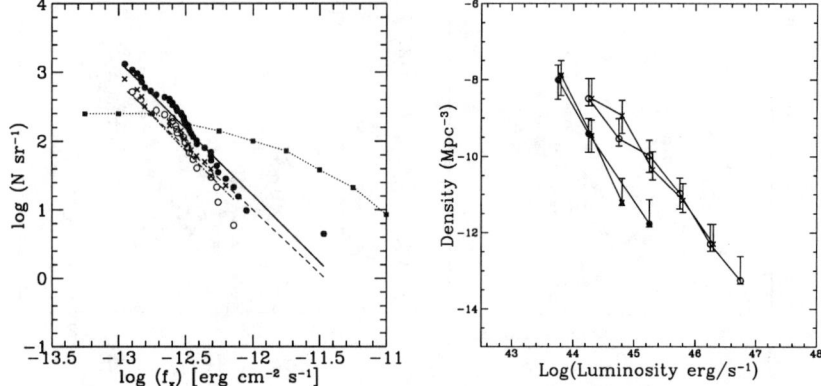

Figure 2. *Left*: Integral number counts for FS (crosses), SS (circles) and all ERL (filled circles). The lines represent the corresponding linear fits: heavy line for ERL; dashed line for FS, and dot-dashed line for SS. For comparison, the number counts of EMSS BL Lacs are reported (from Wolter et al. 1991). *Right*: Differential luminosity functions for the SS and FS sources, in step of $0.5\log L$ (symbols as before; filled circles and stars are de-evolved—exponential form—SS and FS sources respectively). SS points are shifted on the L axis for clarity.

4. Comparison with Other Samples

In Fig. 1a we show the ERL and samples of both FSRQ and radio and X-ray selected BL Lacs. The ERL occupy the transition region (in the "boomerang" shaped blazar distribution) between FSRQ and X-ray selected ($\sim$ HBL) BL Lac objects, roughly overlapping with radio-selected ($\sim$ LBL) ones. Note that the location of LBL and FSRQ in the spectral index plane is due to the dominance in the X-ray band of a flat Compton component.

Even the X-ray selection is unable to detect radio-loud quasars with high peaked synchrotron components, albeit sampling sources with L_r comparable to that of HBL (see Fig. 1b). These findings are therefore in global agreement with the expectation of the blazars sequence scenario, thus re-enforcing the view of a strong connection between the SED and the radio luminosity.

We propose that the concentration of ERL in the LBL region could be at least in part ascribed to the increasing dominance of a quasi-thermal optically-UV component (blue bump) in quasars of increasingly lower (non-thermal) radio power. Indeed the typical $\alpha_{\rm ro}$ and $\alpha_{\rm rx}$ of the low radio power ERL corresponds to SED whose peak is located in the optical-UV region. The presence of such a component would flatten $\alpha_{\rm ro}$ and steepen $\alpha_{\rm rx}$ in sources with decreasing non-thermal continuum. Optical and X-ray spectra will be able to verify such hypothesis.

5. LogN-logS, Evolution, and Luminosity Functions

We plot the number counts of FS and SS sources in Fig. 2a. There is marginal evidence (though over the whole range in flux) for flatter counts of FS (see Table 1.) The count distributions of FS is in complete agreement with what found for all AGN—mostly radio quiet—in the EMSS (1.61 ± 0.06, Della Ceca et al. 1992); SS and all ERL counts are marginally steeper (at 1 σ level) than all AGN, possibly indicating larger amount of evolution.

Table 1. Statistical Properties (in brackets the 1 σ confidence range)

	ERL	FS	SS
Count slope	1.9 [1.7–2.2]	1.8 [1.6–2.1]	2.0 [1.7–2.3]
$\langle z \rangle$	0.92 ± 0.40	0.99 ± 0.40	0.85 ± 0.39
V_e/V_a	0.78 ± 0.05	0.76 ± 0.06	0.81 ± 0.07
C		6.6 [5.6–7.4]	6.4 [5.6–7.1]
LF integral slope	1.6 [1.4–1.7]	1.6 [1.3–1.8]	1.9 [1.6–2.2]

The most interesting aspect is the extension to a factor 100 lower fluxes provided by the X-ray selection. At radio fluxes of a few tens of mJy (which corresponds for an average SED to $\sim 10^{-13}$ erg cm^{-2} s^{-1}) the radio counts of X-ray selected objects are marginally consistent with the Euclidean extrapolation from higher fluxes and thus largely exceed (factor > 10 for FS and ~ 3 for SS) those predicted by the beaming model, which drop below ~ 0.1 Jy (Padovani & Urry 1992). The EMSS BL Lac distribution (see Fig. 2a) is flatter, dominating by a factor ~ 10 at $\sim 10^{-12}$ erg cm^{-2} s^{-1}, and an 'inversion' of the two populations occurs just above $\sim 3 \times 10^{-13}$ erg cm^{-2} s^{-1}.

We have studied the cosmological evolution by applying the V_e/V_a test (Avni & Bachall, 1980)—see Table 1. The derived evolution rate for ERL and FS is consistent with the total EMSS AGN sample (Della Ceca et al. 1992). SS require a slightly higher value for the evolution parameter. The differential luminosity functions are plotted in Fig. 2b as observed, and de-evolved at $z = 0$. Derived slopes are in Table 1.

Acknowledgments. This work has received partial financial support from the Italian MURST under grant COFIN98-02-32.

References

Avni, Y. & Bachall, J. N. 1980, ApJ, 235, 694
Della Ceca, R. et al. 1992, ApJ, 389, 491
Fossati, G., et al. 1998, MNRAS, 299, 433
Ghisellini, G., et al. 1998, MNRAS, 301, 451
Gioia, I. M., et al. 1990, ApJS, 72, 567
Laurent-Muehleisen, S. A., et al. 1999, ApJ, 525, 127
Padovani, P. & Urry, M. C. 1992, ApJ, 387, 449
Perlman, E. S., et al. 1998, AJ, 115, 1253
Stocke, J. T., et al. 1991, ApJS, 76, 813
Wolter, A., et al. 1991, ApJ, 369, 314

Surveys and the Blazar Parameter Space

Eric S. Perlman

Joint Center for Astrophysics, University of Maryland, 1000 Hilltop Circle, Baltimore, MD 21250, USA

Paolo Padovani, Hermine Landt

Space Telescope Science Institute, 3700 San Martin Drive, Baltimore, MD 21218, USA

John T. Stocke

Center for Astrophysics and Space Astronomy, University of Colorado, Campus Box 389, Boulder, CO 80309, USA

Luigi Costamante

Osservatorio di Brera, via Bianchi 46, 23807 Merate (LC), Italy

Travis Rector

NOAO, P. O. Box 26851, Tucson, AZ 85721, USA

Paolo Giommi

SAX Science Data Center, Agenzia Spatiale Italiana, viale Regina Margherita 202, I-00198 Roma, Italy

Jonathan F. Schachter

Smithsonian Astrophysical Observatory, 60 Garden Street, Cambridge, MA 02138, USA

Abstract. The rareness of blazars, combined with the previous history of relatively shallow, single-band surveys, has dramatically colored our perception of these objects. Despite a quarter-century of research, it is not at all clear whether current samples can be combined to give us a relatively unbiased view of blazar properties, or whether they present a view so heavily affected by biases inherent in single-band surveys that a synthesis is impossible. We will use the coverage of X-ray/radio flux space for existing surveys to assess their biases. Only new, deeper blazar surveys approach the level needed in depth and coverage of parameter space to give us a less biased view of blazars. These surveys have drastically increased our knowledge of blazars' properties. We will specifically review the discovery of "blue" blazars, objects with broad emission lines but broadband spectral characteristics similar to HBL BL Lac objects.

1. Myths and Facts about Blazar Surveys

Every survey has its biases, whether imposed by its flux limits or selection techniques—despite the best efforts of the scientists involved. Proper analysis of the results of any survey requires one to understand the impact of these biases on both the range of parameter space to which the survey is sensitive, and also on the broader scheme, including properties which do not form part of the survey definition, but which nevertheless represent characteristics of the class and/or important diagnostics.

Historically, our knowledge of blazar properties began with radio surveys, which tend to be dominated by radio-bright objects. As we discovered in the 1980s, such objects are the most luminous of all AGN, as well as violently variable and radio core dominated. Indications that these properties were not typical of all blazars did not come until the publication of the first large X-ray survey, the *Einstein* EMSS (Stocke et al. 1991). The EMSS BL Lacs are considerably less luminous, less variable, less polarized, and less core dominated than their radio-selected cousins (Perlman & Stocke 1993; Jannuzi, Elston & Smith 1994; Kollgaard et al. 1996; Giommi et al. 1995; Rector et al. 2000). They also have synchrotron peaks in the UV/X-ray rather than IR/optical (Giommi et al. 1995, Sambruna et al. 1996, Fossati et al. 1998). Various explanations for these differences have been proposed (see the review by Urry & Padovani 1995, and also Ghisellini et al. 1998 and Georganopoulos & Marscher 1998). But it was not until the mid-1990s that opinions began to come full circle.

Even if one assumes an unbiased identification process (not always the case, see e.g. Marchã & Browne 1993, 1996; Perlman et al. 1996; Rector et al. 1999) each survey's flux limits affect its sensitivity to parameter space. In his paper, Paolo Padovani presented two figures (Padovani 2000, Figures 1 and 2) which illustrate the sensitivity of various surveys to X-ray/radio parameter space, using their flux limits. For completeness and for the sake of minimum bias, one wants surveys that are: (1) as close as possible to the diagonal lines defined by the inverse-Compton and synchrotron peak limits—which are also the extrema for known HBLs and LBLs respectively; and (2) large enough that all spectral shape classes are represented in statistically significant numbers. Yet even with surveys meeting these requirements, some biases remain, since each flux limit imposes diagonal "completeness contours" on the $(\alpha_{ox}, \alpha_{ro})$ plane. Thus even a survey which is near an extreme line in the (f_x, f_r) plane does not have uniform sensitivity to all regions of $(\alpha_{ox}, \alpha_{ro})$ space. Nevertheless, once samples meeting these criteria are created, it will truly be possible to use their contents to simulate the overall blazar content of the universe.

Figure 1 illustrates these biases for the case of BL Lacs. Due to their small sizes and high flux limits, each of the "classical" surveys (EMSS, 1 Jy and Slew) was sensitive only to a small diagonal swath of $(\alpha_{ox}, \alpha_{ro})$ parameter space. As a result, they presented an almost completely disjoint picture of BL Lac properties (it is an instructive exercise to plot the X-ray, radio and optical fluxes of the EMSS, 1 Jy and Slew BL Lacs on three planes—we omit this here for lack of space). As this diagram shows, the new DXRBS survey fills in the gap very nicely, covering a much wider range of the $(\alpha_{ox}, \alpha_{ro})$ plane much more evenly, and nicely revealing objects in the intermediate range. That intermediate range is also covered by RGB, but not as evenly because of its high optical flux limit.

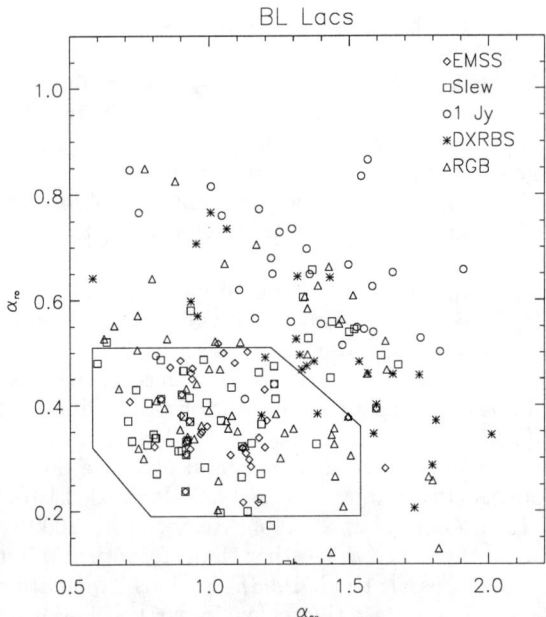

Figure 1. The $(\alpha_{\rm ox}, \alpha_{\rm ro})$ plane for BL Lacs. We have outlined the HBL "box" to guide the eye. The 1 Jy, Slew and EMSS contain very few intermediate objects. The results of the new DXRBS and RGB surveys show this to be a selection effect.

2. The Discovery of X-ray Bright FSRQ: A Case in Point

As shown above, the classical surveys are far too small and shallow to contain significant amounts of objects in all spectral shape classes. Thus, even when lumped together, they present a view that is overwhelmingly colored by biases. For example, by comparing the broadband spectral shapes of the EMSS and 1 Jy BL Lacs, and the S5 FSRQ, Sambruna et al. (1996) predicted that there should be no FSRQ with $\alpha_{rx} \gtrsim 0.78$—the historical dividing line between HBL and LBL type objects. This prediction was proven incorrect by the findings of a newer, deeper radio-limited survey, the DXRBS (Perlman et al. 1998, Landt et al. 2000), which found that approximately 1 in 4 FSRQ had $\alpha_{rx} \gtrsim 0.78$. A subsequent cross-correlation of emission line objects in the RGB (Laurent-Muehleisen et al. 1998) revealed an even higher percentage of these objects ($\sim 40\%$, Padovani et al. in preparation, Perlman 2000).

We will talk more about these objects (called "HFSRQ") below, but first a historical comment is useful. As is well known, the Slew and EMSS both included radio observations at 5 GHz in their identification process, but not radio spectral indices. The biases that imposed were not appreciated for several years, even though both Stocke et al. (1991) and Elvis et al. (1992) noted the presence of radio-loud quasars in both surveys. Only after the results of DXRBS began to come out was this bias considered, and corrected. Cross-correlation of

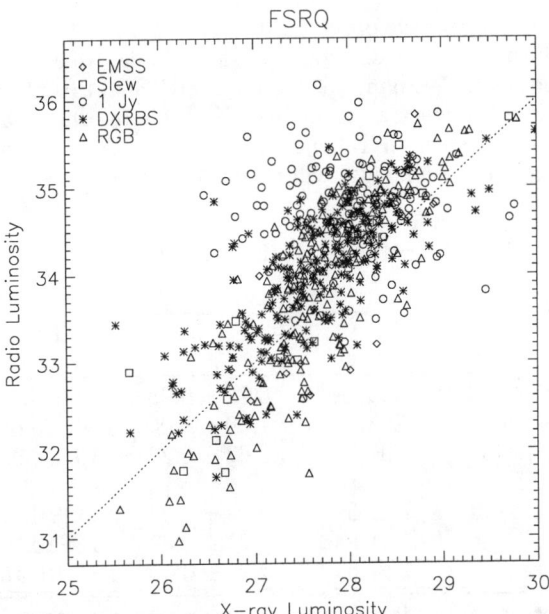

Figure 2. The X-ray and Radio luminosities of FSRQ in the EMSS, Slew, 1 Jy, DXRBS and RGB surveys. The 1 Jy contains only 14 FSRQ, ($\sim$ 5%) of FSRQ to the right of the HBL/LBL dividing line (dotted). By comparison, 25% of DXRBS FSRQ, and 40% of EMSS, Slew and RGB FSRQ, fall into this category.

the Slew and EMSS object lists with lower-frequency radio surveys reveal small, but significant samples of FSRQ in both (22 and 16 objects respectively; Perlman et al. 1999 and in preparation), of which about 40% are HFSRQ. By comparison, the very high radio flux limit of the 1 Jy sample makes it almost completely insensitive to these objects (Perlman et al. 1998). Figure 2 summarizes these findings by comparing the FSRQ content of the new and classical surveys.

As was the case for HBL, HFSRQ cover a different region of (L_x, L_r) parameter space than previously known FSRQ, and dominate increasingly at lower luminosities. As was pointed out by Urry & Padovani (1995), the knee of the FSRQ radio luminosity function sits at $L_r = 10^{33-33.5}$ erg s^{-1} Hz^{-1}. A quick look at Figure 2 is enough to convince the reader that the 1 Jy survey had basically no sensitivity below that luminosity (in fact only 2 of its objects have lower radio luminosities), yet it is precisely at these luminosities where the HFSRQ become increasingly common.

3. X-ray Observations of HFSRQs

In order to understand the full spread of blazar properties, and develop appropriate constraints upon their physics, it is imperative to investigate the properties of HFSRQ in the same depth as has already been done for their BL Lac cousins.

To start the process, we have observed four of the X-ray brightest HFSRQ with SAX to analyze their 0.1–10 keV spectra. In so doing, we aim to investigate the general trend noted in Perlman (2000) that DXRBS HFSRQ tend to have somewhat steeper ROSAT spectra (as defined by their hardness ratios) than their more radio-bright cousins. In Table 1, we give the results of these observations (a full accounting of these results will be presented in a later paper).

Table 1. SAX Observations of HFSRQ—Results

Name	N_H 10^{20} cm^{-2}	α_x	F_{1keV} μJy	χ^2_r/d.o.f.
WGA J0546.6−6415	4.54 fix	0.72 ± 0.08	0.64 ± 0.07	1.07/41
RGB J1629+401	0.852 fix	1.50 ± 0.06	0.66 ± 0.05	0.90/28
RGB J1722+243	4.95 fix	0.62 ± 0.22	0.16 ± 0.05	0.63/18
S5 2116+81:				
.....29/2/98	7.41 fix	0.73 ± 0.04	2.71 ± 0.17	0.96/61
...12/10/98	7.41 fix	0.77 ± 0.07	2.15 ± 0.19	0.98/46
...........sum	7.41 fix	0.73 ± 0.04	2.75 ± 0.16	1.04/62

Note: the errors are at 90% conf. level for one parameter of interest.

As can be seen, we find flat spectra, more similar to LBL BL Lacs and previously known FSRQ (Sambruna et al. 1996) for three of four objects. Only one object (RGBJ1629+401) has a steep, HBL-like spectrum. This represents the first systematic observation of objects with demonstrably HBL-like broadband continua based upon their α_{ox} and α_{ro} values. The ASCA observations of "blue" FSRQ by Sambruna et al. (2000) achieved an outwardly similar result, with all objects having flat hard X-ray spectra. However, the Sambruna et al. result cannot really be considered as an observation of HFSRQ, as their selection criteria was for steep ROSAT spectra and not $(\alpha_{ox}, \alpha_{ro})$ values indicative of a high frequency peak—and indeed, some of the Sambruna et al. objects are *not* in the HBL region of the $(\alpha_{ox}, \alpha_{ro})$ plane.

In this light, it is unclear how we should interpret the broadband spectra of HFSRQ. The most straightforward interpretation would be that in all cases except for RGB J1629+401 we are seeing only inverse-Compton emission at > 0.1 keV. However, extreme caution is required. At the very least these objects have two emission components (both synchrotron and emission line) which can serve as seeds for inverse-Compton emission—and therefore it is likely that the synchrotron to inverse-Compton ratio is considerably higher for HFSRQ than it is for HBL BL Lacs. It is of course unclear where this additional emission might evidence itself, but according to the models of Sikora, Begelman & Rees (1994), inverse-Compton scattered emission-line photons should evidence themselves at energies of order 1–100 keV. Thus it is very likely that for HFSRQ of a given peak frequency the synchrotron to inverse-Compton ratio is considerably higher than for an HBL of the same peak frequency (see Georganopoulos 2000 for an interesting discussion of exactly this point). Thus it is still quite feasible that these objects have peak energies as high as $\sim 10^{16}$ keV. By contrast, for

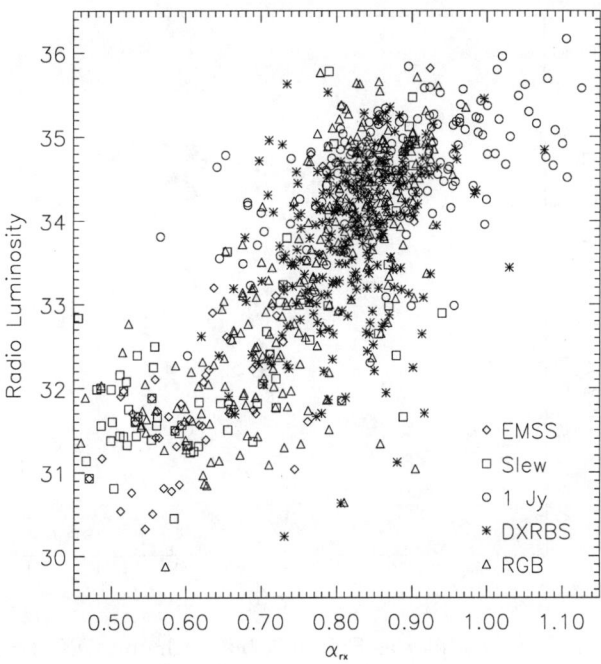

Figure 3. The $(\alpha_{\rm rx}, L_r)$ plane for blazars including the EMSS, Slew, RGB, DXRBS and 1 Jy samples. Note how much broader the correlation becomes after all surveys are included. While the correlation is still > 99.99% significant, the most prominent feature of this graph is the prominent "outlier" regions at the upper left and lower right hand corners, representing extreme objects.

RGB J1629+401 the most consistent explanation of its broadband spectrum would be that we are seeing synchrotron emission all the way up to 10 keV.

4. Discussion

It is now worthwhile to reassess the analysis of Fossati et al. (1998). Figure 3 shows the (α_{rx}, L_r) plane for blazars when all five blazar surveys are added to the diagram. As discussed in Fossati et al. (1998), an object's $\alpha_{\rm rx}$ value is related to the location of its synchrotron peak, with $\alpha_{\rm rx} \approx 1$ representing $\nu_{\rm peak} \sim 10^{12}$ Hz, and $\alpha_{\rm rx} \approx 0.5$ representing $\nu_{\rm peak} \sim 10^{17}$ Hz. Thus, as explained by Fossati et al., the correlation between these two parameters was meant to explore the relationship between total luminosity and peak frequency. As can be seen, this correlation still persists, but when all five samples are added, it grows quite a bit broader—in fact, most of the horizontal extent in the graph is taken up by the "outlier" sections in the lower left and upper right hand corners. These corners represent areas where only one or two surveys had overwhelming sensitivity. If one only examines the middle section of the graph, the correlation is still very

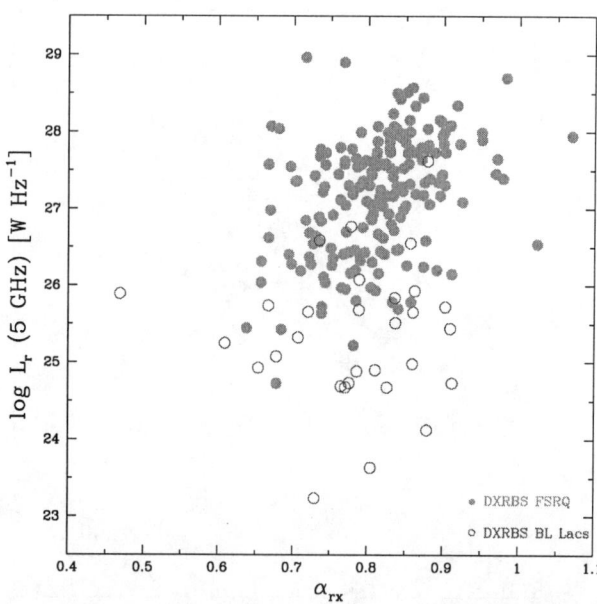

Figure 4. The same plot as Figure 3, but with only DXRBS objects shown. As can be seen, the Fossati et al. correlation does not persist when only single surveys are considered, making it quite conceivable that the observed correlation reflects the parameter space sensitivity of each survey rather than a physically relevant correlation.

significant, but its physical meaningfulness is far less persuasive. And in fact, when *individual* surveys are examined no such correlation is present! This is shown convincingly by Figure 4, which is the same plot with only DXRBS objects shown. It is therefore impossible to tell whether the (α_{rx}, L_r) plot represents a physically important correlation, or is merely reflective of the parameter space sensitivity of each survey.

The comparison of Figures 3 and 4 illustrate quite well the dangers inherent in combining surveys with very different biases and sensitivities, and attempting to derive correlations. These biases will continue to color our perception of blazar properties until either radio and X-ray selected samples become large and deep enough to include significant numbers of objects in every spectral shape class. DXRBS is the first radio-limited survey to accomplish this, but it will take much deeper radio and optical surveys to accomplish this for an X-ray flux limited sample, even with ROSAT data (see Padovani 2000). Thus to fully understand the findings of surveys whose parameter space coverage lies in the middle of the Padovani diagrams (e.g. REX, RGB), significant modeling will be required.

Fossati (2000) has described a promising avenue for such modeling efforts. However, it is instructive to note that the error bars on his models are still quite large, probably because of the almost disjoint parameter space coverage of the EMSS, Slew and 1 Jy surveys. Once Fossati has included deeper surveys such as DXRBS in his seed samples, it will be interesting to see whether the resulting error bars are sufficiently small that the technique can be used to predict the outcomes of future blazar surveys. A similar comment applies to the modeling efforts of Georganopoulos & Marscher.

References

Elvis, M., Plummer, D., Schachter, J., & Fabbiano, G. 1992, ApJS, 80, 2
Fossati, G. 2000, these proceedings
Fossati, G., Maraschi, L., Celotti, A., Comastri, A., & Ghisellini, G. 1998, MNRAS, 299, 433
Georganopoulos, M. & Marscher, A. P. 1998, ApJ, 506, 621
Georganopoulos, M. 2000, ApJ, 543, L15
Giommi, P., Ansari, S. G., & Micol, A. 1995, A&AS, 109, 267
Giommi, P. & Padovani, P. 1994, MNRAS, 268, L51
Ghisellini, G., Celotti, A., Fossati, G., Maraschi, L., & Comastri, A. 1998, MNRAS, 301, 451
Jannuzi, B. T., Elston, R., & Smith, P. A. 1994, ApJ, 428, 130
Kollgaard, R. I., Palma, C., Laurent-Muehleisen, S. A., & Feigelson, E. D. 1996, ApJ, 465, 115
Landt, H., Padovani, P., Perlman, E. S., Giommi, P., Tzioumis, A., & Bignall, H. 2000, MNRAS, in press
Laurent-Muehleisen, S. A., Kollgaard, R. I., Ciardullo, R., Feigelson, E. D., Brinkmann, W., & Seibert, J. 1998, ApJS, 118, 127
Marchã, M. J. M. & Browne, I. W. A. 1993, MNRAS, 261, 795
Marchã, M. J. M. & Browne, I. W. A. 1996, MNRAS, 279, 92
Padovani, P. 2000, these proceedings
Perlman, E. S. 2000, in GeV-TeV Gamma-Ray Astrophysics Workshop, eds. B. L. Dingus, M. H. Salamon & D. B. Kieda (AIP: New York), 53 (astro-ph/9910321)
Perlman, E. S., Padovani, P., Giommi, P., Sambruna, R., Jones, L. R., Tzioumis, A., & Reynolds, J. 1998, AJ, 115, 1253
Perlman, E. S. & Stocke, J. T. 1993, ApJ, 406, 413
Perlman, E. S., Stocke, J. T., & Schachter, J. F., 1999, BAAS, 195, 1601
Perlman, E. S., Stocke, J. T., Wang, Q. D., & Morris, S. L. 1996, ApJ, 456, 451
Rector, T. A., Stocke, J. T., & Perlman, E. S. 1999, ApJ, 516, 145
Rector, T. A., Stocke, J. T., Perlman, E. S., Morris, S. L., & Gioia, I. M. 2000, AJ, 120, 1626
Sambruna, R. M., Maraschi, L., & Urry, C. M. 1996, ApJ, 463, 444
Sambruna, R. M., Chou, L. L., & Urry, C. M. 2000, ApJ, 533, 650
Sikora, M., Begelman, M. C., & Rees, M. J. 1994, ApJ, 421, 153
Stocke, J. T., Morris, S. L., Gioia, I. M., Maccacaro, T., Schild, R., Wolter, A., Fleming, T. A., & Henry, J. P. 1991, ApJS, 76, 813
Urry, C. M. & Padovani, P. 1995, PASP, 107, 803

How Many 'Flavors' Can a BL Lac Have?

Maria J. Marchã

Observatório Astronómico de Lisboa, Portugal

Abstract. Recent years have seen an increase in the number of BL Lacs discovered. Although undoubtedly related to the variety of surveys aiming to find such objects, the result is not alien to the fact that the adopted criteria for BL Lac classification has also changed. In this work the findings about the relationship between BL Lacs and other low radio luminosity flat spectrum sources are presented, and evidence for a continuity, rather than a bimodality in the observed properties, is discussed. Such findings suggest that BL Lacs can appear under several 'flavors,' and stress the need for a multi-frequency classification of sources.

1. Introduction

BL Lacs are low luminosity radio-loud sources which are amongst the most highly polarized and variable AGN known in the universe. Such "violent" behavior is shared only with Flat Radio Spectrum Quasars (FRSQs) which are found predominantly at much higher luminosities. The two types of sources are usually grouped together and termed as blazars, even though their optical properties are remarkably different. While BL Lacs show only weak, if any, emission lines, the spectra of FRSQs are characterized by strong and broad emission features. The differences between the two populations were at the origin of the dichotomy created in the zeroth-order unifying schemes, where orientation/extinction is the most important factor to determine the source's appearance. Due to the lack of strong emission lines in their spectrum and their low radio luminosity, BL Lacs have been unified with FRI radio galaxies, while quasars have been unified with the high luminosity radio galaxies (FRIIs).

This zeroth-order unification, however, was based on samples containing the strongest radio sources for each category, since these were the only ones available from the existing surveys at the time. Nevertheless, recent years have been very successful in producing deeper surveys which have blurred the apparently clear divisions between sources. Curiously enough, BL Lacs are essential to these studies since by going deeper, the recognition of a weak BL Lac nuclei in the core of a luminous galaxy has become subtler, thus enhancing the necessity of a classification based on intrinsic properties rather than on parameters that could be due to orientation or dilution effects. At the same time, the clear separation between BL Lacs and other flat radio spectrum sources (FRSS), namely quasars is becoming less clear since broad lines have been observed in the spectra of the former (Vermeulen et al. 1995; Corbett et al. 1996), and a certain trend is found in the shape of the Spectral Energy Distribution (SED) of the two types of sources (Fossati et al. 1998).

2. The 'Flavors'

In Marchã et al. (1996) the optical definition of a BL Lac was investigated through the study of a sample of optically bright FRSS. In particular, based on the distribution of the sources on the 'contrast-EW' ('C-EW') plane it was suggested that the definition of BL Lac within this plane should not be restricted to C ≤ 0.25 and EW ≤ 5 Å. Instead it was suggested that these criteria should be extended to larger contrasts and EW. Here 4 ranges of optical criteria are presented as slightly different manifestations of the same intrinsic physical phenomena that would have a BL Lac nucleus at its origin.

2.1. Flavor 1: C ≤ 0.25 and EW ≤ 5 Å

These criteria correspond to the classical definition of a BL Lac. The reduced CaII-break contrast (this measure is typically 0.5 in elliptical galaxies) is an indication that there is a strong contribution of the non-thermal continuum even in the optical regime. The fact that the emission features (if present) are so weak could be interpreted as either the lack of material to be ionized (lack of a broad line region—BLR), or the lack of strong enough ionizing radiation, as suggested by Baum, Zirbel, & O'Dea (1995) to explain the absence of strong emission features in FRI radio galaxies, the assumed parent population of BL Lacs.

2.2. Flavor 2: C ≤ 0.4 and EW ≤ 10 Å

These criteria correspond to the BL Lac candidates proposed in Marchã et al. (1996) and were based on two major facts: (1) the lack of sharp separation in the 'C-EW' plot between the classical BL Lacs and other flat radio spectrum sources of the 200 mJy sample with EW roughly less than 10 Å and contrasts below 0.4, and (2) less than ~ 5% of a sample of ellipticals studied by Dressler & Schectman (1987) show contrasts below 0.4. The fact that these sources show larger starlight contribution means that either there is an intrinsically weaker BL Lac nucleus which is being diluted by the host galaxy, or that they are seen at slightly larger angles from the line of sight (l.o.s). To investigate this further, Dennett-Thorpe & Marchã (2000) studied the radio properties of these sources and found that on the one hand, the high frequency radio spectra is indistinguishable from those of BL Lacs, but on the other, their α_{ro} is steeper. Both of these observations are consistent with the dilution scenario. However, the radio core polarization in these BL Lacs of Flavor 2 is systematically lower than that found in 'classical' BL Lacs, something that is consistent with either a larger observing angle from the l.o.s, or that these sources are on average seen through larger amounts of depolarizing medium. In any of the situations, the data strongly supports the claim that these sources are intrinsically similar to the BL Lacs of Flavor 1.

2.3. Flavor 3: $C \geq 0.4$ and $EW \leq 10$ Å

The fact that the break contrast in these sources is above 0.4 means that it is not possible to use this measure to infer the presence of a weak non-thermal component in the spectrum. In principle, however, it is possible that there is a weak BL Lac nucleus in the core of such galaxies which is simply too diluted. In order to investigate whether these apparently 'Passive Elliptical Galaxies' (PEGs) hide a BL Lac nucleus, it is necessary to make use of other diagnostics. With this in mind the radio properties of these sources were investigated by Dennett-Thorpe & Marchã (2000). It was found that these sources show, on average, steeper high-frequency radio spectra and lower percentage radio core polarizations, thus suggesting that, on average, these sources will be observed further from the l.o.s. than 'classical' BL Lacs. The same study, however, has also shown that the PEGs with higher polarized cores are also those that have flatter high radio frequency spectra. These sources constitute good candidates for 'hidden BL Lacs,' i.e. sources physically identical to BL Lacs, seen at the similar angles to the l.o.s but just intrinsically weaker than their Flavor 1 counterparts. Further evidence that these sources are weak BL Lacs comes from preliminary results concerning the luminosity function (LF) for a deep radio selected flux limited sample ($S_{5GHz} \geq 30$ mJy) of FRSS sources—the 'CLASS' sample—(Marchã & Caccianiga, in preparation). The results show that the LF of these Flavor 3 sources is a natural extension (same slope but lower luminosities) of the LF obtained for the BL Lacs.

2.4. Flavor 4: $C \leq 0.4$ and $EW \sim 60$ Å

In Marchã et al. (1996) two sources from the 200 mJy sample were identified as having simultaneously properties of BL Lacs and quasars. In fact these two low z flat radio spectrum sources showed variable optical polarization, a galaxy-like spectrum with a reduced break contrast ($C \leq 0.4$), and broad Hα. Relative to the 'C-EW' plot, these two sources were placed in an intermediate position between the BL Lacs and the strong emission line sources. The 'hybrid' properties of these two sources were further investigated through a spectropolarimetry campaign with the objective of establishing whether the polarization and the presence of broad emission features were connected to each other. The findings established that the polarization level had not only increased since the previous observations five years before, but that it was solely due to the continuum. In other words, the emission features were not the cause of the large polarization measured by Jackson & Marchã (1999). This suggests that the emission features are therefore being observed directly and not through some polarizing medium. The presence of these broad emission lines in such low radio luminosity ($L_{5GHz} = 10^{24}$ W/Hz) sources is difficult to interpret according to the unification schemes, since in this regime the parent population is thought to be the FRI radio galaxies which are themselves thought to lack a BLR.

3. Discussion

There is now good observational evidence to support that the 'classical' definition of BL Lacs (Flavor 1) should be extended beyond the rigid set of criteria: C $\leq$

0.25 and EW ≤ 5 Å. This evidence comes primarily from the systematic study of samples of low z, low radio luminosity, flat radio spectrum sources which are showing that, unlike what is predicted by the unification schemes, the sources in this luminosity regime show a wide diversity of optical spectra. The investigation of the broadband properties of these sources leads to the following conclusions:

- There are intrinsically weak BL Lac nuclei hiding in the core of what appear to be simply luminous ellipticals. These are sources physically similar to the 'classical' BL Lacs of Flavor 1, i.e. relativistically boosted sources seen at similar angles to the l.o.s but which are intrinsically weaker in power. The recognition of these weak nuclei requires complementary diagnostics to the traditional position on the 'C-EW' plot.

- The BL Lacs of Flavors 1 and 2 from the 200 mJy sample show intermediate broadband properties and radio core dominance parameter between the High and Low energy peaked BL Lacs (HBL and LBL, respectively) (Bondi et al. 2000). Such findings revive the importance of orientation in the unification of BL Lacs.

- The detection of broad, even if weak emission lines in some BL Lacs, and moreover, the existence of sources of Flavor 4 require that a BLR co-exist with BL Lac nuclei.

In summary, the evidence being built on larger and deeper samples of BL Lacs reveals that instead of a sharp transition between BL Lacs and the other FRSS, namely quasars, there must be a smooth transition in both intrinsic power and orientation of the sources.

References

Baum, S. A., Zirbel, E. L., & O'Dea, C. P. 1995, ApJ, 451, 88
Bondi, M., Marchã, M. J. M., Dallacasa, D., & Stanghellini, C. 2000, MNRAS, submitted
Corbett, E. A., Robinson, A., Axon, D. J., Hough, J. H., Jeffries, R. D., Thurston, M. R., Young, S. 1996, MNRAS, 281, 737
Dennett-Thorpe, J. & Marchã, M. J. M. 2000, A&A, in press
Dressler, A. & Schectman, S. 1987, AJ, 94, 899
Fossati, G., Maraschi, L., Celotti, A., Comastri, A., & Ghisellini, G. 1998, MNRAS, 299, 433
Jackson, N. & Marchã, M. 1999, MNRAS, 309, 153
Marchã, M. J. M., Browne, I. W. A., Impey, C. D., & Smith, P. S. 1996, MNRAS, 281, 425
Vermeulen, R. C., Ogle, P. M., Tran, H. D., Browne, I. W. B., Cohen, M. H., Readhead, A. C. S., & Taylor, G. B. 1995, ApJ, 452, L5

Discovery of Hidden Blazars inside Quasars

Feng Ma, Beverley J. Wills

Astronomy Department, the University of Texas at Austin, Austin, TX 78712-1083

Abstract. We have carried out a spectroscopic search for hidden blazars in a sample of $z\sim2$ radio-loud quasars, by comparing new spectra with historical ones taken over 10 years ago. The search is motivated by our prediction that in every radio-loud quasar, collisionally excited emission lines such as CIV $\lambda1549$ can vary at $\sim 50\%$ level without variations in most other lines, including Lyα and CIII] $\lambda1909$. Here we report the discovery of large line variations consistent with our prediction.

1. Introduction

Jet-disk systems may exist in many astronomical objects such as gamma-ray bursters, proto-stars and radio-loud quasars. Radio-loud quasars constitute $\sim 10\%$ of the known quasar population. Their strong radio emission arises from two jets shooting away from the center in opposite directions. It is believed that the great diversity in radio structures and optical spectra may be explained by a unified scheme based on viewing angle (see, e.g. Antonucci 1993; Urry & Padovani 1995). If we look directly into the jet, strong beamed synchrotron radiation can be seen over a wide range of wavelengths from radio to γ-rays. These objects are highly variable and are called "blazars." If the line of sight is a few degrees away from the beam, we can still see a bright radio core, which dominates the radio power, and these quasars are classified as "core-dominant." As the viewing angle increases further, the objects are seen as "lobe-dominant" quasars, then as radio galaxies.

If the unified scheme is correct, every radio-loud quasar should harbor a blazar, which we are often not able to observe. It would offer more solid evidence for the unified scheme if we could find signatures of the hidden blazars via their interactions with their otherwise normal host quasars. This would be analogous to the unification of Seyfert 1 and Seyfert 2 galaxies, for which there is solid physical evidence from the polarized broad emission lines in Seyfert 2 galaxies (Miller & Goodrich 1990). The hidden Broad Emission Line Regions (BELRs) in Seyfert 2s reveal their existence through scattered light.

The beamed synchrotron radiation believed to exist in every radio-loud quasar, but absent or much weaker in radio-quiet AGNs, can heat gas in the BELR and greatly enhance collisionally-excited line emission such as CIV $\lambda1549$ and SiIV $\lambda1397$ (hereafter CIV and SiIV). Lyα, CIII] $\lambda1909$ (hereafter CIII]) and all Balmer lines are much less affected because they are more sensitive to hydrogen-ionizing UV photons that are scarce in the beamed steep-spectrum synchrotron radiation. We thus predicted that the CIV line intensity should vary at ~ 50% level in every radio-loud quasar, during the outbursts of the "hidden" blazars (Ma & Wills 1998), and a key signature would be large variations in line ratios such as CIV/Lyα or CIV/CIII]. We have observed and collected from the literature a sample of 62 radio-loud quasars to search for this significant variability, by comparing new spectra with historical ones taken over 10 years ago. The quasars in our sample have z~1–3 and $-31 < M_B < -25$.

Earlier studies of AGN variability have focused on nearby, low-luminosity, radio-weak objects with known continuum variability, such as NGC 5548, for which it is found that Lyα, CIV and CIII] lines vary with similar amplitudes in response to "normal" (non-synchrotron) AGN continuum variations (e.g. Kaspi & Netzer 1999). The amplitude of line variations is usually less than half that of the continuum, which is also true for Balmer lines in a sample of low-redshift (z~0.1) quasars (Maoz et al. 1994). Systematic *spectroscopic* monitoring of luminous high-redshift quasars has not been done before. From *photometric* studies of large samples of quasars it is found that those of higher luminosity are less variable (Hook et al. 1994; Cristiani et al. 1996; Hawkins 2000). For a quasar in the luminosity range of our sample, when measured at two random epochs, the "typical" variation (or the variability index) is less than 0.2 mag (Hook et al. 1994; Cristiani et al. 1996). The expected emission line variations are < 10% in response to a 0.2 mag "normal" continuum variations. The line *ratio* variations are even smaller if not un-measurable (Shields, Ferland, & Peterson 1995; Kaspi & Netzer 1999).

2. Observations

Most new spectra were taken during 1998–2000 using the Large Cassegrain Spectrograph on the 2.7-m Harlan J. Smith telescope at McDonald Observatory. In our observations, we typically use a 2" slit for long exposures and then calibrate the spectra with short exposures using an 8" slit. Many spectra were taken under non-photometric conditions and thus are not on an absolute flux density scale. Observational details and spectra of all the quasars will be presented elsewhere.

Most historical spectra we use, such as those from Barthel, Tytler, & Thomson (1990), were taken using narrow slits and are not absolutely flux calibrated. Hence, we focus on comparing emission line equivalent width (EW) ratio variations. Direct division of two spectra is a useful way of showing the variability when the spectra are not absolutely calibrated (Pérez, Penston, & Moles 1989). A featureless division spectrum is probably an indication of no variations in continuum or emission lines, because clouds cause only a grey suppression of spectra, and wavelength-dependent slit losses due to atmospheric dispersion cause a low order continuum shape difference. We thus "calibrate" one spectrum using a spectrum at a different epoch, i.e. dividing spectrum 1 by spectrum 2, then fitting the division spectrum (excluding regions of emission lines) with a low-order curve, and then multiplying spectrum 2 with this curve. This way, we correct for all the observational causes of the continuum difference between spectra 1 and 2. The information about changes in emission-line ratios is preserved unless there is a real variation in continuum shape.

3. Results

The observed variability can be summarized in Table 1, where we classify the objects into 5 classes, as indicated in column 1. Column 2 gives the number of objects in each class. The number of core-dominant objects is given in parentheses. Examples of spectra are plotted in Fig. 1. For half of the objects in our sample the spectra at two epochs match well. This increases the credibility of our comparison method and is consistent with photometric findings that quasars at this redshift and luminosity range are not highly variable. Also, it is not surprising that most class "D" objects are core-dominant, known to be more variable than lobe-dominant objects (Pérez et al. 1989), and could cause the apparent proportional line variations using our method of comparison. Large line variations are only seen in CIV and SiIV lines, consistent with our prediction.

Table 1. Classification of EW Ratio Variability in the Sample

Class	#(C)	Description	Interpretation
A	3(1)	large (>20%) relative CIV variations	outbursting hidden blazars uncovered
B	9(3)	smaller (10–15%) relative CIV variations	hidden blazars active
C	33(7)	no >10% variations in any lines	high-z and high-L quasars less variable
D	15(10)	all lines vary in proportion	continuum variation at different epochs; emission line response to "normal" AGN continuum
E	2(1)	Lyα shows 10–15% variations but not CIV	continuum more variable in the blue than in the red

MRC 0238+100 has $z = 1.83$ (Barthel et al. 1990) and $M_B = -27.7$ (Véron-Cetty & Véron 1996). Its CIV line EW has increased by $\sim 70\%$. SiIV is even more sensitive to infrared heating and may increase by over 100% in our modeling. It clearly appears in our new spectrum but not in the historical one. This further suggests the cause of the CIV enhancement to be infrared heating from the additional blazar continuum. Due to the relatively low S/N ratio near the CIII] line, we can only constrain the CIII] variation to be $< 30\%$. Accurate broad-band photometric measurements are scarce for these quasars. From the Digitized Sky Survey I & II (red) we find for MRC 0238+100 a differential variation of 0.01±0.20 mag from the year 1954 to 1990. Comparing our new observations at three epochs (1998 November 22, 1999 November 10, and 2000 March 2) reveals no significant continuum or emission line variations during the time interval of 0.45 years (restframe). These results suggest that MRC 0238+100 does not have significant continuum variations. The hidden blazar inside it is in its high state during 1998–2000, maintaining strong CIV and SiIV emission lines, with no changes in observed continuum or CIII] line.

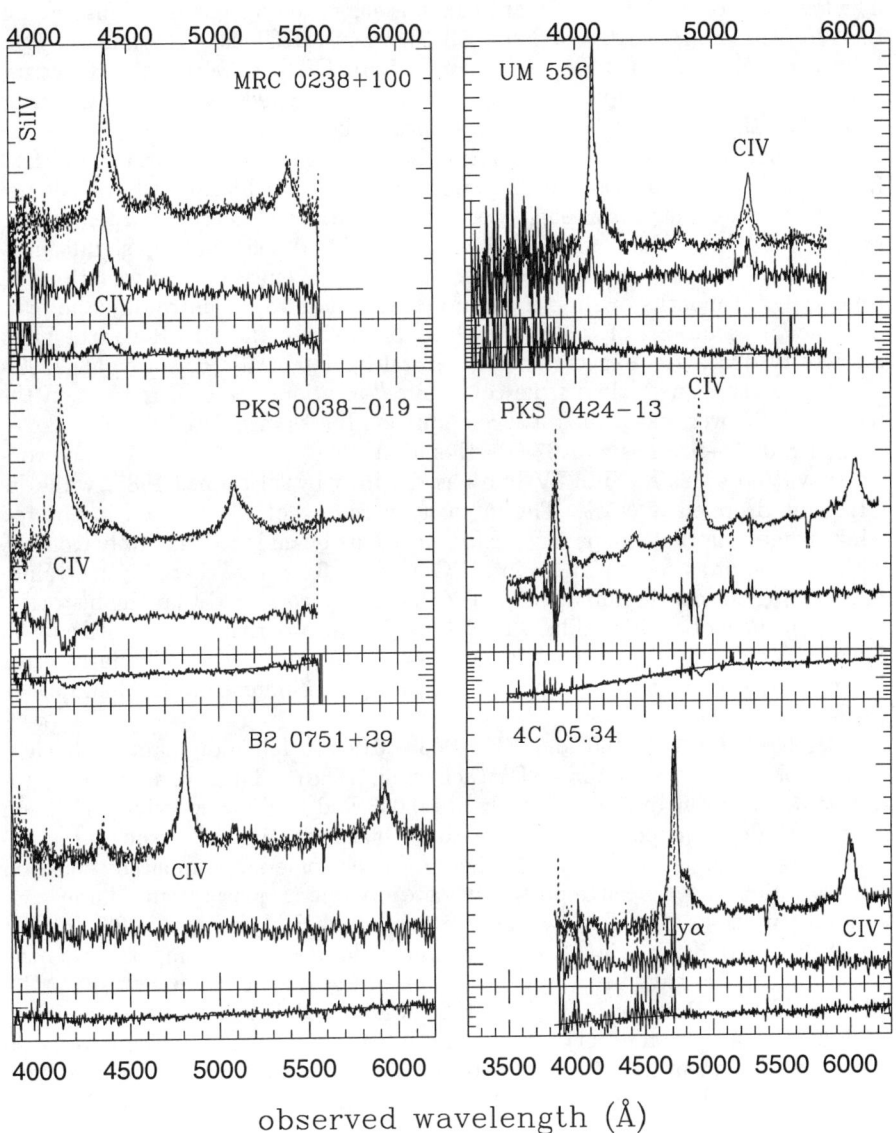

Figure 1. The continua at different epochs have been scaled to the same level. Solid lines are new spectra taken during 1998–2000. Dotted lines are historical spectra. Subtraction spectra are plotted in the same panels. In the lower panels the division spectra and the fitting curve are plotted together. MRC 0238+100, UM 556, PKS 0038−019 and PKS 0424−13 show 30–70% CIV variations with little CIII] or Lyα variation. B2 0751+298 and 4C 05.34 are examples of objects whose spectra taken at a time interval of 10 years are in excellent agreement.

The low state of the hidden blazar in 1986 is suggested by an earlier observation in 1976 (Mitton, Hazard, & Whelan 1977), when MRC 0238+100 had the line-to-continuum ratios of 2.5 for CIV, and 1.5 for CIII] (however, this spectrum is not available for comparison), consistent with the 1986 values, while during 1998–2000 the ratios are 3.8 and 1.6, respectively.

UM 556 also shows a 70% variation in CIV with a $\sim$ 10% variation in Lyα from 1988 March 6 to 2000 June 5. If we assume that the blazar outbursts do not last longer than one year as suggested by observations of nearby blazars, we can predict that the CIV line in MRC 0238+100 and UM 556 will drop significantly within the next three years. It is unlikely to see the emission line variations on time scales of months or shorter despite of the more rapid (intra-day and even intra-hour) variations of the hidden blazars, because the BELR size of high-luminosity quasars is expected to be several light months (Kaspi et al. 2000), and hence the shorter time scale continuum variations are smeared out in the whole BELR over which the observed emission line is integrated.

PKS 0038−019 has z=1.67 (Barthel et al. 1990) and M_{abs}= − 26.9 (Véron-Cetty & Véron 1996). Its CIV line has dropped by 30% since 1986, while its CIII] line dropped by 15%. The intensity variation of CIV is seen in the red wing, suggesting that the part of BELR gas in the beam is moving away from us. This case is not as convincing as in MRC 0238+100 because there is uncertainty in matching the continuum at the blue end of the spectra and that the historical spectrum did not include SiIV λ1397 line. We also derive from Digitized Sky Surveys a differential photometric variation of 0.11±0.45 mag. The large error bar is likely due to the plate quality and the slight difference in the bandpasses at the two epochs.

Both MRC 0238+100 and PKS 0038−019 are lobe-dominant with clear double-lobed radio structures (Barthel et al. 1988). Their core to total radio power ratios at restframe 5 GHz are 0.098 and 0.067, respectively (Nilsson 1998). We do not expect to see the variable blazar continuum at such large implied viewing angles away from the beam. While some large apparent emission line variations have been reported in core–dominant quasars with large continuum variations (e.g. Pérez et al. 1989 reported EW change of 68% in CIV and 82% in CIII] for 3C446), no large line ratio variations comparable to the case of 0238+100 have been reported so far except for the low luminosity AGN NGC 5548 (e.g. Peterson & Ferland 1986, where a HeII λ 4686 flare was interpreted as an accretion event). We also note that since 90% of quasars are radio quiet, some of them have inevitably been observed more than once yet we have not found large line ratio variations from the literature.

4. Discussion

If our observations and interpretations are proved (via repeated observations of the objects of interest and detailed modeling), the results presented here will offer strong support for the unified scheme for radio-loud quasars. In addition, the disk-wind model for the BELR with the winds blowing off the accretion disk by radiation pressure (Murray et al. 1995), will be challenged because in this model there are no winds in the polar regions and hence no beam-BELR interactions are expected. Another widely adopted BELR model, the stellar atmosphere model (Alexander & Netzer 1994), has been challenged by emission line profile studies (Arav et al. 1998). Hence, other explanations for the origin of BELR gas may be needed.

The results from this work suggest that radio-loud quasars should not be used when studying cosmology via the Baldwin effect (Baldwin 1977), as they introduce extra scatter in the inverse correlation between EW(CIV) and continuum luminosity.

The idea of searching for hidden blazars inside quasars may also be applied to observations of misaligned gamma-ray bursts to look for re-processed, more isotropic emission.

Acknowledgments. We thank Jack Baldwin, Peter Barthel, Paul Francis, and Marianne Vestergaard for making their data available in digital form. We are also grateful to Greg Shields for helpful discussions, Derek Wills for help with the manuscript and David Doss for help with the observations. This work makes uses of the ST ScI Digitized Sky Survey and the NED database.

References

Alexander, T. & Netzer, H. 1994, MNRAS, 270, 781
Antonucci, R. 1993, ARA&A, 31, 473
Arav, N., Barlow, T. A., Laor, A., Sargent, W. L. W., & Blandford, R. D. 1998, MNRAS, 297, 990
Baldwin, J. A. 1977, ApJ, 214, 679
Barthel, P. D., Miley, G. K., Schilizzi, R. T., & Lonsdale, C. J. 1988, A&AS, 73, 515
Barthel, P. D., Tytler, D. R., & Thomson, B. 1990, A&AS, 82, 339
Cristiani, S., et al. 1996, A&A, 306, 395
Hawkins, M. R. S. 2000, A&AS, 143, 465
Hook, I. M., McMahon, R. G., Boyle, B. J., & Irwin, M. J. 1994, MNRAS, 268, 305
Kaspi, S. & Netzer, H. 1999, ApJ, 524, 71
Kaspi, S., et al. 2000, ApJ, 533, 631
Ma, F. & Wills, B. J. 1998, ApJ, 504, L65
Maoz, D., Smith, P. S., Jannuzi, B. T., Kaspi, S., & Netzer, H. 1994, ApJ, 421,
Miller, J. S. & Goodrich, B. F. 1990, ApJ, 355, 456
Mitton, S., Hazard. C., & Whelan, J. A. J. 1977, MNRAS, 179, 569
Murray, N., Chiang, J., Grossman, S. A., & Voit, G. M. 1995, ApJ, 451, 498
Nilsson, K. 1998, A&AS, 132, 31
Pérez, E., Penston, M. V., & Moles, M. 1989, MNRAS, 239, 55
Peterson, B. M. & Ferland, G. J. 1986, Nature, 324, 345
Shields, J. C., Ferland, G. J., & Peterson, B. M. 1995, ApJ, 441, 507
Urry, C. M. & Padovani, P. 1995, PASP, 107, 803
Véron-Cetty, M. P. & Véron, P. 1996, A catalogue of quasars and active nuclei (7th Edition), ESO Sci. Rep., 17, 1

Demographics of Blazars

Giovanni Fossati

University of California at San Diego, Center for Astrophysics and Space Sciences, 9500 Gilman Drive, La Jolla, CA 92093-0424, USA

Abstract. We discuss the preliminary results of an extensive effort to address the fundamental, and yet un-answered, question that can be trivialized as: "are there more blue or red blazars?" This problematic is tightly connected with the much debated issue of the unified picture(s) of radio-loud AGNs, which in turn revolves around the existence and the properties of relativistic jets. We address this question by comparing—simultaneously—the properties of the collection of heterogeneously selected samples that are available now, with the predictions of a set of plausible unifications scenarios. We show that it is already possible to make significant progress even by using only the present samples. The important role of selection effects is discussed. For instance we show that the multiple flux selections typical of available surveys could induce some of the correlations found in color-color diagrams. These latter results should apply to any study of flux limited samples.

1. The Factor of 100 Problem

More than 95% of all catalogued blazars have been found in either shallow radio or shallow X-ray surveys (e.g. see Padovani, these proceedings). Because of the range of blazar spectral energy distributions (SED) the two selection methods yield different types, the "red" objects (with the peak of the synchrotron emission at IR-optical wavelengths, LBL) in radio samples, and the "blue" (whose synchrotron emission peaks at UV-X-ray wavelengths, HBL) in X-ray samples. The differences in the SEDs do reflect different physical states but only as the extrema of an underlying continuous population.

The relative space densities of the different types, not to mention their absolute space densities or their evolution in cosmic time still remain indeterminate. Different scenarios predict a difference of two orders of magnitude (!) in the ratio of the "red" and "blue" types, nevertheless the presently available samples are unable to distinguish between them. The blazar demographics are this uncertain essentially because the flux limits of current complete samples are high, so only the tip of the population is sampled. The interpretation of observed phenomenology depends on the complicated sensitivity of diverse surveys to a range of spectral types. Ultimately, this means we do not know which kind of jets nature preferentially makes: those with and high B and γ_e ("blue" blazars) or low B and γ_e ("red" blazars). We also do not know whether they evolve differently and/or if "red" blazars dominate at high redshift and evolve into "blue"

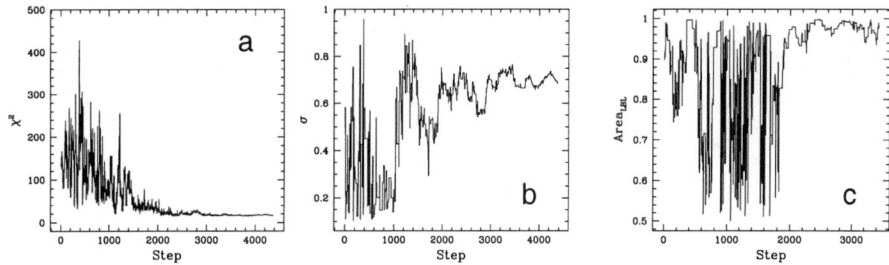

Figure 1. Evolution during the fit of the values of (a) χ^2 and (b) of the width σ of the L–ν_{peak} relationship for the bolometric model. In (c) is plotted the area of the "LBL Gaussian" for the radio-leading model.

blazars at low redshift, and what is the relationship between the "non-thermal" and "thermal" power/components. The implications for understanding jet formation are obvious.

Here we present a concise account of the preliminary results of numerical simulations of a set of unification models, including an actual fit of the model parameters to reproduce the general characteristics of a few reference samples (§2). We also introduce a "concept" experiment, devised to address the role of selection effects (§3), and discuss a couple of issues that are connected to this problem. In §4 we comment on future developments.

2. Testing Unification Scenarios

We compared the existing surveys with a set of three alternative unified schemes, following the discussion developed in recent years after Padovani & Giommi (1995), and Fossati et al. (1997, 1998). They are: i) the "radio-leading," where the primary[1] luminosity is the radio one and $N_{\text{LBL}} > N_{\text{HBL}}$. ii) The "X-ray-leading," where the primary band are the X-ray, and $N_{\text{LBL}} < N_{\text{HBL}}$. iii) The "bolometric," where the SED properties (and in turn the distribution of L_X/L_R, i.e. the balance between LBL and HBL) are determined by the total power of the source, with HBLs being the less powerful objects. In Fossati et al. (1997) the input parameters of each model were pre-set to values based on those of the observed samples. The most interesting results was the success of the new model, the bolometric one.

2.1. The Fit Method, and Results

In this work our approach is different. First we normalize/optimize each unifying scheme by performing an *actual fit* to three reference samples (EMSS, Slew, 1 Jy). We leave free to vary 7–8 variables, such as the normalization and slope

---

[1] Defined as the band where a flux limited selection would be objective with respect to the range of intrinsic properties.

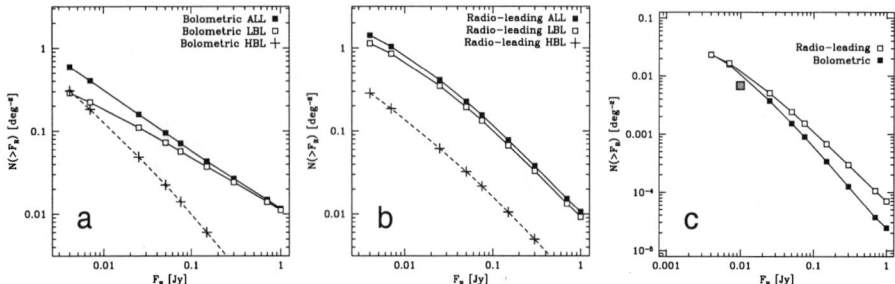

Figure 2. Radio log(N)–log(S) predicted by the (a) bolometric and the (b) radio-leading models for the DXRBS sample. (c) Bolometric and radio-leading predictions for the "sedentary" sample; the grey-filled square represents the observed density.

of the primary luminosity function, and the distribution of the L_X/L_R ratio[2]. For those parameter for which there is a measured value (e.g. the luminosity function) we allowed their values to move within their 2σ interval. The observational quantities to reproduce were the number, and average radio and X-ray luminosities of HBLs and LBLs.

The technique used for the fit is *"simulated annealing"* (e.g. Kirkpatrick et al. 1983), which is based on statistical mechanics, and implemented via MonteCarlo. It is a very robust technique, very well suited for many parameter fits. Moreover the "global" nature of the technique is very effective for cases where there might be multiple secondary local minima in the parameter space.

In Fig. 1 we show examples of the evolution of the fit. We here just point out interesting results concerning two of the "core" issues: i) the best fit of the bolometric model requires a finite width for the L–ν_{peak} relationship (see Fig. 1b). The best fit value is $\sigma \simeq 0.6$, i.e. at any given L the synchrotron peak frequency will be distributed as a Gaussian of width σ centered at the ν_{peak} value determined by the relationship. ii) In both the radio- and X-ray (not shown) leading cases the best fit L_X/L_R distribution is basically a single, broad, Gaussian (see Fig. 1c). For the radio-leading case the LBL Gaussian comprises 98% of the total area, and it is centered at $\simeq -6.3$ with $\sigma \simeq 0.6$.

2.2. Comparison with Real Samples

The next step is to use the results of the fits to predict the properties of samples that have not been used to optimize the parameters of the models. We present here only the integral log(N)–log(S) curves, and we only sub-divide the samples in HBL/LBL (according to the values of F_X/F_R). It is worth noting that the absolute normalizations may not be completely reliable, because of uncertainties on the sky coverage. The uncertainty on the (details of) sky coverage is indeed probably the main one involved in the simulations. The relative fraction of

---

[2]Note on L_X/L_R: for the bolometric scenario we allow for a spread in the relationship between peak frequency and luminosity. For the radio and X-ray leading scenarios we use the combination of two Gaussians, for which we fit the mean, sigma and area.

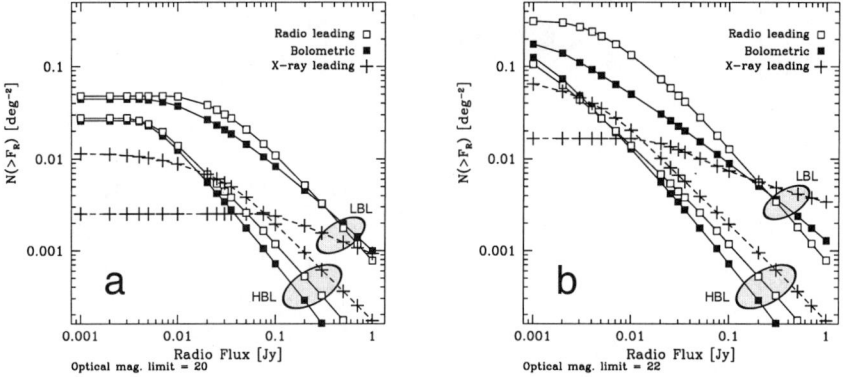

Figure 3. Radio log(N)–log(S) predicted by the bolometric/radio-/X-ray-leading models: HBLs and LBLs for a sample with an additional cut at (a) $m_V <20$, or (b) $m_V <22$.

HBL and LBL may be a more robust parameter, and it is the one more easily amenable to a quick comparison.

DXRBS. The DXRBS sample (Perlman et al. 1998) is still in progress, but an "off record" comparison of the predicted log(N)–log(S) (shown if Fig. 2a,b) with the observed one seems to show that the models are (still) in good agreement with the data. The predictions of the bolometric and radio-leading models become radically different below about 100 mJy, a domain now reachable. The LBL/HBL density ratio at a few radio flux limits are the following:

Flux @5GHz	Bolometric	X-ray leading	Radio-leading
@300 mJy	9.7	6.9	6.6
@150 mJy	6.2	5.7	6.4
@50 mJy	3.2	5.2	6.1

Sedentary survey. The "sedentary" sample (Giommi, Menna & Padovani 1999) comprises only HBLs because of the built-in cut in α_{RX}. The radio log(N)–log(S) is shown in Fig. 2c, where the grey square represent a density @10 mJy between the actual "sedentary" and the EMSS, showing that there is a quite good agreement. Here, as for the DXRBS, we do not plot the predictions of the X-ray leading model because they are not satisfactory. In fact this scenario does not seem to be able to explain the properties of these recent samples, at least with its parameters set at the best fit values.

2.3. Going Deeper

In Fig. 3a,b we show the predictions of the three scenarios for the number densities of HBLs and LBLs in radio surveys with a secondary cut in optical magnitude at $m_V = 20$, and $m_V = 22$. We see that the X-ray leading model is giving a substantially different answer from the two other competing models, which seem to agree over most of the accessible radio flux range. The bolometric and radio-leading models actually start to give different predictions only at very faint

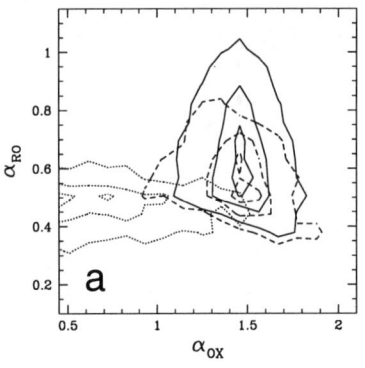 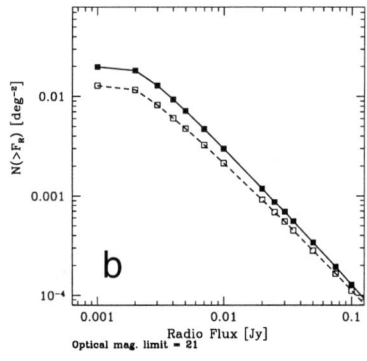

Figure 4. (a) α_{RO} vs. α_{OX} diagram with the density contours predicted by the "cube" for an EMSS-like (dotted), a DXRBS-like (dashed), and a radio (solid) selected samples. (b) "Observed" (empty symbols) and "intrinsic" (filled) log(N)–log(S) of extreme HBLs.

radio fluxes, as seen in Fig. 3b. In the radio-leading model the radio counts of HBLs and LBLs keep a fixed ratio by definition, while in the bolometric picture HBLs are deemed to eventually outnumber the LBLs, but this seems to happen at radio fluxes lower that expected. However, we are not far from the range of radio fluxes that will be the most sensitive to discriminate among the different pictures. Actually there are already a few samples going deep enough.

2.4. The "Cube" (Caveat #1)

To try to assess the problem of selection effects we introduced the "cube." (Fossati & Urry, in preparation), a toy model stripped down of every a priori assumption as to the presumed intrinsic properties of the SED. We assume that the radio/optical/X-ray luminosities are completely un-correlated, and we take simple power law luminosity functions. We then simulate samples of sources that would be selected by a generic flux limited radio or X-ray survey (including the flux dependent sky coverage), with a possible additional cut in another spectral band. An example of the results of this exercise is shown in Fig 4a. It seems to be relatively "natural" to obtain patterns in a color–color diagram which looks like those that are actually observed, and promptly interpreted as tracing intrinsic properties of the sources. Of course the "cube" is not able to reproduce the large variety of patterns and correlations observed in luminosity-luminosity, color–color diagrams, nevertheless we regard it as a very instructive example of how careful we need to be when dealing with selection effects.

2.5. Caveat #2: on Cutting in F_X/F_R color

Figure 4b shows the log(N)–log(S) of *observed* extreme HBL (lower dashed line) and of *intrinsic* extreme HBL (upper solid line), defined as such according to the observed or intrinsic X-ray/radio ratio. Because of the K-correction and their SED shape, blazars systematically shift towards the LBL side when seen at higher redshift, when "classified" on the basis of the *observed* X/radio ratio.

The effect can be sensible when comparing relative populations of HBL and LBL.

3. Conclusions: Blazars Demographics and Not-so-perfect Samples

On the basis of the analysis presented here, we think that there might be already enough information available to proceed to constrain meaningfully the main features of unified scenarios. The comparison of observed samples with simulations performed in a systematic fashion (e.g. by means of simultaneous fit) may provide an extremely powerful and effective tool to address the problem of the intrinsic properties of blazars.

In fact, although there is not a single sample comprising all the desirable characteristics to provide the least possible biased picture of the intrinsic properties of blazars, the F_X/F_R plane is now well covered (see Figs. 1, 2 in Padovani's contribution). Moreover, the quality of the most recent samples will allow to compare the predictions and the data directly by using the distribution of the α_{RX}, an important step forward and past some confusion created by selection effects combined with the "two bins" approach (e.g. §2.5).

If the selection biases of each of surveys can be regarded as being under control (and therefore reliably implemented in the simulations) we may soon be able not only to test a given unified scheme, but even to derive directly from the data what should be the general properties of a successful unified scheme.

Finally, we think that more than ever it is necessary to shift the focus away from the BL Lacs sub-class, because this could still be the source of significant confusion. The best progress could be made by considering the BL Lacs–FSRQs relationship as a whole, also from the observational point of view. The bolometric scenario was meant from the beginning to unify BL Lacs and FSRQs, and it tries to connect some basic physical ideas to the observed phenomenology. On the other hand, we need to figure out how to explain the HBL/LBL ratios assumed by the radio and X-ray leading scenarios, and in turn how to extend these models to include smoothly the FSRQs. There should be a way to tell from "first principles" on which side of the 1/10–10/1 range the real value of N_{HBL}/N_{LBL} ratio is more likely to belong.

Acknowledgments. I'd like to thank the organizers for a great workshop, and for bearing with my request of delaying my talk by one day, and Ilaria Cagnoni for very kindly accepting to swap our talks in the schedule. I also thank *pippol* for the neverending support.

References

Fossati, G., et al. 1997, MNRAS, 289, 136
Fossati, G., et al. 1998, MNRAS, 299, 433
Giommi, P., Menna, M. T., & Padovani, P. 1999, MNRAS, 310, 465
Kirkpatrick, S., et al. 1983, Science, 220, 671
Padovani, P. & Giommi, P. 1995, ApJ, 444, 567
Perlman, E., et al. 1998, AJ, 115, 1253

Part 4
Evolution

The Cosmological Evolution of BL Lacertae Objects

Paolo Giommi, Alberto Pellizzoni

Agenzia Spaziale Italiana, Viale Liegi 26, 00198 Roma, Italy

Matteo Perri

BeppoSAX Science Data Center, Via Corcolle 19, 00131 Roma, Italy

Dipartimento di Fisica, Università di Roma "La Sapienza," P.le A. Moro 2, 00185 Roma, Italy

Paolo Padovani

Space Telescope Science Institute, 3700 San Martin Drive, Baltimore, MD, 21210, USA

Affiliated to the Astrophysics Division Space Science Department, European Space Agency

On leave from Dipartimento di Fisica, II Università di Roma "Tor Vergata," Via della Ricerca Scientifica 1, 00133 Roma, Italy

Abstract. We review the main results from several radio, X-ray and multi-frequency surveys on the topic of cosmological evolution of BL Lacertae objects. Updated findings on BL Lac evolution following the recent identification of many sources in the "Sedentary Multi-Frequency survey" are also discussed. By means of extensive Monte Carlo simulations we test some possible explanations for the peculiar cosmological evolution of BL Lacs. We find that a dependence of the relativistic Doppler factor on radio luminosity (as expected within the beaming scenario) may induce low values of $\langle V/V_\mathrm{m} \rangle$ and that both edge effects at the low luminosity end of the BL Lacs radio luminosity function, and incompleteness at faint optical magnitudes may be the cause of the low $\langle V/V_\mathrm{m} \rangle$ found for extreme HBL sources in X-ray selected samples.

1. Introduction

BL Lacertae objects, like quasars, were discovered in the late sixties as unexpected extragalactic counterparts of bright radio sources. Soon after their discovery the number of known QSOs (radio loud, and especially the radio quiet ones) grew rapidly thanks to the detection of many new objects at optical and radio frequencies. Very few BL Lacs could instead be found due to their peculiar characteristics which include lack of strong emission lines and absence of the "big blue bump." The strong optical variability and polarization properties that are typical of BL Lacs turned out not to be sufficiently strong features to allow the discovery of many new objects of this class at optical frequencies. The number of known BL Lacs remained very low for a long time. When the first generation of X-ray satellites (UHURU, ARIEL V) established that emission line AGN are strong X-ray emitters also BL Lacs were detected as powerful X-ray sources. Shortly afterwards the HEAO1 all sky X-ray surveys discovered many new AGN including a few BL Lacs (Piccinotti et al. 1982). The small number of BL Lacs compared to QSOs (both at radio and X-ray frequencies), together with their extreme properties, nicely fitted into the picture where these objects are the fraction of relativistically beamed sources that happens to be viewed at a small angle with respect to our line of sight (Blandford & Rees 1978). The total number of beamed sources pointing in any direction (the parent population of BL Lacs) would then be much larger.

When the *Einstein* observatory detected X-ray emission form nearly all known AGN and discovered many new ones as serendipitous sources, the number of X-ray discovered BL Lacs was expected to grow in proportion to that of QSOs. Surprisingly instead (although some BL Lacs were indeed found as X-ray serendipitous sources establishing that the X-ray selection is one of the main methods for the selection of new BL Lacs) the fraction of BL Lacs was found to sharply decrease at faint X-ray fluxes. The strong increase of space density or luminosity at earlier cosmic epochs (cosmological evolution) that was found in the radio, optical and X-ray counts of QSOs was clearly not present in the population of BL Lac objects (Stocke et al. 1982; Maccacaro et al. 1983). The *Einstein* Medium Sensitivity Survey (EMSS) convincingly showed that X-ray selected BL Lacs display negative cosmological evolution, that is either there were less objects or they were less luminous in the past (Morris et al. 1991; Wolter et al. 1991; Rector et al. 2000).

In the following paragraphs we review the main results on the cosmological evolution of BL Lacs as obtained from several surveys in different energy bands.

2. BL Lac Surveys

BL Lacs are intrinsically rare objects, therefore it is necessary to select them amongst a much larger number of other 'contaminating' sources. Historically this has been done by means of two main techniques:

1. BL Lacs are searched for in flux limited samples by means of optical spectroscopy of *all* the candidates detected above the survey limit. This approach can be afforded only if the number of candidates is of the order of 1000 or less. In the following we will refer to these surveys as 'classical surveys.'

2. The search makes use of a pre-selection achieved through a cross-correlation of sources detected in surveys carried out in different energy bands, sometimes combined with multi-band spectral restrictions. The candidate BL Lacs can then be identified through optical spectroscopy as in 'classical surveys' or through statistical identification techniques. The big advantage of this method is that the pre-selection can reduce the number of candidates from several thousands or even millions to much smaller samples including only a few hundreds objects or less.

Table 1. Single Band Flux Limited Surveys

Survey	No. of BL Lacs	Radio flux limit (mJy)	X-Ray flux l. erg cm^{-2} s^{-1}	$\langle V/V_m \rangle$
Early radio surveys		>1000	–	–
1Jy complete sample	34	1000	–	0.60 ± 0.05
S4/S5(incomplete)	7,11	500,250	–	–
Einstein EMSS	41	≈ 1	$\approx 2 \times 10^{-13}$	0.427 ± 0.045
Einstein Slew Survey	66	≈ 1	$\approx 5 \times 10^{-12}$	–
RASS HBX BL Lacs	35	≈ 2	8×10^{-13}	0.41 ± 0.05

Table 2. Multi-frequency BL Lac Surveys

Survey	En. Band limits	No. of objects	Vmag limit	Radio fl limit (mJy)	X-Ray flux l. erg cm^{-2} s^{-1}	$\langle V/V_m \rangle$
RGB	R/O/X	127	18	25	$\approx 1 \times 10^{-12}$	–
REX	R/O/X	44	20	5	4×10^{-13}	0.52 ± 0.04
DXRBS	R	30	25	50	2×10^{-14}	0.57 ± 0.05
Sedentary	R	162	21	3.5	$\approx 2 \times 10^{-12}$	0.41 ± 0.02

Table 1 lists the most important classical radio and X-ray surveys together with the main parameters and results on the cosmological evolution of BL Lacs. These surveys select different types of BL Lacs (High energy peaked or HBL, Low energy peaked or LBL, and intermediate objects IBL, see Padovani & Giommi 1995) depending on the energy band and the flux limits. Given the very limited number of objects in each survey it is not possible to study the cosmological evolution comparing luminosity functions built in different redshift shells. The $\langle V/V_m \rangle$ estimator, expected to be 0.5 for a non evolving population of objects (Schmidt 1968), however provides a suitable tool to study cosmological evolution in BL Lacs. Table 1 shows that $\langle V/V_m \rangle$ indicates positive evolution (i.e. $\langle V/V_m \rangle$ > 0.5) for bright radio sources and negative evolution for X-ray selected samples (where radio fluxes are rather low). In all cases the statistical significance is about 2 sigma. In the case of the RASS HBX sample $\langle V/V_m \rangle$ decreases and the significance improves if only the extreme HBL objects are considered (Bade et al. 1998). This effect is also present in the EMSS sample (Rector et al. 2000). In both cases less extreme objects are consistent with no cosmological evolution.

Table 2 lists a number of BL Lac samples selected by means of surveys with flux limits in more than one energy band, or where a multi-frequency preselection has been applied to define a sample that is flux limited in one band. Again the different flux limits tend to select different types of BL Lacs. The $\langle V/V_m \rangle$ results range from consistency with no evolution to strong negative evolution for extreme HBL ($\alpha_{rx} < 0.56$) (Giommi, Menna & Padovani 1999). In the latter survey $\langle V/V_m \rangle$ is significantly below 0.5 only at radio fluxes below 10 mJy.

Strong negative evolution seems to be confined to extreme HBL sources, probably at low radio fluxes.

First results on the REX survey (Caccianiga et al. 2000) however do not seem to confirm this trend, although the radio flux limit (5 mJy) in this case is somewhat higher than that of the other X-ray surveys.

The sedentary survey is the most peculiar among those listed above since it does not rely on direct optical spectroscopy identification of the candidates. In the next paragraph we briefly summarize the main characteristics of this survey and we give an update on the status of the identification process.

3. The Multi-frequency Sedentary Survey: An Update

The Multi-frequency Sedentary Survey (Giommi, Menna & Padovani 1999) makes use of a very efficient multi-frequency selection technique to define a large (radio flux limited) sample of extreme ($\alpha_{rx} < 0.56$) High Energy Peaked BL Lacs (HBL, Padovani & Giommi 1995). In the original paper the sample included 155 objects 58 of which were known BL Lacs and the remaining 97 were still unidentified (but with $\approx 85\%$ chance of being BL Lacs). Using this first version of the sedentary survey Giommi et al. (1999a) found evidence for strong negative evolution at fluxes lower than 10–20 mJy. In order to use the best possible estimate of the cosmological evolution of BL Lacs from this survey we have updated it by adding about 40 new identifications obtained partly through our own optical identification program and (mostly) from the recent publication of the results of massive identification campaigns of X-ray sources discovered in the Rosat All Sky Survey (Bauer et al. 2000; Schwope et al. 2000). As expected a very large fraction of the new identifications are indeed BL Lacs. To give an example, of the 27 candidates in the sedentary survey that are in the list of Schwope et al. (2000) 24 are BL Lacs, 1 is a cluster of galaxies (or a BL Lac in a cluster) and only 2 are emission line AGN.

The updated sample now includes 162 objects resulting from the inclusion of a few extra sources that have been identified as BL Lacs just below the α_{ro} limit that was used to avoid contamination with Seyfert Galaxies (see Giommi et al. 1999a) and the removal of the few candidates identified with emission line AGN. Of the 108 identified sources (67% of the sample) 104 are BL Lacs and 4 have been classified in the literature as clusters of galaxies (although these could well be BL Lacs in a cluster). The remaining 54 candidates are still unidentified but, given the very successful identification rate obtained so far, we are confident that these sources are in very large percentage (85%–90%) BL Lacs.

The application of the V_e/V_a test, a variant of V/V_m for surveys with more than one flux limit (see Avni & Bachall 1980), gives the results reported in Table 3.

The strong indication for negative cosmological evolution ($\langle V_e/V_a \rangle < 0.5$) at low radio fluxes found by Giommi et al. 1999a is confirmed. The updated radio logN–logS (not shown here for reasons of space) is also very similar to that derived with the original sample of 155 objects. The source counts fall sharply

Table 3. The Multi-frequency Sedentary Survey: $\langle V_e/V_a \rangle$ Results

Radio flux limit (mJy)	$\langle V_e/V_a \rangle$	No. of objects in sample
3.5	0.41 ± 0.02	162
5.0	0.43 ± 0.02	140
10.0	0.45 ± 0.03	87
20.0	0.48 ± 0.04	45

below the 'Euclidean' slope at low fluxes consistently with negative cosmological evolution.

4. Possible Interpretations

The most direct explanation of the low $\langle V/V_m \rangle$ values of BL Lacs calls for a cosmological evolution of these objects that is considerably less than that of emission line AGN which have been found to be either more numerous or brighter in the past than now. The space density of BL Lacs (or their luminosity) could instead be approximately constant or even decrease at high redshifts. A model that predicts a lower amount of cosmological evolution in BL Lacs compared to QSOs as a result of changes in the values of fundamental parameters in the fueling mechanism of the central engine in AGN is described by D'Elia & Cavaliere (2000).

In the following we explore a different hypothesis, namely that the low values of $\langle V/V_m \rangle$ might be caused (at least in part) by geometrical and relativistic effects expected in the beaming scenario. These effects influence the position of the synchrotron peak in the broad-band spectral energy distribution of blazars, and hence the composition of flux limited samples selected in energy bands above the position of ν_{peak}.

It is commonly accepted that the broad band electromagnetic spectrum of BL Lacs is dominated by synchrotron emission, up to a peak frequency ν_{peak} (in a ν vs. $L(\nu)$ representation) followed at higher energies by Inverse Compton radiation produced in a relativistically moving plasma seen at a small angle with respect to the line of sight (e.g. Urry & Padovani 1995). Within this scenario the position of the synchrotron peak is determined by the strength of the magnetic field (B), the maximum energy of the electron spectrum (γ_{max}) and by the beaming Doppler factor (δ) which depends on the bulk motion and on the viewing angle:

$$\nu_{peak} \propto B \, \gamma_{max}^2 \, \delta \qquad (1)$$

The observed range of ν_{peak} is extremely wide spanning from 10^{12}–10^{14} Hz for the large majority of BL Lacs selected in radio surveys, to 10^{17}–10^{18} Hz (and possibly higher) for X-ray discovered HBL sources (Pian et al. 1998; Giommi et al. 1999b). Since radio selection is not expected to influence the position of ν_{peak} objects showing low frequency peaks are thought to be much more common

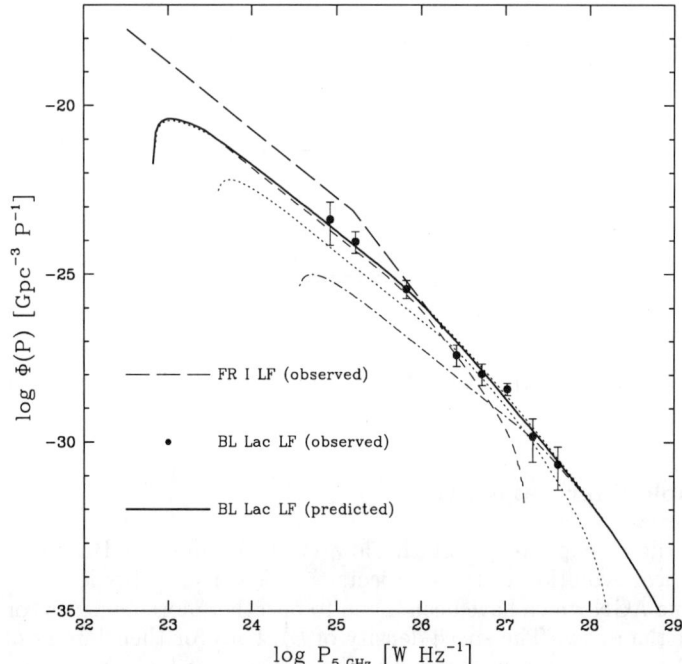

Figure 1. The observed and predicted luminosity function of BL Lacs based on the beaming model described in Urry & Padovani (1995). The contribution to the full luminosity function (solid line) coming from different beaming amplifications are shown as short-dashed ($5 < \delta < 10$), dotted ($10 < \delta < 25$) and dot-dashed thin lines ($25 < \delta < 50$).

than HBLs (Padovani & Giommi 1995). Emission line Blazars (also known as Flat Spectrum Radio Quasars FSRQ) also show a variety of ν_{peak} values but without reaching very high frequencies as BL Lacs (Padovani 2000; Perlman et al. 2000).

BL Lacs (like all Blazars) are thought to be the fraction of intrinsically much fainter sources that happen to have the beaming axis oriented close to our line of sight. In this framework the luminosity function of BL Lacs can be estimated starting from that of their parent population (Urry & Padovani 1995). Figure 1 shows the observed radio luminosity function (LF) together with that obtained by "beaming" the LF of FRI radio galaxies, the assumed parent population of BL Lacs. In the beaming scenario the high luminosity end of the beamed LF is built combining the highest luminosities in the parent population with the highest Doppler amplification factors (δ). Equation (1) then implies that, on average, brighter objects will have higher synchrotron peak frequencies (ν_{peak}). At the other end of the LF the opposite occurs, namely the low luminosity beamed objects will be characterized by lowest δ factors, and consequently by lower ν_{peak} frequencies. The contributions to the beamed LF of low, intermediate

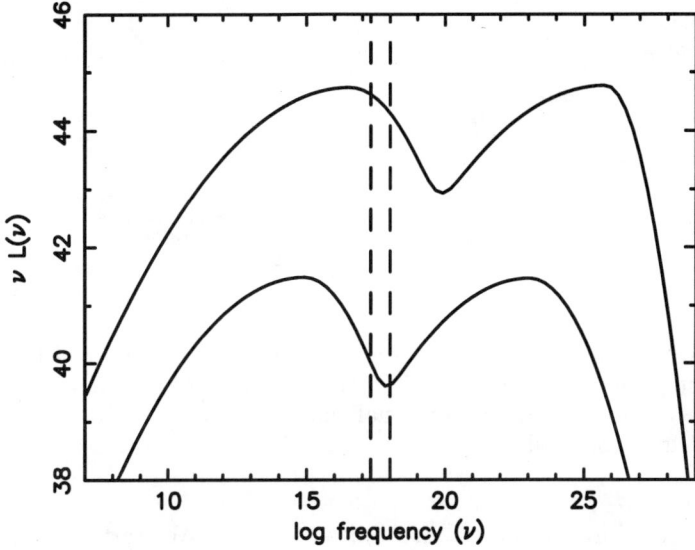

Figure 2. Spectral Energy Distributions of BL Lac objects with different ν_{peak} due to different δ factors at the low and high power ends of the radio luminosity function. The dashed lines indicate the X-ray band. Because of the different ν_{peak} energies the f_x/f_r is much higher in the high radio luminosity object.

and high Doppler factors are shown in Figure 1 as short-dashed, dotted and dashed-dotted lines respectively.

Figure 2 shows the SEDs of two sources at the bright and faint end of the radio luminosity function. Since the δ factor is on average fairly large in high luminosity objects, eq. (1) implies that also ν_{peak} is on average located at higher energies in bright sources. The brightness in the X-ray band (marked by the dashed lines) compared to radio intensity (f_x/f_r) is then larger in high radio luminosity objects than in low power radio sources. This difference favors the inclusion of powerful BL Lacs in surveys of extreme HBLs where the f_x/f_r ratio is by definition very high. Near the flux limit of a deep radio survey (where an increasing fraction of the BL Lacs comes from the low luminosity part of the LF) the number of HBLs does not grow as fast as the rest of the population because ν_{peak} gets smaller and smaller. This effect could flatten the logN–logS of extreme HBLs and bias their $\langle V/V_m \rangle$ towards low values.

5. Monte Carlo Simulations of BL Lac Surveys

In order to study in a quantitative way the effects of the δ–radio luminosity dependence described in the previous section we have carried out extensive simulations of surveys with flux limits at radio, optical and X-ray frequencies. In particular we have reproduced the major surveys listed in Table 1 and 2, and we have applied to them the same V/V_m tests that have been applied to real data.

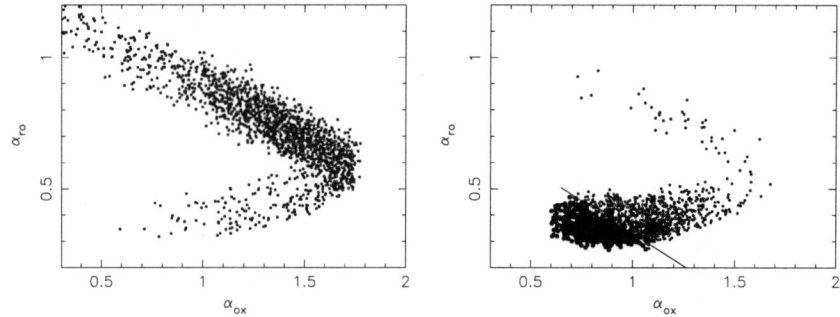

Figure 3. Left: Simulation of a radio flux limited survey of BL Lac objects; Right: Simulation of BL Lacs in the RASS survey. Filled circles to the left of the diagonal line are extreme HBLs as in the multi-frequency sedentary survey.

Our procedure can be summarized as follows. We start by simulating a redshift value assuming a standard Friedman cosmological model with $q_o = 0$ and $H_0 = 50$ km s^{-1} Mpc^{-1}; then we assign a luminosity to our simulated source by drawing it from the radio luminosity function of BL Lacs shown in Figure 1, which was obtained by beaming the luminosity function of FRI galaxies (Urry & Padovani 1995). No cosmological evolution is applied, i.e. the evolution parameter is set to 0. The source intensities at other frequencies (optical and X-ray in this case, but any other energy band can be simulated) are calculated through a synchrotron Self Compton (SSC) model that produces Spectral Energy Distributions similar to that shown in Figure 2. The simulated source is then accepted if its radio, optical and X-ray fluxes are all above the limits chosen for the simulation. In calculating the broad band spectrum from the SSC model the value of the Doppler factor (δ) (which determines the position of $\nu_{\rm peak}$ in the SED, see Figure 2) was correlated to the radio luminosity function so that high luminosity sources have higher probability to have a large δ as predicted by the beaming model (Figure 1). All other SSC parameters (maximum Lorenz factor γ, magnetic field (B), the electron spectral slope etc.) were chosen so that the prediction of the model are consistent with the observed BL Lac SEDs available in the literature.

5.1. Simulations Results

Figure 3 shows the $\alpha_{\rm ox}$ –$\alpha_{\rm ro}$ plane of a simulated radio flux limited survey with $f_r > 1$ mJy (left panel) and of an X-ray survey limited by $f_{\rm x} > 1.0 \times 10^{-12}$ erg cm^{-2} s^{-1} and by radio flux $f_{\rm r} > 2.5$ mJy (right panel). This last simulation reproduces the subset of BL Lacs in the Rosat all sky survey (RASS) that have radio flux high enough to be included in the NVSS survey (Condon et al. 1998). Thick circles to the left of the diagonal line represent those objects with $\alpha_{\rm rx} < 0.56$ and $f_r > 3.5$ mJy that make a radio flux limited sample of extreme HBLs as in the 'sedentary survey.' A sample of 2000 sources was produced in each simulation run. We have then analyzed the $V/V_{\rm m}$ distribution

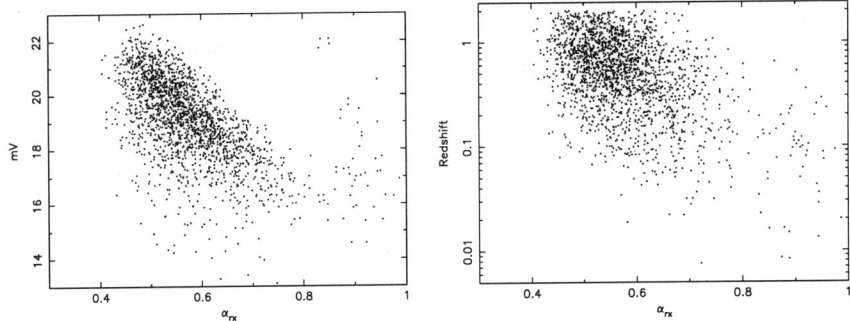

Figure 4. The optical magnitudes (left panel) and redshift (right panel) of all sources in a simulated X-ray flux limited survey are plotted against the radio to X-ray spectral index $\alpha_{\rm rx}$.

in all simulated surveys and compared the results with those of the 'sedentary' and EMSS surveys.

The results of the $V/V_{\rm m}$ analysis applied to the 'sedentary survey' give $\langle V/V_{\rm m}\rangle = 0.45 \pm 0.01$ for the case of a radio flux limit of 3.5 mJy and $\langle V/V_{\rm m}\rangle = 0.50$ for a 20 mJy flux limit. This increase of the $\langle V/V_{\rm m}\rangle$ with radio flux reproduces very well the observational results reported in Table 3 demonstrating that a dependence of the beaming Doppler factor δ on radio luminosity can induce a bias in the estimation of $\langle V/V_{\rm m}\rangle$ at low fluxes. However we note that these results can be obtained only within a narrow range of the parameters that determines how the Doppler factor δ depends on radio luminosity. This mechanism works in BL Lacs and not in emission line blazars because the distribution of synchrotron peak energies is different in the two classes, with $\nu_{\rm peak}$ reaching X-ray and possibly higher frequencies in BL Lacs (Giommi et al. 1999b) and at most the optical/UV band in other blazars (Padovani 2000).

Table 4. Simulated X-ray flux Limited Survey: $\langle V/V_{\rm m}\rangle$ Results

$\alpha_{\rm rx}$ limits	$\langle V/V_{\rm m}\rangle$	No. of objects in sample
All values	0.49(0.47)	2000(1900)
< 0.55	0.47(0.43)	852(749)
< 0.50	0.44(0.37)	348(279)
< 0.48	0.42(0.35)	204(160)

We have next considered the case of a 'classical' X-ray flux limited survey like the EMSS. To keep things simple we have chosen a single X-ray flux limit equal to 5×10^{-13} erg cm^{-2} s^{-1} rather than a sky coverage with multiple flux limits. We do not expect significant differences between the two cases. The $\langle V/V_{\rm m}\rangle$ that we obtained for the total sample is 0.49, only slightly lower than the expected value of 0.5 in case of no dependence between δ and luminosity. We have then verified whether our simulated survey shows the strong dependency of $\langle V/V_m \rangle$ on the f_x/f_r ratio (or, equivalently on α_{rx}) found by Rector et al. 2000. Table 4 shows the $\langle V/V_m \rangle$ values for increasing values of α_{rx}. A clear dependency of the $\langle V/V_m \rangle$ on α_{rx} is confirmed. This dependence, however is still present even if δ is set to be constant for all simulated sources. The decrease of $\langle V/V_m \rangle$ at high f_x/f_r values in our simulation is due to the fact that high f_x/f_r sources close to the X-ray survey limit must be weak radio sources near the low luminosity end of the radio luminosity function. At even higher f_x/f_r values the radio luminosity becomes lower than the low luminosity limit of our LF and the number of sources drops. Another, possibly more important, reason for the observed V/V_m dependency on f_x/f_r could be incompleteness of the sample at very faint optical magnitudes. In a X-ray flux limited sample high f_x/f_r (low α_{rx}) sources tend to have faint optical magnitudes as shown in the left part of Figure 4 where the α_{rx} is plotted against mV. A small incompleteness, possibly arising from the misclassification of optically very faint BL Lacs with cluster of galaxies or radio galaxies, introduces a spurious correlation between f_x/f_r and $\langle V/V_m \rangle$. We have simulated this incompleteness by removing 5% of the sources at the optically faint end of the simulated survey. The results are reported in parenthesis in Table 4 where we can see that the decrease of $\langle V/V_m \rangle$ with f_x/f_r becomes considerably higher than in the complete sample. The right plot in Figure 4 shows a clear correlation between redshift and α_{rx} showing that extreme HBL sources are expected to be on average at higher redshift that less extreme BL Lacs as found by Bade et al. (1998).

6. Conclusions

Various surveys at radio and X-ray frequencies show that the amount of cosmological evolution in BL Lacs is significantly lower than that found in other types of AGN. The results however vary from no evolution (or slightly positive evolution) at high flux values to strong negative evolution at faint fluxes, especially for extreme objects with very high f_x/f_r.

By means of extensive Monte Carlo simulations we have explored various causes that might contribute to explain these findings. The following results were obtained.

- A dependency of the beaming Doppler factor δ on radio luminosity (expected in the relativistic beaming scenario) can at least partially explain the low $\langle V/V_m \rangle$ values found in some X-ray and radio flux limited samples.

- Edge effects due to the flattening at the low luminosity end of the radio luminosity function of BL Lacs can be one of the reasons for the correlation of $\langle V/V_m \rangle$ and X-ray loudness (f_x/f_r).

- Even small amounts of incompleteness at very faint optical fluxes induce a correlation between $\langle V/V_m \rangle$ and f_x/f_r similar to that seen in the EMSS and RASS HBX Surveys.

References

Avni, Y. & Bachall, J. N. 1980, ApJ, 235, 694
Bade, N., Beckmann, V., Douglas, N. G., et al. 1998, A&A, 334, 459
Bauer, F. E., Condon, J. J., Thuan, T. X., & Broderick, J. J. 2000, ApJS, 129, 547
Blandford, R. D. & Rees, M. J. 1978, in Pittsburgh Conference on BL Lac Objects, ed. A. N. Wolfe (Pittsburgh: University of Pittsburg Press), 328
Caccianiga, A., Maccacaro, T., Wolter, A., Della Ceca, R., & Gioia, I. M. 2000, these proceedings
Condon, J. J., Cotton, W. D., Greisen, E. W., Yin, Q. F., Perley, R. A., Taylor, G. B., & Broderick, J. J. 1998, AJ, 115, 1693
D'Elia, V. & Cavaliere, A. 2000, these proceedings
Giommi, P., Menna, M. T., & Padovani, P. 1999a, MNRAS, 310, 465
Giommi, P., Ghisellini, G., Padovani, P. & Tagliaferri, G. 1999b Proc of X-Ray Astronomy 1999, Bologna, in press, astro-ph/0003021
Morris, S. L., Stocke, J. T., Gioia, I. M., Schild, R. E., Wolter, A., Della Ceca, R. 1991, ApJ, 380, 49
Maccacaro, T., et al. 1983, ApJ, 266, 73
Padovani, P. & Giommi, P. 1995, ApJ, 444, 567
Padovani, P. 2000, these proceedings
Perlman, E. et al. 2000, these proceedings
Piccinotti, G., Mushotzky, R. F., Boldt, E. A., et al. 1982, ApJ, 253, 485
Pian, E., et al. 1998 ApJ, 492, L17
Rector, T. A., Stocke, J. T., Perlman, E. S., Morris, S. L., & Gioia, I. M. 2000, AJ, 120, 1626
Schmidt, M. 1968, ApJ, 151, 393
Schwope, A., et al. 2000 Astronomische Nachrichten, 321, 1
Stocke, J., Liebert, J., Stockman, H., Danziger, J., Lub, J., Maccacaro, T., Griffiths, R., Giommi, P. 1982, MNRAS, 200, 27
Urry, C. M. & Padovani, P. 1995, PASP, 107, 803
Wolter, A., Gioia, I. M., Maccacaro, T., Morris, S. L., Stocke, J. T. 1991 ApJ, 369, 314

The Cosmological Evolution of BL Lacs: The REX Point of View

A. Caccianiga

Observatório Astronómico de Lisboa, Lisbon, Portugal

T. Maccacaro, A. Wolter, R. Della Ceca

Osservatorio Astronomico di Brera, Milan, Italy

I. M. Gioia

Istituto di Radioastronomia del CNR, Bologna, Italy

Abstract. We present the results on the cosmological evolution of BL Lac objects as derived from a statistically complete sample of 44 BL Lacs selected from the X-ray bright tail of the REX survey. With this sample, we have investigated the cosmological properties of BL Lacs taking into account the radio, optical and X-ray limits. We infer that no evolution is clearly visible down to the flux limits reached by our sample. On the other hand, deeper samples are probably needed in order to detect the negative evolution found in the EMSS sample. The identification of such deeper sample, extracted from the REX survey, is in progress.

1. Introduction

The first study of the evolutionary behavior of BL Lac objects was based on EMSS data. By studying the surface density of the BL Lacs as a function of the flux (Maccacaro et al. 1989; Wolter et al. 1991) and by applying the V_e/V_a test (Morris et al. 1991; Wolter et al. 1994), evidence for negative evolution (objects less numerous/luminous in the past) has been found. At the same time, the analysis of the 1 Jy radio selected sample of BL Lacs did not show any strong evidence for cosmological evolution (Stickel et al. 1991) in contradiction with what had been found in the X-ray band. These results were based on small samples containing 22 and 34 objects respectively. Because of the difficulty of selecting larger complete samples of BL Lacs a revision of their cosmological evolution is still a difficult task. Nevertheless, the recent availability of large radio and X-ray surveys boosted the attempts of selecting large sample of BL Lacs (e.g. the REX survey, Maccacaro et al. 1998; Caccianiga et al. 1999; the DXRBS, Perlman et al. 1998; the RGB, Laurent-Muehleisen et al. 1998). Using the ROSAT All Sky Survey, Bade et al. (1998) selected a complete sample of 33 BL Lacs with a flux limit of 8×10^{-13} ergs cm^{-2} s^{-1} in the 0.5–2.0 keV band. They have applied the $V/V_{\max}$ test to compute the cosmological evolution of BL Lacs, finding hints of negative evolution ($<V/V_{\max}> = 0.40 \pm 0.06$).

Moreover, they have found that the negative evolution depends on the optical-to-X-ray spectral index (α_{OX}) being more extreme ($<V/V_{max}>= 0.34 \pm 0.06$) for "X-ray" extreme ($\alpha_{OX} \leq 0.91$) objects. A negative evolution has been found also by Giommi, Menna & Padovani (1999) in a sample of HBLs ($\alpha_{RX} \leq 0.56$). The analysis of this sample gives evidence for a negative evolution which depends on the radio flux limit, being stronger for radio flux limit of 3.5 mJy ($<V_e/V_a>= 0.42 \pm 0.02$) and consistent with no-evolution for a flux limit of 20 mJy ($<V_e/V_a>= 0.49 \pm 0.04$). In this paper we present an independent study of the cosmological evolution of BL Lacs based on the REX survey.

2. The X-ray Bright REXs

The REX survey is the result of a positional cross-correlation between the NRAO VLA Sky Survey (NVSS, Condon et al. 1998) at 1.4 GHz and an X-ray catalogue of about 17,000 serendipitous sources detected in ~ 1200 pointed ROSAT PSPC fields. The flux limit in the radio band (at 1.4 GHz) is 5 mJy. In the X-ray band the flux limits range from $\sim 3.5 \times 10^{-14}$ erg s^{-1} cm^{-2} to $\sim 2 \times 10^{-13}$ erg s^{-1} cm^{-2} in the 0.5–2.0 keV band. The area covered at the highest flux limit is about 2200 deg^2. The cross-correlation has produced a catalogue of ~ 1600 Radio Emitting X-ray sources (REXs). The spectroscopical identification of the sample is in progress and, so far, about 36% of the sample has been identified. In this paper, we present a complete sample, the X-ray Bright REX sample (XB-REX), selected from the REX survey with the following additional criteria:
1) X-ray flux (f_X) in the 0.5–2.0 keV band $\geq 4 \times 10^{-13}$ erg s^{-1} cm^{-2};
2) Magnitude (APM O) brighter than 20.4.

The radio flux limit is the same of the whole REX survey (5 mJy). The resulting sample contains 190 objects. About 93% of these objects have been already identified, either from literature or from our own spectroscopy. The classification of a BL Lac is based on the following criteria: 1) Equivalent width of any emission lines ≤ 5 Å; 2) Ca II contrast at 4000 Å (B) $\leq 40\%$. The last criteria is an extension of the "classical" one (B $\leq 25\%$), first used by Stocke et al. (1989). There are many evidences, in fact, that the limit of 25% is too restrictive and can miss some true BL Lac objects.

In total, the XB-REX sample contains 44 BL Lacs. Given its relatively high X-ray flux limit we do not expect to significantly detect the cosmological evolution found in the EMSS sample. In fact, the X-ray flux limit of the XB-REX sample corresponds to a flux limit of 7.2×10^{-13} erg s^{-1} cm^{-2} in the 0.3–3.5 keV band (the Einstein IPC energy band) assuming $\alpha_X=1$, which is higher than the deepest flux limit of the EMSS complete sample of BL Lacs ($f_{(0.3-3.5)} = 5 \times 10^{-13}$ erg s^{-1} cm^{-2}). The flux limit of the XB-REX is in the region where the LogN-LogS produced from the EMSS survey (Wolter et al. 1991) starts to flatten and, thus, we are probably not sampling the region where the negative evolution is more evident.

3. The V_e/V_a Analysis

In order to compute the cosmological evolution of BL Lacs we have applied the $<V_e/V_a>$ method described in Avni & Bahcall (1980) where the V_a is the smallest "avaliable" volume among the V_a computed in the three selection bands (X-ray, optical, radio). For the objects without a redshift, we have assumed $z = 0.27$ (the mean value for the sample). We have also assumed $\alpha_R = 0$, $\alpha_O = 1.5$ and $\alpha_X = 1$ for the spectral indices in the radio, optical and X-ray bands, respectively. The resulting value of the $<V_e/V_a>$ is reported in Tab. 1.

Table 1. Results of the V_e/V_a analysis

Sample	N. of objects	$<V_e/V_a>$ value	Error[a]	KS prob. (%)
Total	44	0.52	0.04	94
"classical" BL	35	0.50	0.05	84
$\alpha_{OX} \leq 0.91$	16	0.52	0.07	91
$\alpha_{RX} \leq 0.62$	20	0.53	0.06	97
EL AGNs	78	0.64	0.03	2

[a] $1/\sqrt{12N}$, where N = number of objects.

The X-ray band is the limiting one in many cases (20) but also the radio and the optical bands play an important role (they are the limiting band in 13 and 11 cases, respectively). The K-S test shows that the $<V_e/V_a>$ values are uniformly distributed between 0 and 1 (probability = 94%). The result, as expected, is not significantly affected by the missing z. If we ignore the objects without z we obtain $<V_e/V_a>= 0.52 \pm 0.05$ while if we assign to all the objects without a measured z the maximum value observed in the sample ($z = 0.6$) we obtain $<V_e/V_a>= 0.54 \pm 0.04$.

We have then divided the sample in different ways and applied the V_e/V_a analysis to these sub-samples. The results are reported in Table 1. First of all, we have investigated whether the "extended" criteria used to define a BL Lac in the XB-REX sample, namely the Ca break below 40% instead of 25%, affects the result of the $<V_e/V_a>$ analysis. To this end, we have computed the $<V_e/V_a>$ by using only the "classical" BL Lacs, i.e. defined with the most restrictive limit on the Ca break (25%). The 35 "classical" BL Lacs still have a $<V_e/V_a>$ consistent with no-evolution. We have then checked if there is any dependence of the $<V_e/V_a>$ value on the "type" of BL Lac. We have thus considered only the 17 BL Lacs with an "extreme" X-ray/optical ratio ($\alpha_{OX} \leq 0.91$). According to the results found by Bade et al. (1998) these sources should show an extremely negative evolution. Instead, we do not see any evidence for negative evolution and the V_e/V_a value is still consistent with 0.5. We have also divided the sample according to the ratio between the X-ray and the radio flux and considered only the typical X-ray selected BL Lacs (HBL type) with $\alpha_{RX} \leq 0.62$. Again, the resulting $<V_e/V_a>$ is consistent with 0.5.

For comparison, we have computed the V_e/V_a for the 78 Emission Line AGNs found in the XB-REX sample. In this case, the most limiting band is the X-ray one (60 objects limited by the X-ray) while the optical and the radio bands play a marginal role. The result is fully consistent with what found in the EMSS sample ($<V_e/V_a>= 0.62 \pm 0.01$, Della Ceca et al. 1992). The distribution of the V_e/V_a values is clearly not uniform (KS probability of 2%).

4. Discussion and Conclusions

The analysis of the complete sample of 44 BL Lacs presented here shows no evidence for cosmological evolution down to the explored flux limits. Even if we restrict the analysis to the most "X-ray extreme" objects (i.e. $\alpha_{OX} \leq 0.91$ or $\alpha_{RX} \leq 0.62$), we do not find any sign of negative evolution. As stated before, the XB-REX sample is probably not deep enough to see the negative evolution detected in the EMSS survey. Instead, the RASS sample has an X-ray limit a factor 2 higher than the XB-REX sample and the negative evolution found by Bade et al. (1998) should have been detected in our analysis. In any case, a deeper sample will be instrumental to assess if a negative evolution affects the population of BL Lacs. At a flux limit 2.8×10^{-13} erg s^{-1} cm^{-2} (0.5–2.0 keV band), which corresponds to the lowest limit in the EMSS-C sample, the REX survey is identified at the 80% level, with the same radio and optical constraints of the XB-REX sample presented here. After the identification of the remaining sources is completed, we should be able to make a firmer statement on the cosmological evolution of the sample.

Acknowledgments. This research was partially supported by the European Commission, TMR Programme XCT96-0034 "CERES," by the FCT under grant PRO15132/1999, by Italian Space Agency (ASI) and the Italian MURST under grant COFIN98-02-32.

References

Avni, Y. & Bahcall, J. N. 1980, ApJ, 235, 694
Bade, N., et al. 1998, A&A, 334, 459
Caccianiga, A., et al. 1999, ApJ, 513, 51
Della Ceca, R., et al. 1992, apj, 389, 491
Giommi, P., Menna, M. T., & Padovani, P. 1999, MNRAS, 310, 465
Laurent-Muehleisen, S. A., et al. 1998, ApJS, 118, 127
Maccacaro, T., et al. 1989, in BL Lac Objects, eds. L. Maraschi, T. Maccacaro, M.-H. Ulrich (Heidelberg: Springer-Verlag), 242
Maccacaro, T., et al. 1998, Astron. Nachr., 319, 15
Morris, S. M., et al. 1991, ApJ, 380, 49
Perlman, E. S., et al. 1998, AJ, 115, 1253
Wolter, A., et al. 1991, ApJ, 369, 314
Wolter, A., et al. 1994, ApJ, 433, 29
Stickel, M. 1991, ApJ, 374, 431

Extragalactic Radio Source Evolution and Unification: Clues to the Demographics of Blazars

C. A. Jackson
Research School of Astronomy & Astrophysics, The Australian National University, Cotter Road, Weston Creek, ACT 2611, Australia

J. V. Wall
Department of Astrophysics, University of Oxford, Nuclear and Astrophysics Laboratory, Keble Road, Oxford OX1 3RH, UK

Abstract. In this paper we discuss the demographics of the radio blazar population: (i) what are their parent ('unbeamed') sources and (ii) what magnitude and/or type of evolution have they undergone? The discussion is based on models of radio source evolution and beaming based on a 'dual population' unification paradigm. These models, developed from radio blazar properties in bright samples, predict blazar demographic trends at the lower flux-density levels; samples from deep mJy-level surveys (e.g. NVSS and FIRST) may now provide direct tests of these predictions.

1. Dual-population Unification

All galaxies are radio galaxies in a sense. The scale of activity ranges from 'normal' galaxies (like our own) whose radio emission emanates from supernova remnants. At higher radio powers are the 'starburst' galaxies, greatly enhanced stellar activity resulting in stronger radio emission. At similar radio powers are the Seyfert galaxies with their small-scale radio core-jet structures and with clear indication of AGN activity. At the top end of the scale are the powerful radio galaxies, together with BL Lacs and quasars, the latter two being collectively termed blazars.

Radio galaxies tend to have steep radio spectra and therefore dominate low radio-frequency ($\nu < 0.5$ GHz) surveys. The structures are predominantly double-lobed; Fanaroff & Riley (1974) compared the distances between central maxima of the radio structures to overall size and discovered a dichotomy. Sources with regions of maximum brightness separated by more than 0.5 times the overall source size are edge-brightened (FRII), while when the regions of maximum brightness lies within this limit the source has diffuse outer lobes (FRI) (centrally concentrated); and the highest radio power objects are *predominantly* FRIIs whilst those of lower radio power are *usually* FRIs. The overlap between the classes at middling radio powers is seen in the local radio luminosity function (Figure 1), and the division is known to be a function of optical luminosity as well as radio (Owen & Ledlow 1993).

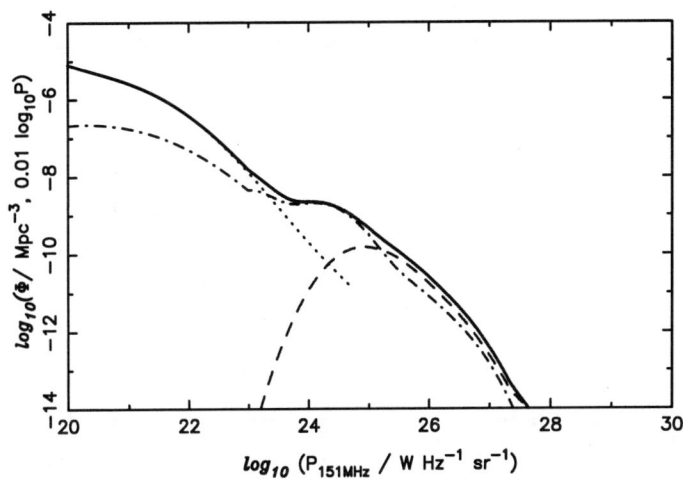

Figure 1. The local radio luminosity function at 151 MHz (dark, solid line) comprises contributions from 3 populations: starburst galaxies (dotted), FRI (dot-dash) and FRII radio galaxies (dashed).

Evidence is accumulating that the powerful radio sources, double-lobed radio galaxies on the one hand and blazars on the other, are 'unified' by projection effects, with a central torus together with relativistic beaming responsible for our classification of objects as radio galaxies or blazars (Scheuer & Readhead 1979; Orr & Browne 1982; Scheuer 1987; Barthel 1989; Urry & Padovani 1995 and references therein). The 'dual-population' unified scheme (Wall & Jackson 1997), summarized in Table 1, posits that FRII radio galaxies are the parent population of *all* quasars and *some* BL Lac type objects. The FRII-quasar sources are galaxies with class 'A' spectra (Hine & Longair 1979), having strong, high-excitation emission lines in their optical/UV spectra. The FRII-BL Lac sources are galaxies with class 'B' spectra (Hine & Longair 1979), having only weak, if any, high-excitation lines in their spectra. The second part of the scheme posits that FRI radio galaxies are the parent population of the remainder of the BL Lac type objects.

2. The Cosmic Evolution of the Parent Populations

Our analysis interprets radio source counts and identification/redshift data for samples at high flux densities, adopting a dual-population unified scheme paradigm. Using a simple parametric model to describe the evolution of the underlying parent populations we first fit low-frequency radio data ($\nu < 0.5$ GHz), where radio samples are unbiased by the effects of Doppler beaming. Then, using these evolution models for the parent populations of FRI and FRII objects, we fit the 5 GHz radio source count using a small number of parameters to describe

Table 1. Unified classes of powerful extragalactic radio sources

Population	UV/optical emission type	Radio spectrum at 5 GHz	Source type
FRII, class A	narrow	steep	RG
	broad	steep	RG
	broad	'flat'	Quasar
FRII, class B	weak/none	steep	RG
	none	'flat'	BL Lac
FRI	weak/none	steep	RG
	none	'flat'	

the Doppler beaming, which gives rise to blazars (i.e. quasars and BL Lacs with $\alpha_{2.7\,\text{GHz}}^{5\,\text{GHz}} > -0.5$, where $S \propto \nu^\alpha$)[1]

The fundamental dataset to our analysis is the differential radio source count, the count at a single observing frequency of source density on the sky as a function of flux density. Compiling data from wide-area and deep pencil-beam surveys yields source counts spanning 2 decades in frequency and 6 decades in flux density, as shown in Figure 2.

In fact radio source counts contain a wealth of cosmological information, in particular providing potent evidence against a steady-state Universe (Ryle & Clarke 1961) and indicating that differential evolution must take place, with the most luminous sources showing the strongest evolution. In addition the *shape* of the source count changes with observing frequency, suggesting that more complex behavior than just number density evolution is taking place—and our analysis shows just what this behavior is.

Identification programs and further studies of radio source samples have revealed distinct regions within the counts. These regions are indicated along the abscissa in Figure 2 and discussed briefly here:

Region 1: At the very highest flux densities the source count is near-Euclidean, although typically less than 20 sources contribute to this region. This region is comprised of a mixture of nearby sources and cosmologically-distant powerful sources. The flat count arises due to the dilution of the powerful evolving sources by the local lower-radio-power sources which are far more abundant at low redshifts.

Region 2: In this region the counts rise more rapidly with decreasing flux density than the Euclidean ($-3/2$ power law) prediction, and then reach a Euclidean plateau. This region of the count is made up of the most powerful radio sources at cosmological distances. The low-frequency counts

---

[1] Note that our definition of blazar is solely a radio definition and says nothing about the optical equivalent width criteria.

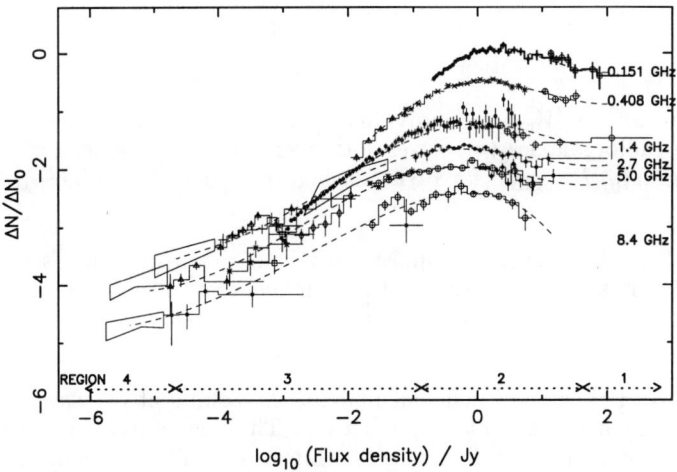

Figure 2. Source counts in relative differential form $\Delta N/\Delta N_0$ where ΔN is the number of sources per sterad with flux density S_ν between S_2 and S_1 and ΔN_0 is the number of sources expected in a uniformly-filled Euclidean universe ($N_0 = K_\nu S_\nu^{-3/2}$ with $K_\nu = 2400$, 2730, 3618, 4247, 5677 and 3738 for the six frequencies shown). The horizontal bars show the flux-density bin width, S_2 to S_1 and the vertical error is the $\sqrt{N}$ error. The curves are polynomial least-square fits to the counts. The counts are compiled from radio survey data described in Wall (1994).

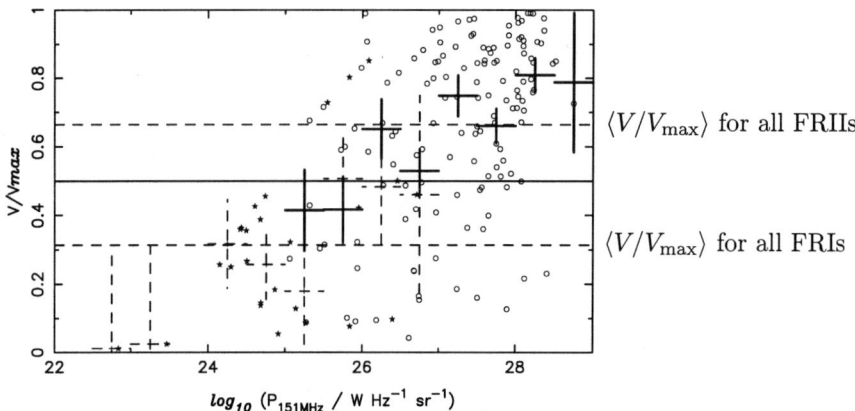

Figure 3. V/V_{max} for individual 3CRR FRI radio galaxies (*) and FRII radio galaxies (o). Overplotted are $\langle V/V_{max}\rangle$ values for bins of 0.5 in $\log_{10}(P_{151\,MHz})$ for FRIs (dashed +) and FRIIs (solid +).

(< 0.5 GHz) comprise almost entirely of steep-spectrum sources whose radio emission is greatest at these frequencies. The width of the plateau increases with increasing frequency, due to an increasing blazar (flat-spectrum) contribution.

Region 3: Below the Euclidean plateau the counts at all wavelengths drop away from the Euclidean prediction. This region extends up to two orders of magnitude in flux density. (Limited) identification data shows that the sources at such flux densities are lower-power objects at intermediate redshifts and not, as might be supposed, powerful radio sources at ever-increasing redshifts. Very little work has been done on the blazar population at these flux density levels.

Region 4: At low flux densities the counts again flatten to near-Euclidean. In this region the dominant populations change dramatically to 'blue' starburst galaxies and 'red' FRI-type galaxies, both relatively local and of low radio powers (e.g. Windhorst et al. 1993; Benn et al. 1993).

Using the spectroscopically-complete 3CRR sample at 178 MHz (Laing, Riley & Longair 1983 and more recently published data collated by R. Laing, private communication) we first determine an appropriate form for the evolution of the FRI and FRII populations using the $\langle V/V_{max}\rangle$ statistic.

Figure 3 shows V/V_{max} values for 137 steep-spectrum FRII radio galaxies and 26 FRI radio galaxies. Both populations show increasing $\langle V/V_{max}\rangle$ with radio power. However, FRIs exhibit negative evolution with $\langle V/V_{max}\rangle$ < 0.5 at $\log_{10}(P_{151\,MHz} \sim 25.0)$ and the FRIIs exhibit strong positive evolution with $\langle V/V_{max}\rangle$ > 0.7 at $\log_{10}(P_{151\,MHz} \sim 27.5)$. Interestingly, there is a trend for $\langle V/V_{max}\rangle$ to continuously increase with $\log_{10}(P_{151\,MHz})$ across the populations, suggesting luminosity-dependent evolution which may be population-independent.

Adopting exponential luminosity-dependent density evolution (LDDE) we fit the source count at 151 MHz comprising the 3CRR and 6C survey (Hales, Baldwin, & Warner 1988), using a minimization routine. The model fit requires strong evolution of the most powerful FRII sources, with little or no evolution of the FRI population (Table 2). Thus the fit reproduces the $\langle V/V_{\max} \rangle$ behavior for the radio galaxy population as a whole, in that only FRIIs with $\log_{10}(P_{151\,\mathrm{MHz}}) > 25.44$ undergo any evolution, the lower-power FRIIs and all FRIs have a constant space density to their cut-off redshift, z_c.

Table 2. Fitted evolution parameter values at 151 MHz for LDDE: $\rho(P, z) = \rho_0(P) \exp M(P)\tau(z)$

Population	Exponential LDDE parameter values	Chi-square test $\chi^2_{\min}$	ν^a
FRI	$M_{\max} = 0.0$, $z_c = 5.0$ P_1 & P_2 not used given $M_{\max} = 0.0$		
FRII, class A & B	$M_{\max} = 10.93$, $z_c = 5.62$ $P_1 = 25.44$, $P_2 = 27.34$		
	best-fit	30.73	33

[a] Degrees of freedom

3. Blazars from Beamed Parent Radio Sources

We fit the 5 GHz source count, which is well-defined over a wide flux-density range, starting from the evolution model fit of Table 2. However at 5 GHz we incorporate the beamed products (blazars) by randomly aligning the sources with respect to our line-of-sight, allowing both different Lorentz factors (γ) and different core-to-extended flux ratios for the two (FRI,FRII) populations. The best-fit beaming parameters are shown in Table 3.

Table 3. Fitted Beaming Parameter Values at 5 GHz

Population	Parameter values	Chi-square test χ^2_{min}	ν^a
FRI	$\gamma = 15.0$, $R_{\mathrm{med}} \propto P_{151\,\mathrm{MHz}}^{-0.55}$		
FRII, class A & B	$\gamma = 8.5$, $R_{\mathrm{med}} = 0.01$ $\theta_c(R_{\mathrm{med}}) = 7°.1$		
	best-fit	32.98	25

The contribution from each population to the 5-GHz source count is shown in Figure 4. The contribution from the quasar population peaks at a higher flux

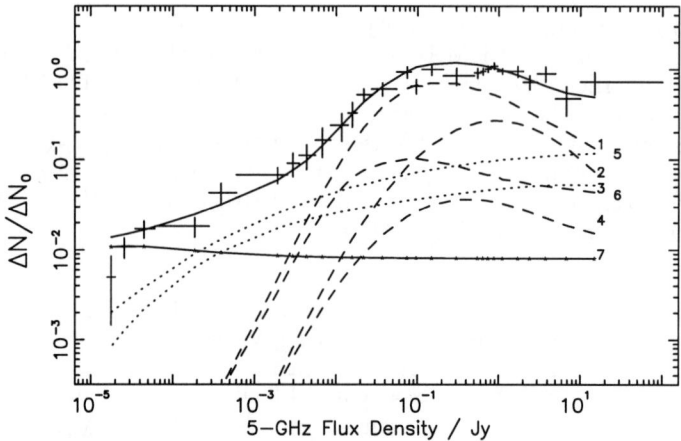

Figure 4. Observed (+) and model (solid line) differential source counts at 5 GHz. The model count comprises 4 contributions from FRII parents (dashed): 1) high-excitation radio galaxies, 2) quasars (from high-excitation parents), 3) low-excitation radio galaxies and 4) BL Lac type objects (from low-excitation parents). There are 2 contributions from FRI parents (dotted): 5) radio galaxies and 6) BL Lac type objects. The contribution from the starburst galaxy population is shown dot-dashed (7).

density limit than its parent population, and a similar behavior is seen for the BL Lac objects.

We can analyze the relative importance of each class of radio source as a function of flux density. Blazars are dominated by quasars (high-excitation FRII parents) at $S_{5\,\rm GHz} > 0.3$ Jy, but dominated by BL Lac-type objects at lower flux densities. The magnitude of the rapidly changing quasar fraction predicted by our model is well-matched by observations from the 0.25 Jy-sample of flat-spectrum sources of Shaver et al. (1996)—see Figure 15 of Jackson & Wall (1999).

In terms of population mix, it is also instructive to consider the integral population count, shown in Figure 5 in which the model fit has been transposed to 1.4 GHz. This figure reveals that complete samples are predicted to comprise $\sim 20\%$ BL Lac objects for flux density limits extending down to $S_{1.4\,\rm GHz} \sim 0.1$ mJy. Optical spectroscopy of objects selected from the FIRST survey (Becker, White, & Helfand 1995) could provide substantial tests of this prediction to $S_{1.4\,\rm GHz} \sim 1$ mJy.

4. Redshift Distributions of Blazars

Transposing our model (evolution plus beaming) to 2.7 GHz we compare our predictions with the limited observational data currently available. Figure 6 shows the mean redshift, $\langle z \rangle$, across a wide flux density limit range for all source types as well as three sub-classes. For complete samples the model predicts that

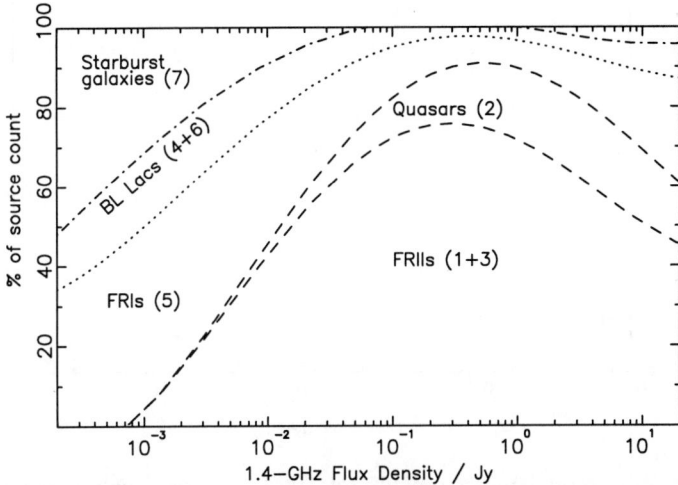

Figure 5. Integral population mix from the model fit at 5 GHz transposed to 1.4 GHz. Numbers in braces refer to the populations described in Figure 4.

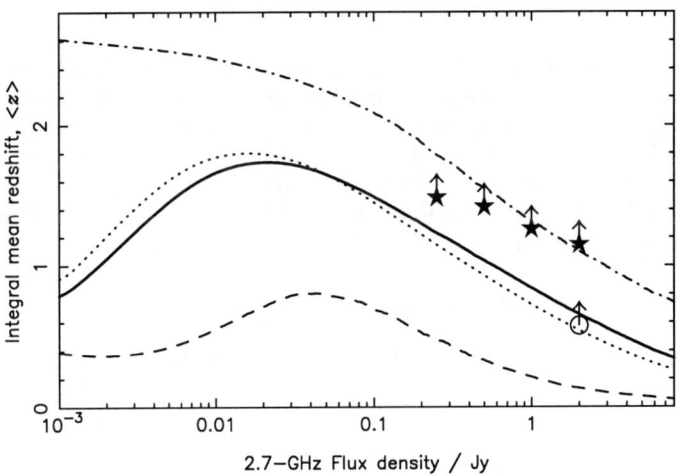

Figure 6. Model $\langle z \rangle$ of all sources to limiting flux density at 1.4 GHz (solid line). Contributions from the steep spectrum FRII, FRI and starburst galaxies (populations 1, 3, 5, and 7) shown dotted, quasars (from high-excitation FRIIs, population 2) dot-dashed and BL Lacs (from low-excitation FRIIs and FRIs, populations 4 and 6) dashed. Data points are from the flat-spectrum quasar sample ($\star$) (Shaver et al. 1996) and the 2 Jy sample (o) (Wall & Peacock 1985, Morganti et al. 1997). The solid curve shows the total.

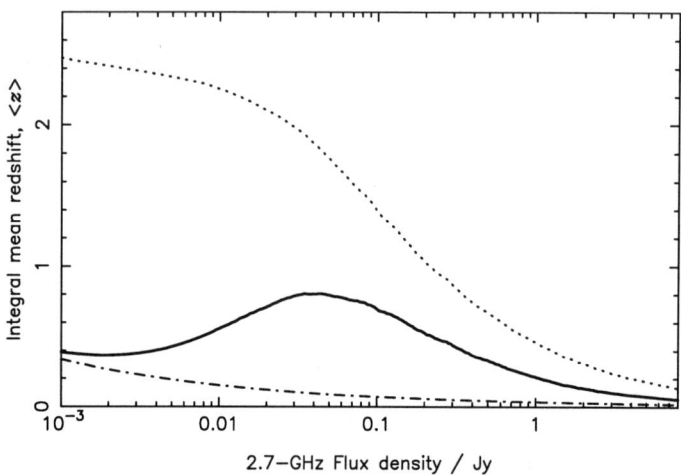

Figure 7. Model mean redshift of BL Lacs from their two parent populations—low-excitation FRIIs (population 4) dotted, FRIs (population 6) dot-dashed.

the value of $\langle z \rangle$ peaks at ~ 20 mJy then drops rapidly. In detail, we find that (i) the contribution from steep-spectrum sources dominates and therefore traces the overall $\langle z \rangle$, (ii) for the quasar population, $\langle z \rangle$ rises continuously as $S_{2.7\,\text{GHz}}$ decreases. However the *number* of quasars is negligible below $S_{2.7\,\text{GHz}} = 20$ mJy so this has little effect on the overall $\langle z \rangle$ value and (iii) the peak $\langle z \rangle$ of the BL Lac sources occurs at a higher flux density than that of the quasars, although the value of $\langle z \rangle$ is much lower.

Comparing the model $\langle z \rangle$ values to the 0.25 Jy flat-spectrum quasar sample of Shaver et al. (1996 and in preparation) we find reasonable agreement. The data points for the 0.25 Jy sample shown on Figure 6 are *lower* limits as this sample selects quasars with $(\alpha_{2.7\,\text{GHz}}^{5\,\text{GHz}} > -0.4)$. Our model, however, defines blazars as having $(\alpha_{2.7\,\text{GHz}}^{5\,\text{GHz}} > -0.5)$. Beamed objects with $\alpha_{2.7\,\text{GHz}}^{5\,\text{GHz}} > -0.4$ have parent sources of lower radio luminosity, which in turn are predicted to have a lower $\langle z \rangle$ distribution than the less-beamed $-0.5 < \alpha_{2.7\,\text{GHz}}^{5\,\text{GHz}} < -0.4$, higher-luminosity objects which are missing from the 0.25 Jy sample.

The $\langle z \rangle$ value for the complete 2-Jy sample (Wall & Peacock 1985; Morganti et al. 1997) agrees well with the overall $\langle z \rangle$ model prediction.

Figure 7 splits the BL Lac $\langle z \rangle$ distribution of Figure 6 into its contributing parent populations. We see that low-excitation-FRII BL Lac sources have a $\langle z \rangle$ distribution which is very similar to the quasar distribution. However, the FRI BL Lacs have a very different distribution, with $\langle z \rangle$ gradually increasing with decreasing flux density. Morphological studies of a sizeable sample of BL Lac objects from e.g. the FIRST survey would provide direct, powerful tests of this prediction.

References

Barthel, P. D. 1989, ApJ, 336, 606
Benn, C. R., Rowan-Robinson, M., McMahon, R. G., Broadhurst, T. J., & Lawrence, A. 1993, MNRAS, 263, 98
Fanaroff, B. L. & Riley, J. M. 1974, MNRAS, 167, 31P
Hales, S. E. G., Baldwin, J. E., & Warner, P. J. 1988, MNRAS, 234, 919
Hine, R. G. & Longair, M. S. 1979, MNRAS, 188, 111
Jackson, C. A. & Wall, J. V. 1999, MNRAS, 304, 160
Laing, R. A., Riley, J. M., & Longair, M. S. 1983, MNRAS, 204, 151
Longair, M. S. 1966, MNRAS, 133, 421
Morganti, R., Oosterloo, T. A., Reynolds, J. E., Tadhunter, C. N., & Migenes, V. 1997, MNRAS, 284, 541
Orr, M. J. L. & Browne, I. W. A. 1982, MNRAS, 200, 1067
Owen, F. N. & Ledlow, M. J. 1994, in The Physics of Active Galaxies, eds. G. V. Bicknell, et al., ASP Conf. Ser., 54, 319
Ryle, M. & Clarke, R. W. 1961, MNRAS122, 349
Scheuer, P. A. G. 1987, in Superluminal Radio Sources, eds. J. A. Zensus & T. J. Pearson, (Cambridge: Cambridge University Press), 104
Scheuer, P. A. G. & Readhead, A. C. S. 1979, Nature, 277, 182
Shaver, P. A., Wall, J. V., Kellermann, K. I., Jackson, C. A., & Hawkins, M. R. S. 1996, Nature, 384, 439
Urry, C. M. & Padovani, P. 1995, PASP, 107, 803
Wall, J. V. & Jackson, C. A. 1997, MNRAS, 290, L17
Wall, J. V. & Peacock, J. A. 1985, MNRAS, 216, 173
Windhorst, R. A., Fomalont, E. B., Partridge, R. B., & Lowenthal, J. D. 1993, ApJ, 405, 494

The Connection between BL Lacs and Flat Spectrum Radio Quasars

V. D'Elia, A. Cavaliere

Astrofisica, Dip. Fisica, Universitá Tor Vergata, Roma, I-00133

Abstract. We discuss the features that mark the Flat Spectrum Radio Quasars from the BL Lacertae objects. We propose that FSRQs exceeding $L \sim 10^{46}$ erg s^{-1} are powered by black hole accreting at rates $\dot{m} \sim 1$; their power is dominated by the disk components, thermal and BZ. Instead, sources accreting at rates $\dot{m} \sim 10^{-2} \div 10^{-3}$ radiate in the BL Lac mode; here the power is mainly non-thermal, and is driven by the rotational energy stored in a Kerr hole which sustains $L \lesssim 10^{46}$ erg s^{-1} in the jet frame for Gyrs. The two populations may be even linked if around the same objects $\dot{m}$ drops in time; then some negative cosmological evolution is expected for the newborn BL Lacs fed by the dying FSRQs. Further implications are discussed.

1. Introduction

Within the AGNs, the Blazars are singled out by their flat-spectrum GHz emission, powerful γ rays into the GeV band and beyond, rapid variability, high and variable optical polarization.

These common properties are widely explained (see Urry & Padovani 1995) in terms of beamed emissions from a relativistic jet, with Lorentz bulk factors $\Gamma \approx 5 \div 20$ (in the following, we will concentrate on the debeamed e.m. power).

Within the Blazars, the Flat Spectrum Radioloud Quasars (FSRQs) differ from the BL Lac objects on the following accounts: 1) optical features 2) integrated power 3) top spectral energies 4) cosmological evolution. The specific features of the two classes are collected in Table 1.

Table 1. Observational features of FSRQs and BL Lacs

	FSRQs	BL Lacs
optical features	em. lines, bump	no lines, no bump
integrated power	$L \sim 10^{46 \div 48}$ erg s^{-1}	$L \lesssim 10^{46}$ erg s^{-1}
evolution	strong	weak if any
top energies	$h\nu \sim 10$ GeV	$h\nu \sim 10$ TeV

These differences stand out of selections and call for explanation. Within the accreting black hole paradigm for the primary power source, we shall order all these features in a *sequence* primarily marked by one parameter, the accretion rate $\dot{m}$ (Eddington units). We propose that the FSRQs, like other quasars, accrete at rates $\dot{m} \sim 1$, while the BL Lacs radiate in conditions of $\dot{m} \ll 1$.

2. The Blazar Luminosities

We begin with the *thermal* luminosities, including the optical-UV bumps; in the standard view these are produced in the accretion *disk* and are given by $L_{th} \approx \dot{m} L_E$. Then the absence or weakness of bumps in the BL Lac spectra can be simply understood in terms of $\dot{m} \ll 1$. If so, low gas densities are also expected around the hole, and these concur to account for the other optical feature of the BL Lacs, namely, the weak or absent emission lines. The FSRQs instead, in spite of their "blazing" non-thermal component, share with other quasars the bumps and the broad emission lines; these features are consistent with values $\dot{m} \sim 1$.

As to the *jets*, we state our guideline: jets are powered by variants of the mechanism originally proposed by Blandford & Znajek (1977) for extraction of rotational energy from a Kerr hole via the Poynting-like flux associated with the surrounding magnetosphere. Variants are necessary in view of the limitations recently discussed to the power extractable from the hole. Since the BZ power scales as $B_h^2 (r_c/r_h)^2$, such variants involve either high strengths of the magnetic fields B_h threading the hole horizon at r_h, or the MHD contribution of the disk from a radius r_c larger than r_h.

Recent discussions (Modersky & Sikora 1996; Ghosh & Abramowicz 1997; Livio, Ogilvie & Pringle 1999) have stressed the continuity of B_h with the field B_d rooted in the inner stable region of the disk; in turn, B_d is bounded after $B_d/8\pi \lesssim P_{max}$ by the maximum pressure. In a standard α-disk the latter scales as $P_{max} \propto (\alpha M_9)^{-9/10} \dot{m}_{-4}^{4/5}$ if it is gas dominated, or as $P_{max} \propto (\alpha M_9)^{-1}$ if it is radiation dominated; in the latter case the *hole* power attains its maximum

$$L_K = 2\, 10^{45}\, M_9\, (J/J_{\max})^2\, \mathrm{ergs}^{-1}\, . \qquad (1)$$

But to have an inner disk region dominated by radiation pressure, accretion rates $\dot{m} \gtrsim 10^{-3}$ are required. This is because the radius r_c bounding the region grows with $\dot{m}^{16/21}$ (Novikov & Thorne 1973), and exceeds the last stable orbit only if $\dot{m} \gtrsim 10^{-3} (\alpha M_9)^{-1/8}$ holds. In such conditions, the hole output can exceed the disk thermal luminosity as given by $L_K/L_{th} = 3.3\, 10^{-2} \dot{m}^{-1} (J/J_{\max})^2$.

With $\dot{m}$ increasing, L_K saturates and its ratio to L_{th} decreases, but the radiation pressure region tends to broaden. Then a larger power component may be extracted from the disk, up to $L_d \sim L_K (r_c/r_h)^2$.

Within this framework, we draw the following implications concerning the Blazar sequence. BL Lac jets can live with accretion rates $\dot{m} \sim 10^{-2}$, since the rotational power levels L_K given by eq. (1) are often adequate even considering the kinetic power remaining in the jets (see Celotti 1999). The most powerful BL Lacs require very massive BHs and/or a larger but still comparable contribution from the dynamically entrained and magnetically connected inner rings of the disk. In all such cases one expects $L_d \sim L_K \gtrsim L_{th}$ to hold.

On the other hand, many FSRQs feature outputs exceeding 10^{46} erg s^{-1} considerably, with specific sources approaching a total of 10^{48} erg s^{-1} (Tavecchio et al. 2000). Such outputs require dominant components $L_d \sim L_{\rm th} \gg L_K$ from a wider disk region dominated by radiation pressure, and so require conditions where $\dot{m} \sim 1$. The hole contribution, though minor, is likely to provide a "high-velocity spine" instrumental for the jet propagation, see Livio 1999.

Alternatively, to account for such huge outputs one needs very strong B_h, as advocated by Meier (1999); fields up to $B_h^2/8\pi \sim \rho c^2$ in the plunging orbit region have been argued by Krolik (1999); Armitage, Reynolds, & Chiang (2000) and Paczynski (2000) discuss how and why such enhancements are unlikely inside a thin disk. In thick disks the status of such enhanced fields is still uncertain; we note they would require high $\dot{m}$ anyway.

In this section we have shown how different values of the key parameter $\dot{m}$ mark thermal and non-thermal features *together* along the Blazar sequence. We summarize our discussion by adding to the previous Table 1 the following lines:

	FSRQs	BL Lacs
Kerr hole vs. disk	$L_K \ll L_d$	$L_K \lesssim L_d$
key parameter: $\dot{m}$	$\dot{m} \sim 1$	$\dot{m} \sim 10^{-3} \div 10^{-2}$

3. The Blazar Evolutions

Strong cosmological evolution is closely shared by the FSRQs with the rest of the quasars (Goldschmidt et al. 1999), and shows up, e.g., in their steep number counts; the BL Lacs, instead, show no signs of a similar behavior (Giommi, Menna, & Padovani 1999; Padovani 2000).

The quasar behavior includes a strong component of luminosity evolution (see Boyle et al. 2000 for the optical and Della Ceca et al. 1994 for the X-ray band). This is widely traced back to the exhaustion in the host galaxy of the gas stockpile usable for accretion, due to previous accretion episodes and to star formation (Cattaneo, Haehnelt, & Rees 1999; Haehnelt & Kauffmann 2000; Cavaliere & Vittorini 2000). With the average rate $\dot{m}$ so decreasing, we expect many objects to switch from being mostly fueled by accretion to being mostly fed by the Kerr hole rotational supply E_K (stored by accretion of angular momentum J along with mass, Bardeen 1970); so we expect many sources to switch from the FSRQ to the BL Lac mode. The moderate BL Lac powers can be sustained for several Gyrs by the coupled system Kerr hole—disk, so the BL Lac luminosity evolution is expected to be slow (Cavaliere & Malquori 1999), with time scales around $\tau_L \lesssim E_K/L_K \approx 8$ Gyr.

In more detail, Cavaliere & Vittorini (2000) trace back the bright quasar evolution to the diminishing rate of the interaction episodes of the host galaxies with their neighbors in a group. These events destabilize the host gas and trigger accretion; they last some 10^{-1} Gyr, a galactic dynamical time, and produce a weak *density* evolution on a time scale $\tau_D \approx 6$ Gyr. But the efficiency of such episodes drops, due to the exaustion of the gas available in the hosts over times $\tau_L \approx 3$ Gyr; this produces a strong *luminosity* evolution.

Our point is that the powerful FSRQ activity based on high $\dot{m}$ will die out over times of some 10^{-1} Gyr after a "last interaction"; but in many instances this will leave behind a maximally spinning hole, and so a long lived BL Lac. Thus the scale τ_D for bright FSRQ deaths is also the scale for BL Lac births.

One sign of evolution is provided by the integrated counts, which at high/medium fluxes may be evaluated from the crude but explicit expression

$$N(> S) \propto S^{-3/2}[1 - C(S_0/S)^{1/2} + 0(S^{-1})] , \qquad (2)$$

with the key time scales directly appearing in the coefficient

$$C = 3 D_0 \langle l^2 \rangle [2(1 + \alpha) - 1/H_0 \tau_D - (\beta - 1)/H_0 \tau_L]/4 R_H \langle l^{3/2} \rangle . \qquad (3)$$

For the BL Lacs counted in the radio band such time scales are: $\tau_D \approx -6$ Gyr (BL Lac births imply *negative* density evolution); $\tau_L \approx 8$ Gyr (marking the slow BL Lac luminosity evolution). Other quantities involved are: the spectral index $\alpha = 0.3$ in the GHz range; the slope $\beta \approx 2.5$ of the radio LF; the normalized moments $\langle l^n \rangle$ of the LF; the distance $D_0 = (L_0/4\pi S_0)^{1/2} \approx 0.05 R_H$ in Hubble units of typical high flux BL Lacs.

For example, in the critical universe with $t_0 \approx 13$ Gyr the result is $C \approx 0.1$. This means $N(> S) \propto S^{-1.5}$ or flatter at high fluxes, consistent with the data by Giommi et al. (1999), see Fig. 1.

In contrast, the values appropriate for the FSRQs, namely: $\tau_D = +6$ Gyr, $\tau_L = 3$ Gyr, $D_0 \approx 0.5 R_H$ and still $\beta = 2.5$, yield $C \approx 4$ and produce radio counts $N(> S)$ steepening well above $S^{-1.5}$.

Note that C includes β and $\langle l^2 \rangle / \langle l^{3/2} \rangle$, which both act to flatten the counts for flatter LFs. In fact, the beaming effect (Urry & Padovani 1995) does flatten the LF at the faint end, which contributes to the flattening of the faint counts; however, the similarly affected FSRQ counts show a bright steep section indicative of intrinsically stronger evolution.

To summarize this discussion, we add to Table 1 the following line:

	FSRQs	BL Lacs
evolution	strong	weak if any

4. The Blazar Spectra: γ Rays

Another feature of the Blazars is their SED that extends into the 10 GeV range for the FSRQs, and into the TeV range for the BL Lac objects. Such high energy photons are likely produced via inverse Compton (Ghisellini 1999) by GeV and by 10^2 GeV electrons, respectively. Can the parameter $\dot{m}$ also explain this difference?

To such energies the particles may be accelerated in two ways: either by weak electric fields ($E \sim 10^{-8}$ cgs units) over large distances ($\sim 10^{16}$ cm) as in the internal shock scenario which, however, falls short of the top energies required in BL Lacs (Ghisellini 1999); or by higher fields (associated with energy transport via "Pointing flux" along the jet) effective over shorter distances.

In pursuing the latter way, we expect the force-free condition $E \bullet B = 0$ governing the BZ magnetosphere to break down at average distances $R \sim 10^{17}$ cm, but to do it inhomogeneously within bubbles or filaments; however, fields with natural values $E \sim 1$ cgs units still would be screened out over distances exceeding some $c/\omega_p \propto (\gamma/n)^{1/2}$. The densities may be estimated

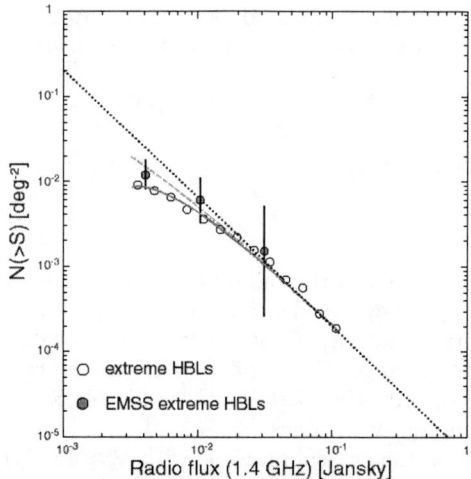

Figure 1. The BL Lac counts evaluated from eqs. (2) and (3) using $\tau_D = -7$ Gyr (dashed line) and $\tau_D = -5$ Gyr (solid line). The dotted line represents the "Euclidean" slope. Data for extreme BL Lacs from Giommi, Menna, & Padovani 1999.

from the emissions which scale as $L \sim \gamma^2 U R^3 n$ with $\gamma^2 U R^3$ roughly constant (Fossati et al. 1999; Ghisellini 1999). So in comparing BL Lacs as a class with the FSRQs, values of n smaller by 10^{-3} obtain; then the top electron (and γ-ray) energies ought to scale like $\mathcal{E}_{\max} \propto (\gamma^3 U R^3)^{1/2} L^{-1/2}$. The scatter of ν_{peak} in the BL Lac class expected from the second parameter L_d/L_K will be discussed elsewhere.

After this discussion we may add to Table 1 the line:

	FSRQs	BL Lacs
top energies	$h\nu \sim 10$ GeV	$h\nu \sim 10$ TeV

5. A Link with Accelerators of UHECR?

If we carry the Blazar sequence to its extreme, we may expect endpoint objects (dead BL Lacs) with very low residual $\dot{m} \lesssim 10^{-4}$, which would feature very faint if any e.m. emission, along with nearly unscreened electric fields. These would accelerate particles including protons, and make ultra high energy cosmic rays up to the long recognized limit given by $\mathcal{E}_{\max} \sim e B_D r_{\text{ms}} (r_{\text{ms}}/R)^p / p \sim 10^{20} M_8 B_4$ eV; after Blandford & Payne (1982), the magnetic fields are assumed to decrease outwards like $r^{-(1+p)}$ with $p = 1/4$.

Several tens of these accelerators could lie within some 100 Mpc (Boldt & Ghosh 1999). As to ultra high energies, these accelerators would evade in the simplest way the GZK cutoff (if it exists); as to accounting for the particle flux, they need to produce only $L \sim 10^{42}$ erg s^{-1}. Note that nG intergalatic fields would blur the geometrical memory of the sources for most but not all UHECRs.

6. Conclusions

The lines added to Table 1 as a summary of each of the Sects. 2–5 lead us to propose the overall Table 2:

Table 2.   The Blazar Sequence

	FSRQs	BL Lacs	$\to$ CR accelerators
optical features	em. lines, bump	no lines, no bump	none
integrated power	$L \sim 10^{47 \div 48}$ erg s^{-1}	$L \lesssim 10^{46}$ erg s^{-1}	$L \lesssim 10^{42}$ erg s^{-1}
evolution	strong	weak if any	negligible
top energies	$h\nu \sim 10$ GeV	$h\nu \sim 10$ TeV	$\mathcal{E}_{max} \sim 10^{21}$ eV
Kerr hole vs. disk	$L_K \ll L_D$	$L_K \lesssim L_D$	very low L_K and L_D
key parameter: $\dot{m}$	$\dot{m} \sim 1$	$\dot{m} \sim 10^{-2}$	$\dot{m} \lesssim 10^{-4}$

Acknowledgments. We thank P. Giommi for discussions of his data and for critical reading, and M. Salvati for helpful exchanges. Grants from ASI and MURST are acknowledged.

References

Armitage, P. J., Reynolds, C. S., & Chiang, J. 2000, astro-ph/0007042
Bardeen, J. 1970, Nature, 226, 64
Blandford, R. D. & Payne, D. G. 1982, MNRAS, 199, 883
Blandford, R. D. & Znajek, R. 1977, MNRAS, 179, 473
Boldt, E. & Ghosh, P. 1999, MNRAS, 307, 491
Boyle, B. J., et al. 2000, astro-ph/0005368
Cattaneo, A., Haehnelt, M. G., & Rees, M. J. 1999, MNRAS, 308, 77
Cavaliere, A. & Malquori, D. 1999, ApJ, 516, L9
Cavaliere, A. & Vittorini, V. 2000, astro-ph/0006194
Celotti, A. 1999, Mem SAIt, 70, 169 and references therein
Della Ceca, R., et al. 1994, ApJ, 430, 533
Fossati, G., et al. 1999, MNRAS, 299, 433
Ghisellini, G. 1999, astro-ph/9906111 and references therein
Ghosh, P. & Abramowicz, M. 1997, MNRAS, 292, 887
Giommi, P., Menna, M. T., & Padovani, P., 1999, astro-ph/9907014
Goldschmidt, P., Kukula, M. J., Miller, L., & Dunlop, J. S. 1998, ApJ, 511, 612
Kauffmann, G. & Haenelt, M. G. 2000, MNRAS, 311, 576
Livio, M. 1999, preprint *Astrophysical Jets: a Phenomenological Examination*
Livio, M., Ogilvie, G., & Pringle, J. 1998, ApJ, 512, 100
Krolik, J. H. 1999, ApJ, 515, L73
Meier, D. L. 1999, astro-ph/9908283
Modersky, R. & Sikora, M. 1996, MNRAS, 283, 854

Novikov, I. D. & Thorne, K. S. 1973 in *Black Holes*, ed. C. De Witt & D. De Witt (New York: Gordon & Breach), 343

Paczynski, B. 2000, astro-ph/0004129

Padovani, P. 2000, these proceedings

Tavecchio, F., et al. 2000, astro-ph/0006443

Urry, C. M. & Padovani, P. 1995, PASP, 107, 83

Author Index

Abraham, Z., **108**
Aller, M., 69
Aller, H., 69
Antón, S., **180**

Bailey, J. A., 140
Balonek, T., 144
Biermann, L., 56
Browne, I., 180

Caccianiga, A., 190, **238**
Cagnoni, I., **176**
Caler, M., 144
Capetti, A., 131
Carini, M. T., 144
Catalano, S., 144
Cavaliere, A., 252
Cavallotti, F., 190
Celotti, A., 50, **105**, 131, 176, 196
Chiaberge, M., **50**, 131
Chiappetti, L., 144
Costamante, L., **135**, 200
Cross, L. L., **140**

D'Alessio, F., 144
D'Elia, V., 252
Della Ceca, R., 190, 238

Edwards, P. G., 18

Falcke, H., **56**
Fodor, S., 18
Fossati, G., 135, **218**
Frasca, A., 144

Georganopoulos, M., **116**
Ghisellini, G., 50, **85**, 135, 144
Giommi, P., 73, 135, 144, 200, **227**
Gioia, I. M., 190, 238
Greyber, H. D., **158**

Heidt, J., 32
Hough, J. H., 140

Jackson, C. A., **242**
Jorstad, S. G., 10, **69**

Kataoka, J., 127
Kato, T., 144
Kirk, J. G., 116
Kurtanidze, O. M., 144

Landt, H., **73**, 200
Lister, M. L., **36**

Ma, F., **212**
Maccacaro, T., 190, 238
Maraschi, L., **40**, 122
Marchã, M. J., **208**
Markoff, S., 56
Marscher, A. P., **10**, 69
Marilli, E., 144
Massaro, E., 144
Mastichiadis, A., 116
Mattox, J., 144
Minoia, M., 190
Montagni, F., 144

Nesci, R., **144**
Nikolashvili, M. G., 144
Nilsson, K., 32
Noble, J. C., 144
Nucciarelli, G., 144

Padovani, P., **3**, 73, 135, **163**, 200, 227
Pellizzoni, A., 227
Perlman, E. S., 73, **200**
Perri, M., 227
Pesce, J. E., 122
Piner, B. G., **18**
Poccecai, D., 176
Pursimo, T., **32**

Ravasio, M., 144
Rector, T., 200

Sambruna, R., **122**
Scarpa, R., **22**, **77**, 122
Schachter, J. F., 200
Sclavi, S., 144
Sikora, M., **95**
Sillanpää, A., 32

Smith, P. S., 36
Stocke, J., **184**, 200

Tagliaferri, G., 135, 144
Takahashi, T., 127
Takalo, L. O., 32
Tanihata, C., **127**
Tavecchio, F., 40, 122
Tosti, G., 144
Tran, H., 150, 154
Tremonti, C., 144
Trussoni, E., **131**

Uemura, M., 144
Urry, C. M., 3, 77, 122, 127

Wagner, S. J., **112**
Wall, J. V., 242
Wills, B. J., 140, 150, 154, 212
Wills, D., 150, 154
Wolter, A., 135, **190**, **196**, 238

Yuan, J., **150**, **154**

ASTRONOMICAL SOCIETY OF THE PACIFIC
CONFERENCE SERIES
(ASP CS) VOLUMES

and

INTERNATIONAL ASTRONOMICAL UNION
(IAU) VOLUMES

Published
by

The Astronomical Society of the Pacific
(ASP)

ASP CONFERENCE SERIES VOLUMES
Published by the Astronomical Society of the Pacific

PUBLISHED: 1988 (* asterisk means OUT OF STOCK)

Vol. CS -1   PROGRESS AND OPPORTUNITIES IN SOUTHERN HEMISPHERE
             OPTICAL ASTRONOMY: CTIO 25TH Anniversary Symposium
             eds. V. M. Blanco and M. M. Phillips
             ISBN 0-937707-18-X

Vol. CS-2   PROCEEDINGS OF A WORKSHOP ON OPTICAL SURVEYS FOR QUASARS
             eds. Patrick S. Osmer, Alain C. Porter, Richard F. Green, and Craig B. Foltz
             ISBN 0-937707-19-8

Vol. CS-3   FIBER OPTICS IN ASTRONOMY
             ed. Samuel C. Barden
             ISBN 0-937707-20-1

Vol. CS-4   THE EXTRAGALACTIC DISTANCE SCALE:
             Proceedings of the ASP 100th Anniversary Symposium
             eds. Sidney van den Bergh and Christopher J. Pritchet
             ISBN 0-937707-21-X

Vol. CS-5   THE MINNESOTA LECTURES ON CLUSTERS OF GALAXIES
             AND LARGE-SCALE STRUCTURE
             ed. John M. Dickey
             ISBN 0-937707-22-8

PUBLISHED: 1989

Vol. CS-6   SYNTHESIS IMAGING IN RADIO ASTRONOMY: A Collection of Lectures
             from the Third NRAO Synthesis Imaging Summer School
             eds. Richard A. Perley, Frederic R. Schwab, and Alan H. Bridle
             ISBN 0-937707-23-6

Vol. CS-7   PROPERTIES OF HOT LUMINOUS STARS: Boulder-Munich Workshop
             ed. Catharine D. Garmany
             ISBN 0-937707-24-4

PUBLISHED: 1990

Vol. CS-8*   CCDs IN ASTRONOMY
             ed. George H. Jacoby
             ISBN 0-937707-25-2

Vol. CS-9   COOL STARS, STELLAR SYSTEMS, AND THE SUN: Sixth Cambridge Workshop
             ed. George Wallerstein
             ISBN 0-937707-27-9

Vol. CS-10* EVOLUTION OF THE UNIVERSE OF GALAXIES:
             Edwin Hubble Centennial Symposium
             ed. Richard G. Kron
             ISBN 0-937707-28-7

Vol. CS-11   CONFRONTATION BETWEEN STELLAR PULSATION AND EVOLUTION
             eds. Carla Cacciari and Gisella Clementini
             ISBN 0-937707-30-9

Vol. CS-12   THE EVOLUTION OF THE INTERSTELLAR MEDIUM
             ed. Leo Blitz
             ISBN 0-937707-31-7

ASP CONFERENCE SERIES VOLUMES
Published by the Astronomical Society of the Pacific

PUBLISHED: 1991 (* asterisk means OUT OF STOCK)

Vol. CS-13   THE FORMATION AND EVOLUTION OF STAR CLUSTERS
ed. Kenneth Janes
ISBN 0-937707-32-5

Vol. CS-14   ASTROPHYSICS WITH INFRARED ARRAYS
ed. Richard Elston
ISBN 0-937707-33-3

Vol. CS-15   LARGE-SCALE STRUCTURES AND PECULIAR MOTIONS IN THE UNIVERSE
eds. David W. Latham and L. A. Nicolaci da Costa
ISBN 0-937707-34-1

Vol. CS-16   Proceedings of the 3rd Haystack Observatory Conference on ATOMS, IONS, AND MOLECULES: NEW RESULTS IN SPECTRAL LINE ASTROPHYSICS
eds. Aubrey D. Haschick and Paul T. P. Ho
ISBN 0-937707-35-X

Vol. CS-17   LIGHT POLLUTION, RADIO INTERFERENCE, AND SPACE DEBRIS
ed. David L. Crawford
ISBN 0-937707-36-8

Vol. CS-18   THE INTERPRETATION OF MODERN SYNTHESIS OBSERVATIONS OF SPIRAL GALAXIES
eds. Nebojsa Duric and Patrick C. Crane
ISBN 0-937707-37-6

Vol. CS-19   RADIO INTERFEROMETRY: THEORY, TECHNIQUES, AND APPLICATIONS, IAU Colloquium 131
eds. T. J. Cornwell and R. A. Perley
ISBN 0-937707-38-4

Vol. CS-20   FRONTIERS OF STELLAR EVOLUTION:
50th Anniversary McDonald Observatory (1939-1989)
ed. David L. Lambert
ISBN 0-937707-39-2

Vol. CS-21   THE SPACE DISTRIBUTION OF QUASARS
ed . David Crampton
ISBN 0-937707-40-6

PUBLISHED: 1992

Vol. CS-22   NONISOTROPIC AND VARIABLE OUTFLOWS FROM STARS
eds. Laurent Drissen, Claus Leitherer, and Antonella Nota
ISBN 0-937707-41-4

Vol CS-23   ASTRONOMICAL CCD OBSERVING AND REDUCTION TECHNIQUES
ed. Steve B. Howell
ISBN 0-937707-42-4

Vol. CS-24   COSMOLOGY AND LARGE-SCALE STRUCTURE IN THE UNIVERSE
ed. Reinaldo R. de Carvalho
ISBN 0-937707-43-0

ASP CONFERENCE SERIES VOLUMES
Published by the Astronomical Society of the Pacific

PUBLISHED: 1992 (asterisk means OUT OF STOCK)

Vol. CS-25 ASTRONOMICAL DATA ANALYSIS, SOFTWARE AND SYSTEMS I - (ADASS I)
eds. Diana M. Worrall, Chris Biemesderfer, and Jeannette Barnes
ISBN 0-937707-44-9

Vol. CS-26 COOL STARS, STELLAR SYSTEMS, AND THE SUN:
Seventh Cambridge Workshop
eds. Mark S. Giampapa and Jay A. Bookbinder
ISBN 0-937707-45-7

Vol. CS-27 THE SOLAR CYCLE: Proceedings of the
National Solar Observatory/Sacramento Peak 12th Summer Workshop
ed. Karen L. Harvey
ISBN 0-937707-46-5

Vol. CS-28 AUTOMATED TELESCOPES FOR PHOTOMETRY AND IMAGING
eds. Saul J. Adelman, Robert J. Dukes, Jr., and Carol J. Adelman
ISBN 0-937707-47-3

Vol. CS-29 Viña del Mar Workshop on CATACLYSMIC VARIABLE STARS
ed. Nikolaus Vogt
ISBN 0-937707-48-1

Vol. CS-30 VARIABLE STARS AND GALAXIES
ed. Brian Warner
ISBN 0-937707-49-X

Vol. CS-31 RELATIONSHIPS BETWEEN ACTIVE GALACTIC NUCLEI
AND STARBURST GALAXIES
ed. Alexei V. Filippenko
ISBN 0-937707-50-3

Vol. CS-32 COMPLEMENTARY APPROACHES TO DOUBLE
AND MULTIPLE STAR RESEARCH, IAU Colloquium 135
eds. Harold A. McAlister and William I. Hartkopf
ISBN 0-937707-51-1

Vol. CS-33 RESEARCH AMATEUR ASTRONOMY
ed. Stephen J. Edberg
ISBN 0-937707-52-X

Vol. CS-34 ROBOTIC TELESCOPES IN THE 1990's
ed. Alexei V. Filippenko
ISBN 0-937707-53-8

PUBLISHED: 1993

Vol. CS-35* MASSIVE STARS: THEIR LIVES IN THE INTERSTELLAR MEDIUM
eds. Joseph P. Cassinelli and Edward B. Churchwell
ISBN 0-937707-54-6

Vol. CS-36 PLANETS AROUND PULSARS
ed. J. A. Phillips, S. E. Thorsett, and S. R. Kulkarni
ISBN 0-937707-55-4

ASP CONFERENCE SERIES VOLUMES
Published by the Astronomical Society of the Pacific

PUBLISHED: 1993 (* asterisk means OUT OF STOCK)

Vol. CS-37   FIBER OPTICS IN ASTRONOMY II
ed. Peter M. Gray
ISBN 0-937707-56-2

Vol. CS-38   NEW FRONTIERS IN BINARY STAR RESEARCH: Pacific Rim Colloquium
eds. K. C. Leung and I.-S. Nha
ISBN 0-937707-57-0

Vol. CS-39   THE MINNESOTA LECTURES ON THE STRUCTURE
AND DYNAMICS OF THE MILKY WAY
ed. Roberta M. Humphreys
ISBN 0-937707-58-9

Vol. CS-40   INSIDE THE STARS, IAU Colloquium 137
eds. Werner W. Weiss and Annie Baglin
ISBN 0-937707-59-7

Vol. CS-41   ASTRONOMICAL INFRARED SPECTROSCOPY:
FUTURE OBSERVATIONAL DIRECTIONS
ed. Sun Kwok
ISBN 0-937707-60-0

Vol. CS-42   GONG 1992: SEISMIC INVESTIGATION OF THE SUN AND STARS
ed. Timothy M. Brown
ISBN 0-937707-61-9

Vol. CS-43   SKY SURVEYS: PROTOSTARS TO PROTOGALAXIES
ed. B. T. Soifer
ISBN 0-937707-62-7

Vol. CS-44   PECULIAR VERSUS NORMAL PHENOMENA IN A-TYPE AND RELATED STARS,
IAU Colloquium 138
eds. M. M. Dworetsky, F. Castelli, and R. Faraggiana
ISBN 0-937707-63-5

Vol. CS-45   LUMINOUS HIGH-LATITUDE STARS
ed. Dimitar D. Sasselov
ISBN 0-937707-64-3

Vol. CS-46   THE MAGNETIC AND VELOCITY FIELDS OF SOLAR ACTIVE REGIONS,
IAU Colloquium 141
eds. Harold Zirin, Guoxiang Ai, and Haimin Wang
ISBN 0-937707-65-1

Vol. CS-47   THIRD DECENNIAL US-USSR CONFERENCE ON SETI --
Santa Cruz, California, USA
ed. G. Seth Shostak
ISBN 0-937707-66-X

Vol. CS-48   THE GLOBULAR CLUSTER-GALAXY CONNECTION
eds. Graeme H. Smith and Jean P. Brodie
ISBN 0-937707-67-8

Vol. CS-49   GALAXY EVOLUTION: THE MILKY WAY PERSPECTIVE
ed. Steven R. Majewski
ISBN 0-937707-68-6

ASP CONFERENCE SERIES VOLUMES
Published by the Astronomical Society of the Pacific

PUBLISHED: 1993 (* asterisk means OUT OF STOCK)

Vol. CS-50   STRUCTURE AND DYNAMICS OF GLOBULAR CLUSTERS
eds. S. G. Djorgovski and G. Meylan
ISBN 0-937707-69-4

Vol. CS-51   OBSERVATIONAL COSMOLOGY
eds. Guido Chincarini, Angela Iovino, Tommaso Maccacaro, and Dario Maccagni
ISBN 0-937707-70-8

Vol. CS-52   ASTRONOMICAL DATA ANALYSIS SOFTWARE AND SYSTEMS II - (ADASS II)
eds. R. J. Hanisch, R. J. V. Brissenden, and Jeannette Barnes
ISBN 0-937707-71-6

Vol. CS-53   BLUE STRAGGLERS
ed. Rex A. Saffer
ISBN 0-937707-72-4

PUBLISHED: 1994

Vol. CS-54*   THE FIRST STROMLO SYMPOSIUM: THE PHYSICS OF ACTIVE GALAXIES
eds. Geoffrey V. Bicknell, Michael A. Dopita, and Peter J. Quinn
ISBN 0-937707-73-2

Vol. CS-55   OPTICAL ASTRONOMY FROM THE EARTH AND MOON
eds. Diane M. Pyper and Ronald J. Angione
ISBN 0-937707-74-0

Vol. CS-56   INTERACTING BINARY STARS
ed. Allen W. Shafter
ISBN 0-937707-75-9

Vol. CS-57   STELLAR AND CIRCUMSTELLAR ASTROPHYSICS
eds. George Wallerstein and Alberto Noriega-Crespo
ISBN 0-937707-76-7

Vol. CS-58*   THE FIRST SYMPOSIUM ON THE INFRARED CIRRUS
AND DIFFUSE INTERSTELLAR CLOUDS
eds. Roc M. Cutri and William B. Latter
ISBN 0-937707-77-5

Vol. CS-59   ASTRONOMY WITH MILLIMETER AND SUBMILLIMETER WAVE
INTERFEROMETRY,
IAU Colloquium 140
eds. M. Ishiguro and Wm. J. Welch
ISBN 0-937707-78-3

Vol. CS-60   THE MK PROCESS AT 50 YEARS: A POWERFUL TOOL FOR ASTROPHYSICAL
INSIGHT, A Workshop of the Vatican Observatory --Tucson, Arizona, USA
eds. C. J. Corbally, R. O. Gray, and R. F. Garrison
ISBN 0-937707-79-1

Vol. CS-61   ASTRONOMICAL DATA ANALYSIS SOFTWARE AND SYSTEMS III - (ADASS III)
eds. Dennis R. Crabtree, R. J. Hanisch, and Jeannette Barnes
ISBN 0-937707-80-5

ASP CONFERENCE SERIES VOLUMES
Published by the Astronomical Society of the Pacific

PUBLISHED: 1994 (* asterisk means OUT OF STOCK)

Vol. CS-62 THE NATURE AND EVOLUTIONARY STATUS OF HERBIG Ae/Be STARS
eds. Pik Sin Thé, Mario R. Pérez, and Ed P. J. van den Heuvel
ISBN 0-9837707-81-3

Vol. CS-63 SEVENTY-FIVE YEARS OF HIRAYAMA ASTEROID FAMILIES:
THE ROLE OF COLLISIONS IN THE SOLAR SYSTEM HISTORY
eds. Yoshihide Kozai, Richard P. Binzel, and Tomohiro Hirayama
ISBN 0-937707-82-1

Vol. CS-64* COOL STARS, STELLAR SYSTEMS, AND THE SUN:
Eighth Cambridge Workshop
ed. Jean-Pierre Caillault
ISBN 0-937707-83-X

Vol. CS-65* CLOUDS, CORES, AND LOW MASS STARS:
The Fourth Haystack Observatory Conference
eds. Dan P. Clemens and Richard Barvainis
ISBN 0-937707-84-8

Vol. CS-66* PHYSICS OF THE GASEOUS AND STELLAR DISKS OF THE GALAXY
ed. Ivan R. King
ISBN 0-937707-85-6

Vol. CS-67 UNVEILING LARGE-SCALE STRUCTURES BEHIND THE MILKY WAY
eds. C. Balkowski and R. C. Kraan-Korteweg
ISBN 0-937707-86-4

Vol. CS-68* SOLAR ACTIVE REGION EVOLUTION:
COMPARING MODELS WITH OBSERVATIONS
eds. K. S. Balasubramaniam and George W. Simon
ISBN 0-937707-87-2

Vol. CS-69 REVERBERATION MAPPING OF THE BROAD-LINE REGION
IN ACTIVE GALACTIC NUCLEI
eds. P. M. Gondhalekar, K. Horne, and B. M. Peterson
ISBN 0-937707-88-0

PUBLISHED: 1995

Vol. CS-70* GROUPS OF GALAXIES
eds. Otto-G. Richter and Kirk Borne
ISBN 0-937707-89-9

Vol. CS-71 TRIDIMENSIONAL OPTICAL SPECTROSCOPIC METHODS IN ASTROPHYSICS,
IAU Colloquium 149
eds. Georges Comte and Michel Marcelin
ISBN 0-937707-90-2

Vol. CS-72 MILLISECOND PULSARS: A DECADE OF SURPRISE
eds. A. S Fruchter, M. Tavani, and D. C. Backer
ISBN 0-937707-91-0

Vol. CS-73 AIRBORNE ASTRONOMY SYMPOSIUM ON THE GALACTIC ECOSYSTEM:
FROM GAS TO STARS TO DUST
eds. Michael R. Haas, Jacqueline A. Davidson, and Edwin F. Erickson
ISBN 0-937707-92-9

ASP CONFERENCE SERIES VOLUMES
Published by the Astronomical Society of the Pacific

PUBLISHED: 1995 (* asterisk means OUT OF STOCK)

Vol. CS-74 PROGRESS IN THE SEARCH FOR EXTRATERRESTRIAL LIFE:
1993 Bioastronomy Symposium
ed. G. Seth Shostak
ISBN 0-937707-93-7

Vol. CS-75 MULTI-FEED SYSTEMS FOR RADIO TELESCOPES
eds. Darrel T. Emerson and John M. Payne
ISBN 0-937707-94-5

Vol. CS-76 GONG '94: HELIO- AND ASTERO-SEISMOLOGY FROM THE EARTH
AND SPACE
eds. Roger K. Ulrich, Edward J. Rhodes, Jr., and Werner Däppen
ISBN 0-937707-95-3

Vol. CS-77 ASTRONOMICAL DATA ANALYSIS SOFTWARE AND SYSTEMS IV - (ADASS IV)
eds. R. A. Shaw, H. E. Payne, and J. J. E. Hayes
ISBN 0-937707-96-1

Vol. CS-78 ASTROPHYSICAL APPLICATIONS OF POWERFUL NEW DATABASES:
Joint Discussion No. 16 of the 22nd General Assembly of the IAU
eds. S. J. Adelman and W. L. Wiese
ISBN 0-937707-97-X

Vol. CS-79* ROBOTIC TELESCOPES: CURRENT CAPABILITIES, PRESENT
DEVELOPMENTS, AND FUTURE PROSPECTS
FOR AUTOMATED ASTRONOMY
eds. Gregory W. Henry and Joel A. Eaton
ISBN 0-937707-98-8

Vol. CS-80* THE PHYSICS OF THE INTERSTELLAR MEDIUM
AND INTERGALACTIC MEDIUM
eds. A. Ferrara, C. F. McKee, C. Heiles, and P. R. Shapiro
ISBN 0-937707-99-6

Vol. CS-81 LABORATORY AND ASTRONOMICAL HIGH RESOLUTION SPECTRA
eds. A. J. Sauval, R. Blomme, and N. Grevesse
ISBN 1-886733-01-5

Vol. CS-82* VERY LONG BASELINE INTERFEROMETRY AND THE VLBA
eds. J. A. Zensus, P. J. Diamond, and P. J. Napier
ISBN 1-886733-02-3

Vol. CS-83* ASTROPHYSICAL APPLICATIONS OF STELLAR PULSATION,
IAU Colloquium 155
eds. R. S. Stobie and P. A. Whitelock
ISBN 1-886733-03-1

ATLAS INFRARED ATLAS OF THE ARCTURUS SPECTRUM, 0.9 - 5.3 μm
eds. Kenneth Hinkle, Lloyd Wallace, and William Livingston
ISBN: 1-886733-04-X

Vol. CS-84 THE FUTURE UTILIZATION OF SCHMIDT TELESCOPES, IAU Colloquium 148
eds. Jessica Chapman, Russell Cannon, Sandra Harrison, and Bambang Hidayat
ISBN 1-886733-05-8

ASP CONFERENCE SERIES VOLUMES
Published by the Astronomical Society of the Pacific

PUBLISHED: 1995 (* asterisk means OUT OF STOCK)

Vol. CS-85* CAPE WORKSHOP ON MAGNETIC CATACLYSMIC VARIABLES
eds. D. A. H. Buckley and B. Warner
ISBN 1-886733-06-6

Vol. CS-86 FRESH VIEWS OF ELLIPTICAL GALAXIES
eds. Alberto Buzzoni, Alvio Renzini, and Alfonso Serrano
ISBN 1-886733-07-4

PUBLISHED: 1996

Vol. CS-87 NEW OBSERVING MODES FOR THE NEXT CENTURY
eds. Todd Boroson, John Davies, and Ian Robson
ISBN 1-886733-08-2

Vol. CS-88* CLUSTERS, LENSING, AND THE FUTURE OF THE UNIVERSE
eds. Virginia Trimble and Andreas Reisenegger
ISBN 1-886733-09-0

Vol. CS-89 ASTRONOMY EDUCATION: CURRENT DEVELOPMENTS, FUTURE COORDINATION
ed. John R. Percy
ISBN 1-886733-10-4

Vol. CS-90 THE ORIGINS, EVOLUTION, AND DESTINIES OF BINARY STARS IN CLUSTERS
eds. E. F. Milone and J. -C. Mermilliod
ISBN 1-886733-11-2

Vol. CS-91 BARRED GALAXIES, IAU Colloquium 157
eds. R. Buta, D. A. Crocker, and B. G. Elmegreen
ISBN 1-886733-12-0

Vol. CS-92* FORMATION OF THE GALACTIC HALO .... INSIDE AND OUT
eds. Heather L. Morrison and Ata Sarajedini
ISBN 1-886733-13-9

Vol. CS-93 RADIO EMISSION FROM THE STARS AND THE SUN
eds. A. R. Taylor and J. M. Paredes
ISBN 1-886733-14-7

Vol. CS-94 MAPPING, MEASURING, AND MODELING THE UNIVERSE
eds. Peter Coles, Vicent J. Martinez, and Maria-Jesus Pons-Borderia
ISBN 1-886733-15-5

Vol. CS-95 SOLAR DRIVERS OF INTERPLANETARY AND TERRESTRIAL DISTURBANCES:
Proceedings of 16th International Workshop National Solar Observatory/Sacramento Peak
eds. K. S. Balasubramaniam, Stephen L. Keil, and Raymond N. Smartt
ISBN 1-886733-16-3

Vol. CS-96 HYDROGEN-DEFICIENT STARS
eds. C. S. Jeffery and U. Heber
ISBN 1-886733-17-1

ASP CONFERENCE SERIES VOLUMES
Published by the Astronomical Society of the Pacific

PUBLISHED: 1996 (* asterisk means OUT OF STOCK)

Vol. CS-97 POLARIMETRY OF THE INTERSTELLAR MEDIUM
eds. W. G. Roberge and D. C. B. Whittet
ISBN 1-886733-18-X

Vol. CS-98 FROM STARS TO GALAXIES: THE IMPACT OF STELLAR PHYSICS ON GALAXY EVOLUTION
eds. Claus Leitherer, Uta Fritze-von Alvensleben, and John Huchra
ISBN 1-886733-19-8

Vol. CS-99 COSMIC ABUNDANCES:
Proceedings of the 6th Annual October Astrophysics Conference
eds. Stephen S. Holt and George Sonneborn
ISBN 1-886733-20-1

Vol. CS-100 ENERGY TRANSPORT IN RADIO GALAXIES AND QUASARS
eds. P. E. Hardee, A. H. Bridle, and J. A. Zensus
ISBN 1-886733-21-X

Vol. CS-101 ASTRONOMICAL DATA ANALYSIS SOFTWARE AND SYSTEMS V – (ADASS V)
eds. George H. Jacoby and Jeannette Barnes
ISBN 1080-7926

Vol. CS-102 THE GALACTIC CENTER, 4th ESO/CTIO Workshop
ed. Roland Gredel
ISBN 1-886733-22-8

Vol. CS-103 THE PHYSICS OF LINERS IN VIEW OF RECENT OBSERVATIONS
eds. M. Eracleous, A. Koratkar, C. Leitherer, and L. Ho
ISBN 1-886733-23-6

Vol. CS-104 PHYSICS, CHEMISTRY, AND DYNAMICS OF INTERPLANETARY DUST, IAU Colloquium 150
eds. Bo Å. S. Gustafson and Martha S. Hanner
ISBN 1-886733-24-4

Vol. CS-105 PULSARS: PROBLEMS AND PROGRESS, IAU Colloquium 160
ed. S. Johnston, M. A. Walker, and M. Bailes
ISBN 1-886733-25-2

Vol. CS-106 THE MINNESOTA LECTURES ON EXTRAGALACTIC NEUTRAL HYDROGEN
ed. Evan D. Skillman
ISBN 1-886733-26-0

Vol. CS-107 COMPLETING THE INVENTORY OF THE SOLAR SYSTEM:
A Symposium held in conjunction with the 106th Annual Meeting of the ASP
eds. Terrence W. Rettig and Joseph M. Hahn
ISBN 1-886733-27-9

Vol. CS-108 M.A.S.S. -- MODEL ATMOSPHERES AND SPECTRUM SYNTHESIS:
5th Vienna - Workshop
eds. Saul J. Adelman, Friedrich Kupka, and Werner W. Weiss
ISBN 1-886733-28-7

Vol. CS-109 COOL STARS, STELLAR SYSTEMS, AND THE SUN: Ninth Cambridge Workshop
eds. Roberto Pallavicini and Andrea K. Dupree
ISBN 1-886733-29-5

ASP CONFERENCE SERIES VOLUMES
Published by the Astronomical Society of the Pacific

PUBLISHED: 1996 (* asterisk means OUT OF STOCK)

Vol. CS-110   BLAZAR CONTINUUM VARIABILITY
eds. H. R. Miller, J. R. Webb, and J. C. Noble
ISBN 1-886733-30-9

Vol. CS-111   MAGNETIC RECONNECTION IN THE SOLAR ATMOSPHERE:
Proceedings of a Yohkoh Conference
eds. R. D. Bentley and J. T. Mariska
ISBN 1-886733-31-7

Vol. CS-112   THE HISTORY OF THE MILKY WAY AND ITS SATELLITE SYSTEM
eds. Andreas Burkert, Dieter H. Hartmann, and Steven R. Majewski
ISBN 1-886733-32-5

PUBLISHED: 1997

Vol. CS-113   EMISSION LINES IN ACTIVE GALAXIES: NEW METHODS AND TECHNIQUES,
IAU Colloquium 159
eds. B. M. Peterson, F.-Z. Cheng, and A. S. Wilson
ISBN 1-886733-33-3

Vol. CS-114   YOUNG GALAXIES AND QSO ABSORPTION-LINE SYSTEMS
eds. Sueli M. Viegas, Ruth Gruenwald, and Reinaldo R. de Carvalho
ISBN 1-886733-34-1

Vol. CS-115   GALACTIC CLUSTER COOLING FLOWS
ed. Noam Soker
ISBN 1-886733-35-X

Vol. CS-116   THE SECOND STROMLO SYMPOSIUM:
THE NATURE OF ELLIPTICAL GALAXIES
eds. M. Arnaboldi, G. S. Da Costa, and P. Saha
ISBN 1-886733-36-8

Vol. CS-117   DARK AND VISIBLE MATTER IN GALAXIES
eds. Massimo Persic and Paolo Salucci
ISBN-1-886733-37-6

Vol. CS-118   FIRST ADVANCES IN SOLAR PHYSICS EUROCONFERENCE:
ADVANCES IN THE PHYSICS OF SUNSPOTS
eds. B. Schmieder. J. C. del Toro Iniesta, and M. Vázquez
ISBN 1-886733-38-4

Vol. CS-119   PLANETS BEYOND THE SOLAR SYSTEM
AND THE NEXT GENERATION OF SPACE MISSIONS
ed. David R. Soderblom
ISBN 1-886733-39-2

Vol. CS-120   LUMINOUS BLUE VARIABLES: MASSIVE STARS IN TRANSITION
eds. Antonella Nota and Henny J. G. L. M. Lamers
ISBN 1-886733-40-6

Vol. CS-121   ACCRETION PHENOMENA AND RELATED OUTFLOWS, IAU Colloquium 163
eds. D. T. Wickramasinghe, G. V. Bicknell, and L. Ferrario
ISBN 1-886733-41-4

ASP CONFERENCE SERIES VOLUMES
Published by the Astronomical Society of the Pacific

PUBLISHED: 1997 (* asterisk means OUT OF STOCK)

Vol. CS-122   FROM STARDUST TO PLANETESIMALS:
              Symposium held as part of the 108th Annual Meeting of the ASP
              eds. Yvonne J. Pendleton and A. G. G. M. Tielens
              ISBN 1-886733-42-2

Vol. CS-123   THE 12th 'KINGSTON MEETING': COMPUTATIONAL ASTROPHYSICS
              eds. David A. Clarke and Michael J. West
              ISBN 1-886733-43-0

Vol. CS-124   DIFFUSE INFRARED RADIATION AND THE IRTS
              eds. Haruyuki Okuda, Toshio Matsumoto, and Thomas Roellig
              ISBN 1-886733-44-9

Vol. CS-125   ASTRONOMICAL DATA ANALYSIS SOFTWARE AND SYSTEMS VI – (ADASS VI)
              eds. Gareth Hunt and H. E. Payne
              ISBN 1-886733-45-7

Vol. CS-126   FROM QUANTUM FLUCTUATIONS TO COSMOLOGICAL STRUCTURES
              eds. David Valls-Gabaud, Martin A. Hendry, Paolo Molaro, and Khalil Chamcham
              ISBN 1-886733-46-5

Vol. CS-127   PROPER MOTIONS AND GALACTIC ASTRONOMY
              ed. Roberta M. Humphreys
              ISBN 1-886733-47-3

Vol. CS-128   MASS EJECTION FROM AGN (Active Galactic Nuclei)
              eds. N. Arav, I. Shlosman, and R. J. Weymann
              ISBN 1-886733-48-1

Vol. CS-129   THE GEORGE GAMOW SYMPOSIUM
              eds. E. Harper, W. C. Parke, and G. D. Anderson
              ISBN 1-886733-49-X

Vol. CS-130   THE THIRD PACIFIC RIM CONFERENCE ON
              RECENT DEVELOPMENT ON BINARY STAR RESEARCH
              eds. Kam-Ching Leung
              ISBN 1-886733-50-3

PUBLISHED: 1998

Vol. CS-131   BOULDER-MUNICH II: PROPERTIES OF HOT, LUMINOUS STARS
              ed. Ian D. Howarth
              ISBN 1-886733-51-1

Vol. CS-132   STAR FORMATION WITH THE INFRARED SPACE OBSERVATORY (ISO)
              eds. João L. Yun and René Liseau
              ISBN 1-886733-52-X

Vol. CS-133   SCIENCE WITH THE NGST (Next Generation Space Telescope)
              eds. Eric P. Smith and Anuradha Koratkar
              ISBN 1-886733-53-8

Vol. CS-134   BROWN DWARFS AND EXTRASOLAR PLANETS
              eds. Rafael Rebolo, Eduardo L. Martin, and Maria Rosa Zapatero Osorio
              ISBN 1-886733-54-6

ASP CONFERENCE SERIES VOLUMES
Published by the Astronomical Society of the Pacific

PUBLISHED: 1998 (* asterisk means OUT OF STOCK)

Vol. CS-135   A HALF CENTURY OF STELLAR PULSATION INTERPRETATIONS:
A TRIBUTE TO ARTHUR N. COX
eds. P. A. Bradley and J. A. Guzik
ISBN 1-886733-55-4

Vol. CS-136   GALACTIC HALOS: A UC SANTA CRUZ WORKSHOP
ed. Dennis Zaritsky
ISBN 1-886733-56-2

Vol. CS-137   WILD STARS IN THE OLD WEST: PROCEEDINGS OF THE 13th NORTH
AMERICAN WORKSHOP ON CATACLYSMIC VARIABLES
AND RELATED OBJECTS
eds. S. Howell, E. Kuulkers, and C. Woodward
ISBN 1-886733-57-0

Vol. CS-138   1997 PACIFIC RIM CONFERENCE ON STELLAR ASTROPHYSICS
eds. Kwing Lam Chan, K. S. Cheng, and H. P. Singh
ISBN 1-886733-58-9

Vol. CS-139   PRESERVING THE ASTRONOMICAL WINDOWS:
Proceedings of Joint Discussion No. 5 of the 23rd General Assembly of the IAU
eds. Syuzo Isobe and Tomohiro Hirayama
ISBN 1-886733-59-7

Vol. CS-140   SYNOPTIC SOLAR PHYSICS --18th NSO/Sacramento Peak Summer Workshop
eds. K. S. Balasubramaniam, J. W. Harvey, and D. M. Rabin
ISBN 1-886733-60-0

Vol. CS-141   ASTROPHYSICS FROM ANTARCTICA:
A Symposium held as a part of the 109th Annual Meeting of the ASP
eds. Giles Novak and Randall H. Landsberg
ISBN 1-886733-61-9

Vol. CS-142   THE STELLAR INITIAL MASS FUNCTION: 38th Herstmonceux Conference
eds. Gerry Gilmore and Debbie Howell
ISBN 1-886733-62-7

Vol. CS-143*   THE SCIENTIFIC IMPACT OF THE GODDARD HIGH RESOLUTION
SPECTROGRAPH (GHRS)
eds. John C. Brandt, Thomas B. Ake III, and Carolyn Collins Petersen
ISBN 1-886733-63-5

Vol. CS-144   RADIO EMISSION FROM GALACTIC AND EXTRAGALACTIC COMPACT
SOURCES, IAU Colloquium 164
eds. J. Anton Zensus, G. B. Taylor, and J. M. Wrobel
ISBN 1-886733-64-3

Vol. CS-145   ASTRONOMICAL DATA ANALYSIS SOFTWARE AND SYSTEMS VII – (ADASS VII)
eds. Rudolf Albrecht, Richard N. Hook, and Howard A. Bushouse
ISBN 1-886733-65-1

Vol. CS-146   THE YOUNG UNIVERSE GALAXY FORMATION
AND EVOLUTION AT INTERMEDIATE AND HIGH REDSHIFT
eds. S. D'Odorico, A. Fontana, and E. Giallongo
ISBN 1-886733-66-X

ASP CONFERENCE SERIES VOLUMES
Published by the Astronomical Society of the Pacific

PUBLISHED: 1998 (* asterisk means OUT OF STOCK)

Vol. CS-147 ABUNDANCE PROFILES: DIAGNOSTIC TOOLS FOR GALAXY HISTORY
eds. Daniel Friedli, Mike Edmunds, Carmelle Robert, and Laurent Drissen
ISBN 1-886733-67-8

Vol. CS-148 ORIGINS
eds. Charles E. Woodward, J. Michael Shull, and Harley A. Thronson, Jr.
ISBN 1-886733-68-6

Vol. CS-149 SOLAR SYSTEM FORMATION AND EVOLUTION
eds. D. Lazzaro, R. Vieira Martins, S. Ferraz-Mello, J. Fernández, and C. Beaugé
ISBN 1-886733-69-4

Vol. CS-150 NEW PERSPECTIVES ON SOLAR PROMINENCES, IAU Colloquium 167
eds. David Webb, David Rust, and Brigitte Schmieder
ISBN 1-886733-70-8

Vol. CS-151 COSMIC MICROWAVE BACKGROUND
AND LARGE SCALE STRUCTURES OF THE UNIVERSE
eds. Yong-Ik Byun and Kin-Wang Ng
ISBN 1-886733-71-6

Vol. CS-152 FIBER OPTICS IN ASTRONOMY III
eds. S. Arribas, E. Mediavilla, and F. Watson
ISBN 1-886733-72-4

Vol. CS-153 LIBRARY AND INFORMATION SERVICES IN ASTRONOMY III -- (LISA III)
eds. Uta Grothkopf, Heinz Andernach, Sarah Stevens-Rayburn,
and Monique Gomez
ISBN 1-886733-73-2

Vol. CS-154 COOL STARS, STELLAR SYSTEMS AND THE SUN: Tenth Cambridge Workshop
eds. Robert A. Donahue and Jay A. Bookbinder
ISBN 1-886733-74-0

Vol. CS-155 SECOND ADVANCES IN SOLAR PHYSICS EUROCONFERENCE:
THREE-DIMENSIONAL STRUCTURE OF SOLAR ACTIVE REGIONS
eds. Costas E. Alissandrakis and Brigitte Schmieder
ISBN 1-886733-75-9

PUBLISHED: 1999

Vol. CS-156 HIGHLY REDSHIFTED RADIO LINES
eds. C. L. Carilli, S. J. E. Radford, K. M. Menten, and G. I. Langston
ISBN 1-886733-76-7

Vol. CS-157 ANNAPOLIS WORKSHOP ON MAGNETIC CATACLYSMIC VARIABLES
eds. Coel Hellier and Koji Mukai
ISBN 1-886733-77-5

Vol. CS-158 SOLAR AND STELLAR ACTIVITY: SIMILARITIES AND DIFFERENCES
eds. C. J. Butler and J. G. Doyle
ISBN 1-886733-78-3

Vol. CS-159 BL LAC PHENOMENON
eds. Leo O. Takalo and Aimo Sillanpää
ISBN 1-886733-79-1

ASP CONFERENCE SERIES VOLUMES
Published by the Astronomical Society of the Pacific

PUBLISHED: 1999 (* asterisk means OUT OF STOCK)

Vol. CS-160   ASTROPHYSICAL DISCS: An EC Summer School
eds. J. A. Sellwood and Jeremy Goodman
ISBN 1-886733-80-5

Vol. CS-161   HIGH ENERGY PROCESSES IN ACCRETING BLACK HOLES
eds. Juri Poutanen and Roland Svensson
ISBN 1-886733-81-3

Vol. CS-162   QUASARS AND COSMOLOGY
eds. Gary Ferland and Jack Baldwin
ISBN 1-886733-83-X

Vol. CS-163   STAR FORMATION IN EARLY-TYPE GALAXIES
eds. Jordi Cepa and Patricia Carral
ISBN 1-886733-84-8

Vol. CS-164   ULTRAVIOLET–OPTICAL SPACE ASTRONOMY BEYOND HST
eds. Jon A. Morse, J. Michael Shull, and Anne L. Kinney
ISBN 1-886733-85-6

Vol. CS-165   THE THIRD STROMLO SYMPOSIUM: THE GALACTIC HALO
eds. Brad K. Gibson, Tim S. Axelrod, and Mary E. Putman
ISBN 1-886733-86-4

Vol. CS-166   STROMLO WORKSHOP ON HIGH-VELOCITY CLOUDS
eds. Brad K. Gibson and Mary E. Putman
ISBN 1-886733-87-2

Vol. CS-167   HARMONIZING COSMIC DISTANCE SCALES IN A POST-HIPPARCOS ERA
eds. Daniel Egret and André Heck
ISBN 1-886733-88-0

Vol. CS-168   NEW PERSPECTIVES ON THE INTERSTELLAR MEDIUM
eds. A. R. Taylor, T. L. Landecker, and G. Joncas
ISBN 1-886733-89-9

Vol. CS-169   11th EUROPEAN WORKSHOP ON WHITE DWARFS
eds. J.-E. Solheim and E. G. Meištas
ISBN 1-886733-91-0

Vol. CS-170   THE LOW SURFACE BRIGHTNESS UNIVERSE, IAU Colloquium 171
eds. J. I. Davies, C. Impey, and S. Phillipps
ISBN 1-886733-92-9

Vol. CS-171   LiBeB, COSMIC RAYS, AND RELATED X- AND GAMMA-RAYS
eds. Reuven Ramaty, Elisabeth Vangioni-Flam, Michel Cassé, and Keith Olive
ISBN 1-886733-93-7

Vol. CS-172   ASTRONOMICAL DATA ANALYSIS SOFTWARE AND SYSTEMS VIII – (ADASS VIII)
eds. David M. Mehringer, Raymond L. Plante, and Douglas A. Roberts
ISBN 1-886733-94-5

Vol. CS-173   THEORY AND TESTS OF CONVECTION IN STELLAR STRUCTURE:
First Granada Workshop
ed. Álvaro Giménez, Edward F. Guinan, and Benjamín Montesinos
ISBN 1-886733-95-3

ASP CONFERENCE SERIES VOLUMES
Published by the Astronomical Society of the Pacific

PUBLISHED: 1999 (* asterisk means OUT OF STOCK)

Vol. CS-174   CATCHING THE PERFECT WAVE: ADAPTIVE OPTICS AND INTERFEROMETRY IN THE 21st CENTURY,
A Symposium held as a part of the 110th Annual Meeting of the ASP
eds. Sergio R. Restaino, William Junor, and Nebojsa Duric
ISBN 1-886733-96-1

Vol. CS-175   STRUCTURE AND KINEMATICS OF QUASAR BROAD LINE REGIONS
eds. C. M. Gaskell, W. N. Brandt, M. Dietrich, D. Dultzin-Hacyan, and M. Eracleous
ISBN 1-886733-97-X

Vol. CS-176   OBSERVATIONAL COSMOLOGY: THE DEVELOPMENT OF GALAXY SYSTEMS
eds. Giuliano Giuricin, Marino Mezzetti, and Paolo Salucci
ISBN 1-58381-000-5

Vol. CS-177   ASTROPHYSICS WITH INFRARED SURVEYS: A Prelude to SIRTF
eds. Michael D. Bicay, Chas A. Beichman, Roc M. Cutri, and Barry F. Madore
ISBN 1-58381-001-3

Vol. CS-178   STELLAR DYNAMOS: NONLINEARITY AND CHAOTIC FLOWS
eds. Manuel Núñez and Antonio Ferriz-Mas
ISBN 1-58381-002-1

Vol. CS-179   ETA CARINAE AT THE MILLENNIUM
eds. Jon A. Morse, Roberta M. Humphreys, and Augusto Damineli
ISBN 1-58381-003-X

Vol. CS-180   SYNTHESIS IMAGING IN RADIO ASTRONOMY II
eds. G. B. Taylor, C. L. Carilli, and R. A. Perley
ISBN 1-58381-005-6

Vol. CS-181   MICROWAVE FOREGROUNDS
eds. Angelica de Oliveira-Costa and Max Tegmark
ISBN 1-58381-006-4

Vol. CS-182   GALAXY DYNAMICS: A Rutgers Symposium
eds. David Merritt, J. A. Sellwood, and Monica Valluri
ISBN 1-58381-007-2

Vol. CS-183   HIGH RESOLUTION SOLAR PHYSICS: THEORY, OBSERVATIONS, AND TECHNIQUES
eds. T. R. Rimmele, K. S. Balasubramaniam, and R. R. Radick
ISBN 1-58381-009-9

Vol. CS-184   THIRD ADVANCES IN SOLAR PHYSICS EUROCONFERENCE: MAGNETIC FIELDS AND OSCILLATIONS
eds. B. Schmieder, A. Hofmann, and J. Staude
ISBN 1-58381-010-2

Vol. CS-185   PRECISE STELLAR RADIAL VELOCITIES, IAU Colloquium 170
eds. J. B. Hearnshaw and C. D. Scarfe
ISBN 1-58381-011-0

ASP CONFERENCE SERIES VOLUMES
Published by the Astronomical Society of the Pacific

PUBLISHED: 1999 (* asterisk means OUT OF STOCK)

Vol. CS-186   THE CENTRAL PARSECS OF THE GALAXY
              eds. Heino Falcke, Angela Cotera, Wolfgang J. Duschl, Fulvio Melia,
              and Marcia J. Rieke
              ISBN 1-58381-012-9

Vol. CS-187   THE EVOLUTION OF GALAXIES ON COSMOLOGICAL TIMESCALES
              eds. J. E. Beckman and T. J. Mahoney
              ISBN 1-58381-013-7

Vol. CS-188   OPTICAL AND INFRARED SPECTROSCOPY OF CIRCUMSTELLAR MATTER
              eds. Eike W. Guenther, Bringfried Stecklum, and Sylvio Klose
              ISBN 1-58381-014-5

Vol. CS-189   CCD PRECISION PHOTOMETRY WORKSHOP
              eds. Eric R. Craine, Roy A. Tucker, and Jeannette Barnes
              ISBN 1-58381-015-3

Vol. CS-190   GAMMA-RAY BURSTS: THE FIRST THREE MINUTES
              eds. Juri Poutanen and Roland Svensson
              ISBN 1-58381-016-1

Vol. CS-191   PHOTOMETRIC REDSHIFTS AND HIGH REDSHIFT GALAXIES
              eds. Ray J. Weymann, Lisa J. Storrie-Lombardi, Marcin Sawicki,
              and Robert J. Brunner
              ISBN 1-58381-017-X

Vol. CS-192   SPECTROPHOTOMETRIC DATING OF STARS AND GALAXIES
              ed. I. Hubeny, S. R. Heap, and R. H. Cornett
              ISBN 1-58381-018-8

Vol. CS-193   THE HY-REDSHIFT UNIVERSE:
              GALAXY FORMATION AND EVOLUTION AT HIGH REDSHIFT
              eds. Andrew J. Bunker and Wil J. M. van Breugel
              ISBN 1-58381-019-6

Vol. CS-194   WORKING ON THE FRINGE:
              OPTICAL AND IR INTERFEROMETRY FROM GROUND AND SPACE
              eds. Stephen Unwin and Robert Stachnik
              ISBN 1-58381-020-X

PUBLISHED: 2000

Vol. CS-195   IMAGING THE UNIVERSE IN THREE DIMENSIONS:
              Astrophysics with Advanced Multi-Wavelength Imaging Devices
              eds. W. van Breugel and J. Bland-Hawthorn
              ISBN 1-58381-022-6

Vol. CS-196   THERMAL EMISSION SPECTROSCOPY AND ANALYSIS OF DUST,
              DISKS, AND REGOLITHS
              eds. Michael L. Sitko, Ann L. Sprague, and David K. Lynch
              ISBN: 1-58381-023-4

ASP CONFERENCE SERIES VOLUMES
Published by the Astronomical Society of the Pacific

PUBLISHED: 2000 (* asterisk means OUT OF STOCK)

Vol. CS-197   XVth IAP MEETING DYNAMICS OF GALAXIES:
FROM THE EARLY UNIVERSE TO THE PRESENT
eds. F. Combes, G. A. Mamon, and V. Charmandaris
ISBN: 1-58381-24-2

Vol. CS-198   EUROCONFERENCE ON "STELLAR CLUSTERS AND ASSOCIATIONS:
CONVECTION, ROTATION, AND DYNAMOS"
eds. R. Pallavicini, G. Micela, and S. Sciortino
ISBN: 1-58381-25-0

Vol. CS-199   ASYMMETRICAL PLANETARY NEBULAE II:
FROM ORIGINS TO MICROSTRUCTURES
eds. J. H. Kastner, N. Soker, and S. Rappaport
ISBN: 1-58381-026-9

Vol. CS-200   CLUSTERING AT HIGH REDSHIFT
eds. A. Mazure, O. Le Fèvre, and V. Le Brun
ISBN: 1-58381-027-7

Vol. CS-201   COSMIC FLOWS 1999: TOWARDS AN UNDERSTANDING
OF LARGE-SCALE STRUCTURES
eds. Stéphane Courteau, Michael A. Strauss, and Jeffrey A. Willick
ISBN: 1-58381-028-5

Vol. CS-202   PULSAR ASTRONOMY – 2000 AND BEYOND, IAU Colloquium 177
eds. M. Kramer, N. Wex, and R. Wielebinski
ISBN: 1-58381-029-3

Vol. CS-203   THE IMPACT OF LARGE-SCALE SURVEYS ON PULSATING STAR RESEARCH,
IAU Colloquium 176
eds. L. Szabados and D. W. Kurtz
ISBN: 1-58381-030-7

Vol. CS-204   THERMAL AND IONIZATION ASPECTS OF FLOWS FROM HOT STARS:
OBSERVATIONS AND THEORY
eds. Henny J. G. L. M. Lamers and Arved Sapar
ISBN: 1-58381-031-5

Vol. CS-205   THE LAST TOTAL SOLAR ECLIPSE OF THE MILLENNIUM IN TURKEY
eds. W. C. Livingston and A. Özgüç
ISBN: 1-58381-032-3

Vol. CS-206   HIGH ENERGY SOLAR PHYSICS – *ANTICIPATING HESSI*
eds. Reuven Ramaty and Natalie Mandzhavidze
ISBN: 1-58381-033-1

Vol. CS-207   NGST SCIENCE AND TECHNOLOGY EXPOSITION
eds. Eric P. Smith and Knox S. Long
ISBN: 1-58381-036-6

ATLAS   VISIBLE AND NEAR INFRARED ATLAS OF THE
ARCTURUS SPECTRUM 3727-9300 Å
eds. Kenneth Hinkle, Lloyd Wallace, Jeff Valenti, and Dianne Harmer
ISBN: 1-58381-037-4

ASP CONFERENCE SERIES VOLUMES
Published by the Astronomical Society of the Pacific

PUBLISHED: 2000 (* asterisk means OUT OF STOCK)

Vol. CS-208   POLAR MOTION: HISTORICAL AND SCIENTIFIC PROBLEMS,
IAU Colloquium 178
eds. Steven Dick, Dennis McCarthy, and Brian Luzum
ISBN: 1-58381-039-0

Vol. CS-209   SMALL GALAXY GROUPS, IAU Colloquium 174
eds. Mauri J. Valtonen and Chris Flynn
ISBN: 1-58381-040-4

Vol. CS-210   DELTA SCUTI AND RELATED STARS: Reference Handbook
and Proceedings of the 6th Vienna Workshop in Astrophysics
eds. Michel Breger and Michael Houston Montgomery
ISBN: 1-58381-043-9

Vol. CS-211   MASSIVE STELLAR CLUSTERS
eds. Ariane Lançon and Christian M. Boily
ISBN: 1-58381-042-0

Vol. CS-212   FROM GIANT PLANETS TO COOL STARS
eds. Caitlin A. Griffith and Mark S. Marley
ISBN: 1-58381-041-2

Vol. CS-213   BIOASTRONOMY `99: A NEW ERA IN BIOASTRONOMY
eds. Guillermo A. Lemarchand and Karen J. Meech
ISBN: 1-58381-044-7

Vol. CS-214   THE Be PHENOMENON IN EARLY-TYPE STARS, IAU Colloquium 175
eds. Myron A. Smith, Huib F. Henrichs and Juan Fabregat
ISBN: 1-58381-045-5

Vol. CS-215   COSMIC EVOLUTION AND GALAXY FORMATION:
STRUCTURE, INTERACTIONS AND FEEDBACK
The 3rd Guillermo Haro Astrophysics Conference
eds. José Franco, Elena Terlevich, Omar López-Cruz, and Itziar Aretxaga
ISBN: 1-58381-046-3

Vol. CS-216   ASTRONOMICAL DATA ANALYSIS SOFTWARE AND SYSTEMS IX -- (ADASS IX)
eds. Nadine Manset, Christian Veillet, and Dennis Crabtree
ISBN: 1-58381-047-1   ISSN: 1080-7926

Vol. CS-217   IMAGING AT RADIO THROUGH SUBMILLIMETER WAVELENGTHS
eds. Jeffrey G. Mangum and Simon J. E. Radford
ISBN: 1-58381-049-8

Vol. CS-218   MAPPING THE HIDDEN UNIVERSE: THE UNIVERSE BEHIND THE MILKYWAY
THE UNIVERSE IN HI
eds. Renée C. Kraan-Korteweg, Patricia A. Henning, and Heinz Andernach
ISBN: 1-58381-050-1

Vol. CS-219   DISKS, PLANETESIMALS, AND PLANETS
eds. F. Garzón, C. Eiroa, D. de Winter, and T. J. Mahoney
ISBN: 1-58381-051-X

ASP CONFERENCE SERIES VOLUMES
Published by the Astronomical Society of the Pacific

PUBLISHED: 2001 (* asterisk means OUT OF STOCK)

Vol. CS-220   AMATEUR - PROFESSIONAL PARTNERSHIPS IN ASTRONOMY:
The 111th Annual Meeting of the ASP
eds. John R. Percy and Joseph B. Wilson
ISBN: 1-58381-052-8

Vol. CS-221   STARS, GAS AND DUST IN GALAXIES: EXPLORING THE LINKS
eds. Danielle Alloin, Knut Olsen, and Gaspar Galaz
ISBN: 1-58381-053-6

Vol. CS-222   THE PHYSICS OF GALAXY FORMATION
eds. M. Umemura and H. Susa
ISBN: 1-58381-054-4

Vol. CS-223   COOL STARS, STELLAR SYSTEMS AND THE SUN:
Eleventh Cambridge Workshop
eds. Ramón J. García López, Rafael Rebolo, and María Zapatero Osorio
ISBN: 1-58381-056-0

Vol. CS-224   PROBING THE PHYSICS OF ACTIVE GALACTIC NUCLEI
BY MULTIWAVELENGTH MONITORING
eds. Bradley M. Peterson, Ronald S. Polidan, and Richard W. Pogge
ISBN: 1-58381-055-2

Vol. CS-225   VIRTUAL OBSERVATORIES OF THE FUTURE
eds. Robert J. Brunner, S. George Djorgovski, and Alex S. Szalay
ISBN: 1-58381-057-9

Vol. CS-226   12th EUROPEAN WORKSHOP ON WHITE DWARFS
eds. J. L. Provencal, H. L. Shipman, J. MacDonald, and S. Goodchild
ISBN: 1-58381-058-7

Vol. CS-227   BLAZAR DEMOGRAPHICS AND PHYSICS
eds. Paolo Padovani and C. Megan Urry
ISBN: 1-58381-059-5

All book orders or inquiries concerning ASP or IAU volumes listed should be directed to the:

The Astronomical Society of the Pacific Conference Series
390 Ashton Avenue
San Francisco CA 94112-1722   USA

Phone: 415-337-2126
Fax: 415-337-5205
E-mail: catalog@aspsky.org
Web Site: http://www.aspsky.org

INTERNATIONAL ASTRONOMICAL UNION (IAU) VOLUMES
Published by the Astronomical Society of the Pacific

PUBLISHED: 1999

Vol. No. 190 NEW VIEWS OF THE MAGELLANIC CLOUDS
eds. You-Hua Chu, Nicholas B. Suntzeff, James E. Hesser,
and David A. Bohlender
ISBN: 1-58381-021-8

Vol. No. 191 ASYMPTOTIC GIANT BRANCH STARS
eds. T. Le Bertre, A. Lèbre, and C. Waelkens
ISBN: 1-886733-90-2

Vol. No. 192 THE STELLAR CONTENT OF LOCAL GROUP GALAXIES
eds. Patricia Whitelock and Russell Cannon
ISBN: 1-886733-82-1

Vol. No. 193 WOLF-RAYET PHENOMENA IN MASSIVE STARS AND STARBURST GALAXIES
eds. Karel A. van der Hucht, Gloria Koenigsberger, and Philippe R. J. Eenens
ISBN: 1-58381-004-8

Vol. No. 194 ACTIVE GALACTIC NUCLEI AND RELATED PHENOMENA
eds. Yervant Terzian, Daniel Weedman, and Edward Khachikian
ISBN: 1-58381-008-0

PUBLISHED: 2000

Vol. No. 195 HIGHLY ENERGETIC PHYSICAL PROCESSES AND MECHANISMS FOR
EMISSION FROM ASTROPHYSICAL PLASMAS
eds. P. C. H. Martens, S. Tsuruta, and M. A. Weber
ISBN: 1-58381-038-2

Vol. No. 197 ASTROCHEMISTRY: FROM MOLECULAR CLOUDS TO PLANETARY SYSTEMS
eds. Y. C. Minh and E. F. van Dishoeck
ISBN: 1-58381-034-X

Vol. No. 198 THE LIGHT ELEMENTS AND THEIR EVOLUTION
eds. L. da Silva, M. Spite, and J. R. de Medeiros
ISBN: 1-58381-048-X

****

Vol. XXIV TRANSACTIONS OF THE INTERNATIONAL ASTRONOMICAL UNION
REPORTS ON ASTRONOMY 1996-1999
ed. Johannes Andersen
ISBN: 1-58381-035-8

****

Complete lists of proceedings of past IAU Meetings are maintained at the
IAU Web site at the URL: http://www.iau.org/publicat.html

Volumes 32 - 189 in the IAU Symposia Series may be ordered from
Kluwer Academic Publishers
P. O. Box 117
NL 3300 AA Dordrecht
The Netherlands